Nutritional Toxicology

Lishi Zhang
Editor

# Nutritional Toxicology

Springer

*Editor*
Lishi Zhang
Department of Nutrition
and Food Safety, West China
School of Public Health
Sichuan University
Chengdu, China

ISBN 978-981-19-0870-5  ISBN 978-981-19-0872-9 (eBook)
https://doi.org/10.1007/978-981-19-0872-9



This Springer imprint is published by the registered company Springer Nature Singapore Pte Ltd.
The registered company address is: 152 Beach Road, #21-01/04 Gateway East, Singapore 189721, Singapore

# Preface

Nutritional toxicology is a developing interdisciplinary integration of science of nutrition and toxicology. It is a discipline to investigate the influence of nutrients/ nutrition on the metabolic processes and toxic effects of toxicants, the impacts of food-borne toxic/harmful substances on the ADME (Absorption, Distribution, Metabolism, Excretion) of nutrients and nutritional processes, as well as the adverse health outcomes of excessive nutrients intake on the human body (and other organisms) as well as its mechanisms of action, and then to provide the corresponding prevention and control measures.

In 1982, the book of *Nutritional Toxicology* by John Hathcock was published, which proposed the concept of "nutritional toxicology" for the first time [1]. The conception, research approaches, and scope of nutritional toxicology were further elaborated in the second edition of *Nutritional Toxicology* published in 2002 and the *Food and Nutritional Toxicology* published in 2004, and gradually, the discipline framework of nutritional toxicology is formed [2, 3]. The research scope of "nutritional toxicology" has substantially overlapped with the discipline of "food toxicology," but they each have their own emphasis. The objects of "food toxicology" are mainly focused on the detrimental outcomes of toxic/harmful substances detected in food (including environmental contaminants, natural toxicants and processing induced toxicants, etc.) on human health and their mode of action, while nutritional toxicology mainly studies the interaction between nutrition/nutrients and toxicants and the adverse biological effects and its assessment of nutrients excessive intake.

The normal human body organism has its certain self-regulation mechanism to maintain nutrients homeostasis when one or more nutrients are mildly deficient or excessive, which means there would observe no adverse health outcomes within a certain range of nutrients intake. However, some certain adverse health outcomes will kick in if the nutrient deficiency/excess progressed to exceed the self-regulating capability. Therefore, the main research scope of nutritional toxicology involves three main aspects: (1) Adverse health outcome evaluation and risk assessment of nutrient excesses on the human body and the development of safety intake limits, for example, the Tolerable Upper Level (UL); (2) Impacts of nutrients on the metabolic processes and toxic effects of food-borne toxicants, for instance, nutrients and their metabolic processes can affect the absorption, metabolism, and toxicity of exogenous toxicants by altering the function of the gastrointestinal tract (such as pH, gastric emptying time, intestinal peristalsis, mesenteric and hepatic portal vein blood flow and bile flow, some digestive enzyme activities in the intestinal tract) and the enzyme system in vivo, etc.; (3) Influence of food-borne toxic/harmful substances on nutrients and nutrition processes, for example, exogenous chemicals can affect the absorption, metabolism, and function of nutrients through direct and indirect mechanisms, thereby affecting the nutritional and health status of human body [1–3]. The main research scope was elaborated as follows:

1. Adverse health outcome and toxicological study of nutrients

   With the increasing capacity of food production and supply worldwide, the level of food consumption has increased significantly. In recent years, with continuous increasing of variety and quantity of nutrients supplements, other dietary components supplements, fortified foods, health foods, and the healthcare consciousness of consumers, the issue of excessive intake of nutrients and other bioactive substances may also arouse. Excessive intake of nutrients may not only cause acute poisoning, but also is closely related to the development of many chronic diseases, such as carcinogenic diseases, osteoporosis, cardiovascular diseases, and diabetes. Therefore, government agencies, academia, and international organizations are now more oriented on the toxicity and safety problems caused by excessive intake of nutrients.

   The experimental research on adverse health effects of nutrients is more difficult than similar toxicological studies of general xenobiotics. Because of the specific toxicity of nutrients and the more apparent difference in metabolism and sensitivity between humans and animals, the specific toxic manifestation of certain organs or tissues caused by long-term low-dose excessive nutrients intake may first be shown in some specific age-gender groups and can be difficult to find in the routine animal toxicity assessment experiments. Moreover, the adverse health outcomes of nutrients may change when the body is in certain specific physiological status or simultaneously exposed to other substances such as other nutrients, drugs, or contaminants. Furthermore, some specific nutrients or phytochemicals have similar structures and maybe overlapping functions; therefore, antagonistic or synergistic effects might happen when exposed to various substances.
2. Dose–response relationship assessment of adverse health outcome of nutrients excessive intake and the development of tolerable upper intake level

   Health risk caused by nutrients has dual characteristics. There are at least two different intake-response curves for both deficiency and excessive intake of nutrients. The two curves are independent of each other and have different toxicological mechanisms and mode of actions, rather than a simple U-shaped curve with the same effect in the case of one xenobiotic. Also some nutrients might show different effects for specific target populations, the shape, and the steepness of the two curves with certain sub-population may also vary a lot. The area between the two curves is the "safe intake range" or "acceptable intake range." In this range, the organism can maintain the homeostasis of the nutrients, but it should be noted that this range is not necessarily recommended nutrients intake range.

The UL of nutrients refers to the maximum daily intake of certain nutrients of a specific group. Under this level of intake, no harmful effects would be observed among almost all healthy individuals in this group. The concept of UL is widely used in China, the United States, and other countries. UL is a limit level which is proposed based on crowd effect, and the risk of adverse effects will increase if the intake level is higher than this limit. In most cases, the consideration of UL intake should include nutrients from all intake sources such as normal diets, fortified food, food additives, and even drinking water. If the toxic and side effects of nutrients are mainly related to the intake of fortified food and nutritional supplements, these intake sources should be taken into special consideration in the establishment of UL value. But in some cases, experimental data about the toxicological characteristic and human reports of many nutrients is not sufficient to develop their UL values. However, with the nutrients without UL values, it could not be considered that excessive intake of the nutrients would bring no harm.

The development of UL value should be based on the specific risk assessment of the detrimental health outcomes with different exposure scenarios. The key steps include: (1) description of the key adverse health outcomes (qualitative and quantitative); (2) derivation of the No Observable Adverse Effect Level/Lowest Observable Adverse Effect Level (NOAEL/LOAEL) or the Benchmark dose (BMD); (3) determination of the uncertainty factor; (4) establishment of a UL value for people of specific age/sex/physical status (such as pregnancy and lactation); (5) deduce the UL values to other population groups [4].

3. Interactions between xenobiotics and nutrition/nutrients

   The dietary components and the nutritional status of human body can significantly affect the pharmacokinetics of exogenous chemicals and their toxic effects in vivo. Nutrients and their metabolic processes can affect the pH value of the stomach, gastric emptying time, intestinal peristalsis, mesenteric and hepatic portal vein flow or bile flow, and it can also affect the activity of some digestive enzymes in the intestines, thus in turn affecting the absorption, metabolism, distribution, and excretion of exogenous chemicals in vivo.

   In recent years, nutritional toxicology has been further extended to other related research fields, for example, the effects of nutrients on genetic profile and tumor development, the interactions between nutrients and drugs, biological and toxic effects of phytochemicals and other non-traditional nutrients. Risk-benefit assessment of nutrients, food, and diet has also become an important research direction of nutritional toxicology.

Chengdu, China
Lishi Zhang

## References

1. Hathcock J. Nutritional toxicology. New York: Academic Press; 1982.
2. Kotsonis F, Mackey M. Nutritional toxicology. 2nd ed. Abingdon: Taylor & Francis; 2002.
3. Omaye S. Food and nutritional toxicology. Boca Raton: CRC Press; 2004.
4. The Food and Agriculture Organization of United Nations (FAO), The World Health Organization (WHO). A model for establishing upper levels of intake for nutrients and related substances: report of a joint FAO/WHO technical workshop on nutrient risk assessment, 2–6 May 2005. 2006. Available from http://www.who.int/ipcs/methods/nra/en/

# Contents

# Chapter 1
# Effects of Nutrients/Nutrition on Toxicants/Toxicity

**Yan Zhao, Lishi Zhang, Jie Shen, Lingyu Ma, and Li Wang**

**Abstract** There is no doubt that food components can interact with mammal's biological functions. Nutrients or nutritional metabolism in the body can influence the cytochrome P450, leading to biochemical changes. The capacity of converting toxicants of the body is also vulnerable to food deprivation. Furthermore, it should not be ignored that dietary fat may have influence on the synthesis and induction of the constitutive mixed-function oxidases. Several examples are given to explore the expanding field that involves the effects of the constituents in plant food on the metabolic process of organisms. Xenobiotics can be metabolized or biotransformed in all major organisms for their normal functioning, and biotransformation can dominate toxicokinetics. A multitude of dietary factors can affect the metabolism of xenobiotics, including nutritional status of the subject. Here, we give a brief introduction of metabolism and toxicity of toxicants influenced by macronutrients and micronutrients.

**Keywords** Xenobiotics · Biotransformation · Nutrients · Toxicity

## 1.1 Introduction

A multitude of substances are metabolized by our body every day. The substances are divided into endogenous and exogenous ones. Endogenous substances include all kinds of bioactive materials, such as hormones and neurotransmitters. Exogenous substances are the xenobiotics that the human body is unavoidably composed to daily life, such as drugs, toxicants, food additives, environmental chemicals, and intestinal spoilage products. These materials may be interacted with nutrients, positively or

Y. Zhao (✉) · J. Shen · L. Ma · L. Wang
Department of Nutrition and Food Hygiene, School of Public Health, Harbin Medical University, Harbin, China
e-mail: yanzhao@hrbmu.edu.cn

L. Zhang
Department of Nutrition and Food Safety, West China School of Public Health, Sichuan University, Chengdu, China



L. Zhang (ed.), *Nutritional Toxicology*,
https://doi.org/10.1007/978-981-19-0872-9_1

negatively. Therefore, the interaction between nutrients and xenobiotics will be deeply interpreted in this chapter.

## 1.2 Nutrients in Detoxification and Metabolic System for Xenobiotics

Xenobiotics are any chemical compounds that are found in a living organism, but not generated in organisms. Several chemicals that occur naturally might be recognized as xenobiotics when present at excessive levels. The "xeno" in "xenobiotics" derives from the Greek word xenos and has the meaning of guest, stranger, or foreigner. It has been recognized that proper nutrition might build the body as a better fortress against invasion of xenobiotics. However, the effects of certain nutrients on metabolism of xenobiotics might be positive or negative. And understanding the basic metabolism system would help understand the potential mechanisms of the interaction.

## 1.3 The Basic Chemical Detoxification and Metabolism System

Metabolism refers to the processes of compounds' decomposition in an organism, mainly through the role of enzymes, to produce energy and synthesize the components of biological molecules needed to maintain the life activities [1]. The process of xenobiotic metabolism is often expressed as the detoxification. The description, however, means that metabolic transformations is associated with decreased toxicity. Therefore, the expression is incorrect to some extent. In fact, metabolism might involve the production of new toxins or enhance the toxicity of parent compounds, and therefore, it is not always a detoxification process. This claim has been verified in some chemical carcinogens, organophosphates, and necrosis-inducing compounds. Since the same metabolic pathways can enhance or attenuate the toxicity of different xenobiotics, it is generally suggested that all these reactions involving structure transforming of the parent compound be referred to as biotransformation.

Nutrients and foreign compounds may share the same pathways in living organisms when considering metabolism and detoxification processes. The gastrointestinal tract (GI) is essentially the place where food, water, and related pollutants are all ingested [2]. However, our body may also be exposed to environmental and occupational toxicants via the respiratory tract and skin. Generally speaking, the skin, lungs, or gastrointestinal tract tend to absorb the compounds with strong lipophilicity [3]. Then, these lipophilic chemicals may accumulate in the body to toxic levels by continuous or even intermittent exposure unless effective elimination methods exist.

The possibility of biotransformation of xenobiotic compounds in vivo depends on the chemical properties [4]. It is difficult to enter the human body for highly polar compounds, such as relatively ionizable carboxylic acids; even though they enter, they are usually discharged rapidly. So these compounds are not available or can only be metabolized by enzymes in a very short time. On the contrary, volatile compounds such as dichloromethane or diethyl ether are excreted from the lungs so quickly that enzymatic metabolism can appear with little possibility.

Excretability is a vital feature for the toxicity of foreign compounds. The kidneys of vertebrates can induce electrolyte ionization better than nonelectrolytes. For example, the more completely organic acids are ionized at physiological pH, the more easily they are excreted by the kidney.

An extensive range of physique tissues and organs can metabolize the foreign compounds. As section of the body's protection in opposition to the entry of xenobiotics, xenobiotic metabolism mainly occurs in the sites associated with the entry into the body. In this regard, the liver is especially important for xenobiotic compound metabolism, although the skin, lungs, and intestine wall are also the sites through which xenobiotic species enter the body [5–7]. The liver is rich in the enzymes or enzyme systems that catalyze the xenobiotic biotransformation. In addition, the liver is also the first organ for the substances to enter the systemic circulation from the gastroenterol tract. Therefore, these materials can be easily extracted from the blood and chemically modified by the liver before their storage or distribution.

The metabolism of xenobiotics generally takes place in the microsomal fraction (smooth endoplasmic reticulum) of hepatocytes [8]. Non-microsomal reactions can be visible in some biotransformations such as the redox reactions related to alcohols, aldehydes, and ketones. Some activities that promote the metabolism of xenogeneic organisms are limited for the overall biotransformation, although virtually every tissue of the human body exhibits such activities. However, the biotransformation of a chemical outside a hepatic tissue can have a vital toxicological significance for the specific tissue [4].

### *1.3.1 Biotransformation Reactions*

The process of biotransformation is mediated by enzymes, resulting either in changes of parent molecules or in the formation of compounds associated with the combinations of normally occurring agents and the parent molecules. Several predominant enzyme systems are usually related to these reactions in the liver [9]. These enzymes can be roughly divided into phase I and phase II enzymes [10]. The phase I enzymes generally regulate the oxidation, reduction, or hydrolysis and mediate reactive, polar functional groups onto lipophilic toxicant molecules, resulting in the formation of more water-soluble products than the parent xenobiotic species. However, the water solubility and polarity of some substances do not change significantly after the phase I reactions, and substances or groups with stronger

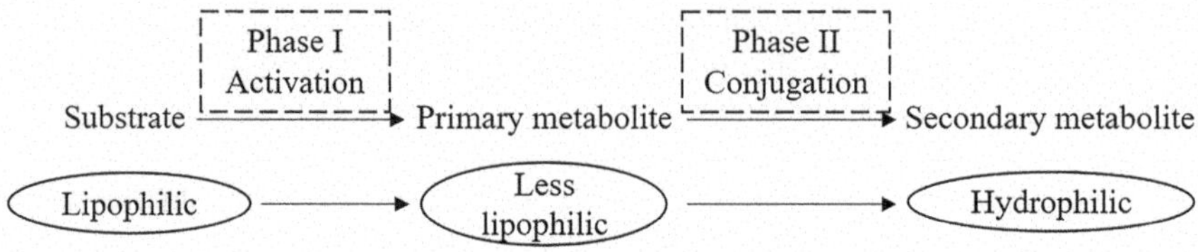


**Fig. 1.1** Two-phase scheme of the biotransformation

polarity need to be combined to further increase their polarity and promote excretion. These conjugation reactions belong to the phase II enzymatic reactions. The phase II reactions thus produce a conjugation product, which is combined between a functional group either on the parent compound or on the introducing group from a phase I reaction and easily excreted from the body (Fig. 1.1).

It is noteworthy that the biotransformation of many substances is complex. Not all xenobiotics go through both phase I and II reactions. Some products from phase I reaction can be directly excreted from the body. Also, a compound with a suitable conjugated functional group can also directly enter phase II reaction without undergoing phase I backward reaction.

### 1.3.2 *Phase I Enzymes*

The phase I enzymes mainly lie in the endoplasmic reticulum, which is a related channel network in the cytoplasm of most cells. The filamentous structure can be observed under the electron microscope, which can be divided into smooth-surface filaments and rough-surface filaments. The tubular endoplasmic reticulum can be broken following that the liver is homogenized into hepatocytes. Microvesicles, that is, microsomes, are formed through the seal of the membrane fragments. Microsomes, the main location of the phase I enzymes, are precipitated from smooth reticular endothelial cells. These enzymes are bound to the cell membrane because the endoplasmic reticulum is essentially a structure of conterminous layer consisting of lipids and proteins. The lipophilic chemicals mainly enter into the lipid layers, the sites of biotransformation; therefore, the enzymes within the lipoprotein matrix are very important. The centrifugation generates many kinds of soluble enzymes involved in phase II biotransformation responses in the supernatant or cytosol. Thus, a number of the critical enzymes involved in the biotransformation are regularly named as cytosolic or microsomal to mark their subcellular sites.

Monooxygenations refer to the combination of one atom of an oxygen molecule with the substrate, but the other atom is reduced to water by nicotinamide adenine dinucleotide phosphate (NADPH). Monooxygenation of xenobiotics, also known as mixed-function oxidations, is either catalyzed by two monooxygenase systems, including cytochrome P-450-dependent monooxygenase system (also known as multi-substrate monooxygenase system) and flavin-containing adenine dinucleotide

(FAD) monooxygenase (also called mixed-function amine oxidase) [11, 12]. These enzyme systems are mainly located in the cellular endoplasmic reticulum and put a hydroxyl group to the xenobiotics.

Reduced NADPH may provide the electrons associated with the reduction of cytochrome P-450 or FAD. Then, a component of the microsomal enzymes that react with molecular oxygen is reduced to form an active oxygen intermediate, leading to the oxidation of the xenobiotic compound [13]. So monooxygenases are known as mixed-function oxygenase (MFO) or hydroxylase too.

The phase I enzymes encompass a series of hydrolases, esterases, and amidases besides the two monooxygenase systems. Regardless of the remaining chemical structure, the cleavage of the ester or amide bond produces two functional groups for further biotransformation: a carboxylic acid plus either an amine (from an amide) or an alcohol (from an ester). Furthermore, the phase I enzymes also include a series of oxidation-reduction enzymes whose functions are to turn over the oxidation state of a carbon, leading to easier excretion or biotransformation by the phase II enzymes.

### *1.3.3 Phase II Enzymes*

In phase II processes, the biotransformation is regarded as biosynthetic reactions in nature, and then, energy is required to trigger. The activation of the cofactors or the production of high-energy intermediates is usually important for the biotransformation. In view of either direct or indirect activation of the cofactors with adenosine triphosphate (ATP), the energy status in the organ is necessary to determine the availability of cofactors [14].

### *1.3.4 Factors Affecting Xenobiotic Metabolism*

Although the biotransformation reactions are complex, the following general conclusions may be proposed [15]:

1. A functional group is usually introduced into a xenobiotic during the phase I reaction, enabling it to bind to endogenous compounds in the process of the phase II reaction.
2. The phase II reactions generally produce more water-soluble and easily excreted conjugates than the parent compound or the phase I metabolites.
3. During the metabolic process, especially in phase I metabolism, more toxic reactive intermediates than parent compounds can be generated. Therefore, xenobiotic metabolism may be a process of detoxification or intoxication.
4. Since metabolic attack may initiate a large number of enzymes involved in phase I and II reactions and there are a number of sites located on organic molecules, the

number of potential metabolites and intermediates available from a single substrate is frequently very large.

5. Because there're qualitative and quantitative differences among species, strains, individual organs, and cell types, a specific exogenous chemical may acquire quite different fates in different scenarios.
6. Since toxicants may be used as inducers and/or inhibitors of the enzymes involved in xenobiotic metabolism and exogenous chemicals are substrates of exogenous metabolic enzymes, they may interact and lead to toxic sequelae, which may be different from the expectation of only using any exogenous chemical.
7. Because the influence of exogenous metabolic enzymes by endogenous factors also affects the enzymes of xenobiotic metabolism, the toxic sequelae from specific toxicants are expected to vary depending on developmental stage, nutritional status, physiological status, gender, and stressors.
8. It has been demonstrated that most of the xenobiotic-metabolizing enzymes appear in the form of several isozymes, which coexist within the same individual and often within the same subcellular organelle. An understanding of the biochemical and molecular genetic processes of these isozymes may help to illustrate the variation among species, individuals, sexes, organs, and developmental stages more clearly.

## 1.4 Nutritional Modulation of Metabolic Enzymes

The relationship between nutrition and exogenous metabolism has always been a subject of widespread concern in academic circles. Effects of nutrients and biological substances, not only on detoxification of xenobiotics but also on pathological process of chronic diseases, have been the hot spot of intensive research. Nutrients, often referred to as xenobiotics from normal food intake, regardless of their nutritional value, mostly have effects on metabolic process of xenobiotics. Due to the fact that most toxicants make enzymes inactivated, the effect of nutrition on drug metabolism is considered to be the effect of nutrition on toxicants. For example, grapefruit juice contains naringin that influences the inactivation of the calcium channel blocker nitrendipine to increase its plasma concentration. This section focuses on two basic mechanisms of nutrients' reactions to xenobiotic metabolizing enzymes, namely, induction and inhibition.

The enzymatic mechanisms of detoxification are involved in two metabolic networks, phase I and phase II detoxification reactions. The cytochrome P450-containing monooxygenases are the major enzyme systems in phase I reactions. About 150 different genes regulate the enzyme, which can be divided into various families and subordinates. It is well accepted that the cytochrome P450 isoenzymes are named and classified. And other oxidative enzymes include microsomal flavin adenine dinucleotide (FAD)-containing monooxygenases, epoxide hydrolase, alcohol dehydrogenases, aldehyde dehydrogenase, esterases, and amidases. The critical enzyme systems involved in phase II biotransformation processes are glucuronosyltransferases, glutathione-*S*-transferases, sulfotransferases, methyl transferases, *N*-acetyl transferases, and so on. There are lots of enzymes and their systems in the liver that catalyze the biotransformation of xenobiotics. Several nutrients may enhance the activity of hepatic microsomal mixed-function oxidases, such as phenobarbital and 3-methylcholanthrene.

The mechanism by which nutrients interact with detoxification (and toxification) enzymes can occur at different levels. At present, there are a lot of discussions about the influences of diet, caloric restriction, and nutrients on metabolic enzyme cytochrome P450, but the nutritional regulation mechanism of cytochrome P450 patterns has not in all cases been extensively studied and elucidated. In the following part, the effects of nutrients on cytochrome P450 and other major metabolic enzymes will be described mainly from the following aspects: starvation diet, dietary fat, protein, sodium, magnesium, and plant constituents.

## 1.5 Effects of Nutrients/Nutrition on Metabolism and Toxicity of Toxicants

Nutrition is involved in the alterations in the body composition, physiological and biochemical functions, and nutritional status of the population. Metabolizing systems require essential macronutrients and micronutrients. A multitude of dietary factors can have obvious influences on the metabolism and toxicity of foreign compounds. Anyhow, the lack of nutrient may result in destroyed health. For example, the decrease in caloric intakes may enhance the toxicity of caffeine and dichlorodiphenyltrichloroethane (DDT) in rats [16]. Some enzymes involved in the detoxification can be affected by diet through different pathways.

Dietary deficiencies in any nutrients can influence protein synthesis, leading to the alteration of cellular membrane, damage of the cellular integrity, changes of the membrane permeability transition and the functions of various macromolecules. Consequently, the ability of the organism to metabolize foreign toxicants may be influenced. The concentrations of critical enzymes required for toxicant metabolism can be disturbed by the bioavailability of nutrients, which interferes with the metabolic reactions involved in toxicant activation, inactivation, and excretion, thus altering toxicity. Cytochrome P450 and flavin proteases are the main enzymes that take part in the reduction of xenobiotics in the body. There is higher oxidoreductase activity in intestinal microflora, which acts a pivotal part in the reduction of xenobiotics. It is known that specific deficiencies in diet may enhance the toxicity of various pesticides, heavy metals, and atmospheric contaminants such as ozone. The levels of important chemicals that participated in the detoxification may also be changed with dietary changes, in that many nutrients, such as protein, lipid, carbohydrate, vitamin C, folate, iodine, and selenium, can change the levels of cytochrome P450 enzymes. Therefore, the effects of nutrients and nutritional status on the metabolism and toxicity of xenobiotics are mainly described in this section, including starvation, dietary macronutrients, micronutrients, and plant constituents.

### *1.5.1 Starvation*

Starvation is a special physiological state; changing the inherent dietary pattern will also increase the susceptibility to liver toxicity injury. Fasting increases catabolic effects and decreases the glycogen stores in the liver, which interferes with the preparation of microsomal enzyme fractions and improves gastric absorption, and thus compromises resistance against cytotoxic agents. However, a set of evidence shows that the mechanisms of metabolism, toxicity, and toxicokinetics of a drug may be influenced via fasting. Eight-hour fasting may result in hypoglycemia and change the activities of some enzymes that metabolize the toxicants. In addition, the activities of cytochrome P450 can also be triggered by fasting in the liver and kidneys of rats, leading to the depletion of glutathione and the production of reactive oxygen species (ROS), oxidative stress, and lipid peroxidation and the reduction in glucuronide conjugation and detoxification [17–19]. Fasting can also increase the amount of certain P450 isoenzymes. Cytochrome P450 increases by 60% and 116% after 24 and 48 h of fasting in rats, respectively. The vital point is that both its activity and the amount of protein itself are increased, possibly by better access of substrate to the enzyme. Immunoblotting and complementary DNA (cDNA) probing have shown an increase in the respective mRNA ensuing from multiplied gene expression. The biosynthesis of cytochrome P450 is affected by the state of starvation (i.e., glycogen depletion and acetonemia). Streptozotocin-induced diabetes mellitus had an effect very comparable to that of fasting on hepatic cytochrome levels, and the acetone/ethanol-inducible structure improved four- to fivefold in diabetic and in fasting rats. In contrast, different isoenzymes were reduced very notably in diabetic rats and multiplied in fasting rats. Cytochrome P450 PCN-e (cyanopregnenolone-inducible) decreases under the conditions in the opposite treatment, namely, hyperalimentation by parenteral infusion of glucose and crystalline amino acids. It is interesting that oral administration of the same nutrient solution does not induce these effects, assuming this reverse effect on blood protein synthesis may be due to the difference in blood composition in the portal vein, similar to the "first-pass effect."

NADPH-producing enzymes have been investigated in an aquatic animal, the American eel, under the influence of fasting and a diet of worms or cow liver. There were no differences in several NADPH-producing enzymes between eels fed with the liver diet and fasting. However, the worm diet may enhance glucose-6-phosphate dehydrogenase (G6PDH) activity. In rats, feed restriction caused a significant increase in the activities of NADPH-generating enzymes in the liver; the enzymes of drug metabolism also were increased by feed restriction.

### 1.5.2 Effects of Macronutrients on Metabolism and Toxicity of Toxicants

Macronutrients, including proteins, lipids, and carbohydrates, are needed in large amounts as energy sources. Excess of macronutrients may exaggerate the toxicity or the results of other macronutrient deficiencies or imbalance. For example, an increase in protein may lead to a decrease in one or both of lipids and carbohydrate.

#### 1.5.2.1 Protein

Protein, an essential nutrient for the human body, is the major structural component to support the growth and tissue repair. Protein is also involved in the synthesis of important molecules like nucleic acid, enzymes, hormones, antibodies and in the maintenance of fluid and acid-base balance. Protein can provide fuel for the body's energy need. Amino acids are the basis for forming a protein. Protein has wide variety and characteristics because 20 different amino acids can be infinitely combined and arranged in living organisms. Dietary protein consists of animal protein and plant protein, and animal protein plus soybean protein is usually considered as good-quality one. Sufficient protein can be easily acquired through a normal dietary intake. In fact, protein-energy malnutrition is one of the major nutritional problems worldwide. Possibly due to the differences in the consumptions of animal food and the economic levels, dietary intakes of protein differ extensively in a variety of components of the world [20]. As a major constituent of enzymes, protein is required in a variety of metabolic reactions. Lack of protein affects the enzymes involved in the metabolism of toxicants. For example, protein-deficient diets and specific changes in dietary protein type can decrease the concentrations of cytochrome P450 in liver microsomes, but not the changes in the amounts of protein [21].

The enzymes may influence the reactions that they catalyze because protein synthesis may be determined by the quality or quantity of amino acids. Protein deficiencies can result in the alterations in amino acid composition of the enzymes, the disturbance of substrate binding, or interaction with the enzyme, in turn the reduction of NADPH-cytochrome P450 (CYP450) reductase and some CYP450 isoenzymes [22]. Studies have shown that feeding rats with a low-protein diet of 6% casein will induce a very significant change in the amino acid composition of the purified form of hepatocyte cytochrome P450, with increased valine, isoleucine, and phenylalanine and decreased glutamine and tyrosine. These kinds of alteration may result in changes in, for example, substrate binding or catalytic activity.

Low-protein diet may also result in the interference with uridine diphosphate glucuronic acid (UDPGA)-glucuronyl transferases (UGT), glutathione-*S*-transferases (GST), and numerous antioxidant enzymes. UGTs are responsible for the conjugation of drugs with UDPGA. Dietary protein acts a pivotal part in the glutathione biosynthesis and the protection of cells against the injury by ROS. Some small molecules, such as glycine, glutamine, cysteine, and taurine, are also associated with the conjugation of xenobiotics. The capacity of detoxication is closely related to enzyme activities. For example, the increase in the activities of mixed-function oxidase led to an obvious decrease of sleeping time triggered by barbiturate in rats. The variation of dietary protein contents from 0% to 50% is correlated with the reduction of serum pentobarbital. With a low-protein diet in rats, it will be noted that the increase in GST-C and glucarate transferases is mainly in the intestines. Also, an increase in urinary recovery of oxazepam glucuronide was observed when changing from a high-protein/low-carbohydrate diet to a low-protein/high-carbohydrate diet.

Protein can influence the carcinogenesis. Male Wistar rats were paired with an isocaloric feed containing 5%, 15%, or 40% casein, and a single dose of aflatoxin B1 was used to induce tumorigenesis. Tumor initiation was achieved by 2-acetylaminofluorene. Significant differences in the total microsomal P450 could not be found, but steroid 16-α-hydroxylation was noted in a high rate. However, α-glutamyl transferase and glutathione-*S*-transferase in precancerous foci were increased in proportion to the dietary protein. The toxicity of several pesticides and other toxic agents can be enhanced by low-protein diets whereas protects against the hepatotoxicity of carbon tetrachloride and dimethylnitrosamine exposure in rats.

Low-protein diets may likely cause a decrease in microsomal enzyme systems and a suppression of protein availability for enzyme synthesis in protein-deficient animals [23]. This is probably the basis that low-protein diets exert protective roles in the toxicity of toxicants. On the other hand, some compounds may increase toxicity through conversion to a more toxic metabolite while exhibit decreased toxicity in malnourished animals. Heptachlor, a chlorinated hydrocarbon insecticide, is an example of such compounds. Heptachlor is converted to heptachlor epoxide, a more toxic metabolite, by mixed-function oxidases of liver microsomes. But this conversion is diminished in rats fed with low-protein diets. Protein deficiency may increase the LD50 for heptachlor by threefold in rats but has no significant effect on the LD50 for heptachlor epoxide, approximately equivalent in rats with pair-fed deficient, adequate, or high levels of protein (Fig. 1.2) [24]. Dietary protein intake also influences other pesticide toxicity: marked increase in the toxicity of carbamate carbaryl, parathion, and phthalidimide captan; the decrease in the toxicity of heptachlor; and no change of the toxicity of dimethoate. In summary, low-protein diets suppress CYP450 and finally may result in the elevation or depression of the toxicity of foreign substances, relying on the toxicity of the products.

High-protein diets may promote oxidative drug metabolism and hepatic mixed-function oxidase activities and reduce the toxicity of chemicals, though excess protein intake may be harmful to health. It was shown that high dietary protein fed to rats decreased the incidences of breast cancer induced by 7,12-dimethyl-benz(*a*) anthracene and the incidence of gastric cancer induced by *N*-methyl-*N*-nitro-*N*-nitrosoguanidine (MNNG) [25, 26]. Sulfur-containing amino acids can provide the protection against the toxicity of some metals. Besides, the toxic effect of cadmium can be antagonized by zinc because of its competition role with −SH. As a free radical scavenger, the special peptides in protein can remove free radicals, protect

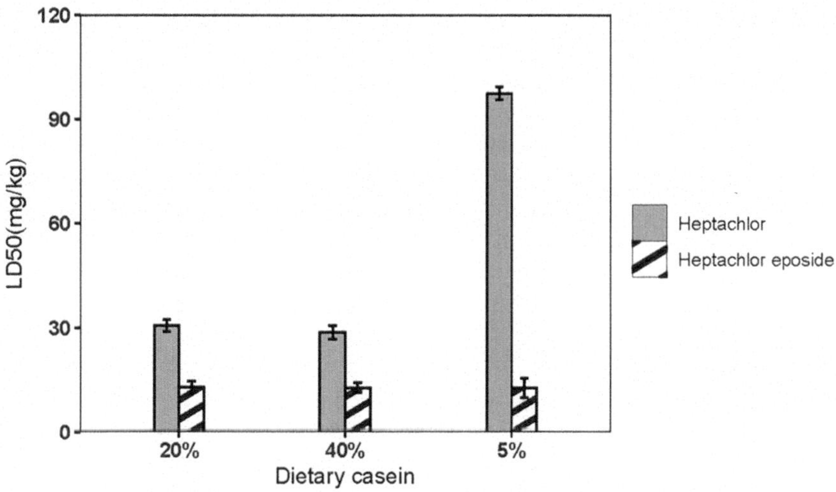


**Fig. 1.2** Effects of diets with different amounts of protein on LD50 values for heptachlor and heptachlor epoxide in rats

the cell membrane from suffering oxidative damage, prevent the erythrocyte hemolysis, and promote the reduction of methemoglobin.

#### 1.5.2.2 Lipids

Lipid comprises a group of naturally occurring molecules soluble in organic solvents, including triglycerides (fats and oils), phospholipids, and sterols. Triglyceride is the chief form of lipid in foods and in the human body, accounting for 95%. A triglyceride molecule is made up of one unit of glycerol and three units of fatty acids. The fatty acids can be classified into saturated, monounsaturated, or polyunsaturated fatty acids according to the degree of saturation. Fatty acids in food may influence the composition of fats in the body. Lipids in the body mainly function as energy fuel and energy stores, cushion of vital organs as shock absorber through fat pad, insulation against extreme temperature, provision of structural materials for cell membranes, and participation in cell signaling. They are also sources of essential unsaturated fats and lipid-soluble vitamins. The primary issue on health problems for lipids is their function in chronic diseases, particularly in cardiovascular diseases and obesity.

Additionally, lipids are involved in the metabolism of toxins. The cytochrome P450 system is actually a coupled enzyme system composed of two enzymes: NADPH-cytochrome P450 reductase and a heme-containing enzyme, cytochrome P450. These enzymes are embedded in the phospholipid matrix of the endoplasmic reticulum. The lipid composition of the endoplasmic membranes not only extensively influences the activity and function of cytochromes P450 but, on the other hand, is susceptible to the number of the dietary fat. Phosphatidylcholine is important since the metabolism of toxins and the combination of substrates with mixed-function oxidases are enzymatically degraded by phospholipase C [27]. Phosphatidylcholine may play roles in maintaining membrane integrity. Steroids and fatty acids are located in CYP450-binding sites and then lead to the substitution of foreign substrates and the interference with their metabolism. The activities of certain toxicant-metabolizing enzymes are suppressed via feeding diet deficient in linoleic fatty acids and essential fatty acids. Dietary lipotropes include choline, methionine, glycine, folate, vitamin B6, vitamin B12, polyunsaturated fatty acids, and phosphate. They are required for the synthesis of phospholipids and membranes, which is essential for the constituents of the microsomal mixed-function oxidase system. The reduction in certain CYP450 isomers and the rise of tumorigenesis by chemical carcinogens may result from the deficiency of choline and methionine in the diet.

The administration of a ketogenic diet containing corn oil for 4 days increased hepatic microsomal P450 2E1 compared with that of rats with lower fat/carbohydrate ratios. Unsaturated fats increase hemoprotein and enzyme activity by twofold compared to olive oil and lard. The induction of hemoprotein of dietary fat has been studied by comparing the effects of a fat free with a 20% corn oil diet. The corn oil admixture did not affect the synthetic level of P450 2B, but increased the constitutive level of P450 2E (ethanol inducible). The decrease in the biosynthesis efficiency of cytochrome P450 heme protein in a fat-free diet may be due to the following two phenomena: insensitivity to inducers and lack of fatty acids needed to synthesize a phospholipid matrix that correctly locates the active protein. Dietary corn oil can increase the activities of other isoenzymes, such as testosterone 6-hydroxylase. Similarly, the phenomenon that aryl hydrocarbon hydroxylase and 7-ethoxycoumarin *O*-deethylase activities was increased in the lung of food-restricted rats and decreased after high-fat diet, but not in liver. In addition, a 20% corn oil diet increases one form of glutathione-*S*-transferase, namely, GST-B, but not GST-A.

Unsaturated essential fatty acids are necessary to maintain membrane function. Therefore, the liver microsomal mixed-function oxidase was stimulated by high intake of polyunsaturated fatty acids. Menhaden oil can not only increase the liver cytochrome P450 but also increase the catalytic ability and affinity to some substrates. In particular, menhaden oil produces higher P4502E1 activity, which significantly increases the metabolic rate.

Both polyunsaturated fatty acids and dietary fat peroxides play an important role in oxidative demethylation as well as phenobarbital inducibility. Briefly, fats may have an effect on the exercise of P450 2B while enhancing the constitutive degree of the ethanol-inducible P450 2E, and it confirmed exclusive consequences on glutathione-*S*-transferases. Also, it appears transferases are influenced in a way similar to P450 in terms of activation by either food restriction or fat-containing diets.

Diet-induced adiposity may influence the availability of circulating toxicants via toxicant sequestration in adipose tissues. The levels of persistent organic pollutants (POPs) in plasma are negatively associated with fat composition in the body [28]. However, POPs exhibit high lipid solubility and bioaccumulation in fatty tissues. POPs may negatively influence the environment via long-distance transport and bioaccumulation. The first process makes these chemicals transfer far from their original source, and the second reconcentrates them to reach potentially dangerous levels for long periods of time. Therefore, these chemical compounds can persistently exist in the environment and also elevate their concentration and toxicity in the environment in view to the bioaccumulation when they are taken in by animals.

#### 1.5.2.3 Carbohydrates

Carbohydrate is a massive and assorted type of organic substances found in nature, consisting of carbon, hydrogen, and oxygen in the ratio of 1:2:1. Carbohydrate is often divided into monosaccharides, disaccharides, and polysaccharides. Carbohydrates are widely located in plants and animals, where they play both structural and metabolic roles. They are essential for providing glucose for metabolic fuels and energy stores. Carbohydrates are also structural components of cell walls and nucleic acid (DNA and RNA) and integral linkers of many proteins and lipids as glycoproteins and glycolipids. There are relatively few reports on toxic effects related to carbohydrate intake in healthy people.

Few evidences are identified concerning the specific function of carbohydrates in biotransformation. Dietary carbohydrate may have generalized effects on intermediary metabolism, such as caloric effects and hormonal alteration effects. A large amount of sugar intake including glucose, sucrose, or fructose may prolong the sleeping time induced by phenobarbital in mice. The longer duration of sleep is negatively connected with the metabolism of the barbiturate. Compared with starch, high-sucrose diets potentiate the lethal reaction to benzylpenicillin because of lower conversion rates of its toxic products. Additionally, the levels of biphenyl 4-hydroxylase activity are lower in rats fed with high sucrose or glucose plus fructose, which has more to do with lower levels of cytochrome P450. Glucose is the precursor of glucuronic acid and closely correlated with phase II detoxification reactions [29]. It is shown that the activities of hepatic CYP450 are decreased and the activation and binding to DNA of carcinogen aflatoxin B1 are increased, given unrestricted feeding, which lead to the reduction of aflatoxin B1 detoxification in the body.

The incidence of spontaneous and chemical-triggered tumors was much lower in rodents and other animals when fed with calorie-restricted diet than fed ad libitum. For example, mammary tumors could not be induced by 7,12-dimethylbenz(alpha)-anthracene in rats whose calories were intermittently restricted [30]. Furthermore, calorie restriction significantly elevated the activities of several drug-metabolizing enzymes (such as CYP2E1), compared with ad libitum feeding in animals [31]. Diet restriction may lead to the decrease in oxygen consumption, the increase in insulin binding, and the changes in energy metabolism through enzymatic alteration involved in glycolysis, gluconeogenesis, and lipid metabolism. Therefore, the activation of some phase II enzymes, such as UDP-glucuronyltransferases, glutathione-*S*-transferases, and *N*-acetyltransferase by calorie restriction, might suppress the incidence of chemical-induced tumors. It is speculated that calorie restriction could also reduce enzyme degradation associated with aging.

### *1.5.3 Effects of Micronutrients on Metabolism and Toxicity of Toxicants*

Micronutrients, which are made up of vitamins and minerals, are essential to life, though the amount needed to meet the body's needs is much less than macronutrients. Many vitamins and minerals have been detected to affect the metabolism of foreign chemicals through the microsome system. Moreover, most studies have shown that micronutrient deficiencies do not have a significant effect on microsomal enzymes as protein deficiencies. But the activities of poison-metabolizing enzymes will be influenced to some extent when micronutrients are severely deficient.

#### 1.5.3.1 Vitamins

Vitamins are divided into water-soluble vitamins (B and C) and fat-soluble vitamins (A, D, E, and K) and play essential roles in health. Insufficient vitamin intakes can result in the classical deficiency diseases. So it is sometimes believed to consume vitamins in large concentrations, even 100 times the recommended dietary levels.

Large amounts of vitamin supplements or the misuse of specific foods can easily lead to toxic reactions.

##### 1.5.3.1.1 Fat-Soluble Vitamins

Fat-soluble vitamins can accumulate in the body, if their intake exceeds the body's requirements, leading to potential toxic effects. It is known that excessive intake of vitamin A and D can cause typical symptoms of toxicity.

Vitamin A may inhibit the bioactivation of carcinogens or bind foreign compounds to microsomal proteins; therefore, it has a function of antitumor. Retinol reduces the mutagenicity of heterocyclic amines by inhibiting their activity. Carotenoids, as provitamin A compounds, may inhibit the mutagenicity of aflatoxin B1 independent of vitamin A. Excessive intake of vitamin A compounds can cause permanent liver damage and stunting. Normally, when excessive intake is stopped, the symptoms of poisoning will be reversed. Common supplements do not easily cause acute toxicity in adults, but there are risks for children. Acute toxicity will appear symptoms within a few hours, such as nausea, vomiting, headache, dizziness, blurred vision, infant fontanelle protuberance, lethargy, and anorexia. Chronic toxicity may take weeks to months to develop clinical symptoms including headache, loss of appetite, hair loss, large liver, muscle pain and stiffness, dry and itchy skin, double vision, bleeding, vomiting, and coma.

Excessive intake of vitamin D will cause the body's absorption of calcium and phosphorus to increase, resulting in hypercalcemia and hypercalciuria, thereby increasing the product of blood calcium and phosphorus in the body and abnormal calcification after reaching saturation. Due to the large amount of calcium excreted by the kidneys, kidney calcification is most obvious, followed by the heart, blood vessels, thyroid, and pancreas. The effects on the skeletal system are mainly thickening and widening of the calcification zone of the long bone metaphysis, thickening of some cortical bone, and bone sclerosis. The symptoms of acute toxicity are mainly hypercalcemia, nausea, vomiting, restlessness, low fever, diarrhea, acidosis, and severe cases include convulsions, coma, and even acute death. The symptoms of chronic toxicity include systemic fatigue, anorexia, polyuria, and constipation. Due to abnormal calcification, there may be different organ damages, such as renal calcification, renal tubular necrosis and proteinuria, hematuria, chronic renal insufficiency, or even kidney decline. Pulmonary calcification has localized epithelial cell necrosis, which can easily lead to repeated infections. Withdrawing the source of the vitamin can reverse the symptoms in general.

Vitamin D may have interactions with diverse inorganic ingredients such as essential minerals and toxic metals [32]. $25(OH)D_3$ is the dominant form of vitamin D in the blood and a precursor of $1,25\text{-}(OH)_2D_3$ (the active form of vitamin D). Adequate amount of $25(OH)D_3$ can adjust the body's absorption, metabolism, and excretion of mineral elements such as calcium, magnesium, iron, phosphorus, zinc, and copper. However, high levels of $25(OH)D_3$ increase the absorption of toxic elements such as aluminum, cadmium, cobalt, lead, and radioisotopes. Conversely, increased absorption of cadmium and lead may lead to decreased levels of active vitamin D in the kidney. In children, due to the increase in intestinal absorption in summer, the level of blood $25(OH)D_3$ will increase seasonally. In turn, the biological accumulation of these toxic metals can disrupt the body's physiological function of vitamin D.

Dietary vitamin E is readily absorbed and transported to cell membranes and to intracellular sites. Vitamin E can scavenge free radicals, prevent oxidative damage to crucial biomolecules, and protect microsomal membranes against lipid peroxidation. A lack of tocopherols, the most abundant form of vitamin E, leads to decreased cytochrome P450 and drug metabolism [33]. Vitamin E can prevent chronic liver injury induced by carbon tetrachloride, increase the levels of CYP2C11 and CYP3A2 in the liver, and inhibit dimethyl benzanthracene mammary tumors. Vitamin E is a scavenger of free radicals. Lack of vitamin E increases free radicals in the body, leading to the body being more sensitive to various oxidative damage, such as ozone, nitrogen dioxide, and other air pollutants. High intakes of vitamin E and vitamin K seem to have relatively less toxicity. No mutagenic, carcinogenic, or teratogenic effects are reported. However, when the intake dose of vitamin E reaches ten times of dietary reference intakes, the immune system may be damaged, such as white cells. Excessive intake of vitamin E may affect platelet aggregation. It can influence blood coagulation system in people with low vitamin K. Although vitamin K is fat-soluble, it is readily excreted from the body and produces scarce toxicity.

#### 1.5.3.1.2 Water-Soluble Vitamins

Water-soluble vitamins tend to produce lower toxic effects than fat-soluble vitamins, mainly due to the fact that they are easily dissolved in water and excreted outside the body. Accordingly, large or excessive consumption can be rapidly eliminated in the urine and sweat, and discernible harm to the individual can be avoided. Therefore, it should be ignored for some water-soluble vitamins that they may have an influence on the body in excessive doses.

Vitamin C, also known as ascorbic acid, is a potent reductant and plays critical roles in hydroxylation and redox reaction. One of its essential functions is to depress the microsomal oxygenation of many xenobiotics. Furthermore, vitamin C can be combined with glucuronides through UDPGA to promote the clearance of the phase I products [34]. Vitamin C deficiency has been shown to decrease the toxicant metabolisms in guinea pigs and also affects the contents of CYP1A and CYP2E in liver. Cytochrome P450 plays a vital role in activating carcinogens such as aflatoxin B1 and heterocyclic amines in foods. Vitamin C is an important antioxidant and free radical scavenger, which has the functions of antiaging, anticancer, reducing serum cholesterol, and preventing atherosclerosis. Vitamin C deficiency is also associated with decreased immunity, skeletal muscle atrophy, neurological disorders, poor wound healing, and impaired capillary integrity. On the other hand, there exist considerable controversies over the efficacy and toxicity of megadoses of vitamin C.

Vitamin $B_1$, that is, thiamin, acts a pivotal part in the carbohydrate metabolism by means of forming thiamine pyrophosphate (TPP), its coenzyme derivative. Dietary deficiency of thiamin can increase aminopyrine and ethylmorphine metabolism. Thiamine deficiency is negatively correlated with increased cytochrome P450 activity, ROS induction, lipid peroxidation, and cytochrome P450 destruction [34]. The clinical manifestation of thiamine deficiency is anorexia, weight loss, cardiac involvement, and neurological involvement.

Vitamin $B_2$, also known as riboflavin, is an important element of NADPH cytochrome P450 reductase and a cofactor for FAD and FMN. Vitamin $B_2$ is involved in biological oxidation and energy metabolism in vivo and is related to carbohydrate, protein, nucleic acid, and fat metabolism. Its deficiency can significantly affect the metabolism of glucose and lipids in tissues and cells, leading to incomplete oxidation, reduced energy utilization, and ultrastructure of mitochondria. And this will also have an adverse effect on poison metabolism. The main clinical symptoms of vitamin $B_2$ deficiency are the inflammations of the eye (blepharitis, photophobia, blurred vision, tears), mouth (stomatitis, glossitis, map tongue), skin (seborrheic dermatitis), and genital (scrotitis, vaginitis).

Niacin is the precursor of nicotinamide adenine dinucleotide (NAD) and nicotinamide adenine dinucleotide phosphate (NADP), two coenzymes involved in almost all redox reactions. Besides, the two coenzymes formed by niacin are also the hydrogen transporters in the first biochemical reaction of the pentose phosphate pathway of glucose, which are related to the synthesis of nucleic acids. In the treatment of niacin deficiency, high doses of niacin (100–300 mg oral or 20 mg intravenous) can cause a variety of side effects including diarrhea, dizziness, fatigue, dry skin, itching, dry eyes, nausea, vomiting, high uric acid, arrhythmias, and hepatotoxic reactions. The treatment dose of niacin deficiency disease can be used to dilate blood vessels and reduce serum cholesterol, but it may be accompanied by some reversible adverse reactions such as pigmentation, dry skin, and ulcers. However, nicotinamide can not lower serum cholesterol, and high doses of this have no side effects.

Vitamin $B_6$, also known as pyridoxol, when taken orally at 2 g/kg or 20 times the dietary reference intake for peripheral neuritis, can cause oral toxicity, resulting in convulsive disorders and inhibition of prolactin secretion. The remaining water-soluble B vitamins, such as pantothenic acid and thiamine, from dietary sources, show few side effects.

The important physiological function of folate is to participate in metabolism as a carrier of carbon monoxide and act a pivotal part in cell division and proliferation, and it is necessary for the growth of many organisms and microorganisms. Folate deficiency can lead to megakaryocytic anemia, hyperhomocysteinemia, preeclampsia, placental abruption, and fetal neural tube abnormalities and is associated with some cancers. Therefore, it is recommended that women of childbearing age increase their intake of folate. Folate is necessary to increase the conversion of toxicant-metabolizing enzymes during chronic drug administration in rats. High doses of folate can also have side effects, affecting zinc absorption, leading to zinc deficiency, delaying fetal growth, increasing low birth weight, inducing convulsions by interfering with the effects of anticonvulsants, masking vitamin $B_{12}$ deficiency symptoms, and interfering with the diagnosis of the disease.

#### 1.5.3.2 Minerals

##### 1.5.3.2.1 Iron

The synthesis of heme, which is vital to the biosynthesis of cytochrome P450 and to regulate mixed-function oxidase activity, depends on iron. Iron deficiency causes anemia but rarely affects cytochrome P450. The levels of cytochrome *c* and myoglobin are reduced in anemic rats with iron-deficient diets [35]. Competitive low intake of iron promotes the absorption of lead, while high intake of iron may inhibit lead absorption and toxicity via the competition with intestinal mucosal carrier protein and other transport carriers. Iron protects against mercury toxicity, and mercury exposure may result in iron deficiency.

The iron content of the body is mainly controlled by the absorption mechanism, but there is no regulatory mechanism to discharge excess iron out of the body. Once the mechanism of iron absorption is damaged due to some reason, such as hereditary hemochromatosis, iron agent treatment, and repeated blood transfusion, the iron transferred into the plasma will increase, which will lead to excess iron in the body [36]. In hemochromatosis, iron is absorbed in excess, and parenchymal cells are loaded with too much iron, resulting in clinical complications of diabetes mellitus, endocrine abnormalities, cardiomyopathy, arthritis, liver cirrhosis, and even hepatic cancer. Furthermore, the main target organ of iron overload injury is the liver, which can cause liver fibrosis and hepatoma. Excessive iron can also lead to excessive production of reactive oxygen species and free radicals, which can cause mitochondrial DNA damage and induce mutations, which are related to colon cancer, rectal cancer, lung cancer, esophageal cancer, bladder cancer, etc. Iron also catalyzes the production of free radicals and promotes lipid peroxidation, which in excess increases the risk of cardiovascular disease and atherosclerosis.

#### 1.5.3.2.2 Selenium

Selenium found in all cells and tissues is an essential trace element, but too much of it can be toxic. Seafood and animal offal are good sources of selenium. The selenium content of plant foods is related to the level of selenium in the surface soil layer. Drinking water contains inorganic forms of selenium, which are not as readily absorbed as the selenoamino acids present in vegetables. Selenium, an important ingredient of some antioxidant enzymes such as glutathione peroxidase and thioredoxin reductase, has the function of anti-oxidation, which can block the damage of reactive oxygen species and free radicals to the body so as to maintain the normal function of cells. Selenium protects cardiovascular and cardiac health. It also has immune-boosting and growth-promoting properties. Selenium and metal have a strong affinity and play a role in detoxification. Selenium is involved in the synthesis of deiodinase, which is necessary for the transformation of thyroxine (T-4) into triiodothyronine (T-3).

According to the epidemiological investigation of the population, it was found that the incidence rate of tumors in selenium-deficient areas increased significantly, suggesting that selenium may have a protective effect on a variety of human tumors. The anticancer mechanisms may include the protection of DNA from oxidative degradation, the promotion of carcinogen metabolism, the enhancement of immune function, the regulation of cell cycle, and the changes in apoptosis [37, 38]. Two large-scale trials (prevention of Cancer by Intervention with Seleniumm pilot study (PRECISE) in Europe and The Selenium and Vitamin E Cancer Prevention Trial (SELECT) in the United States) on the anticancer effect of selenium, which were recruited by over 30,000 participants, respectively, can play an important role in the future.

Selenium deficiency can cause Keshan disease characterized by myocardial injury, which may be related to enhanced lipid peroxidation caused by selenium deficiency, resulting in myocardial fibrosis and necrosis, myocardial arteriole, and capillary injury. Selenium deficiency also leads to growth retardation, neurological visual impairment, and reduced immune function. In addition, selenium deficiency is believed to be an important cause of Kashin-Beck disease. However, long-term excessive dietary selenium can result in growth inhibition, liver damage, and chronic selenium poisoning, with symptoms including hair and nail loss, skin damage and neurological abnormalities, numbness and convulsions in the extremities, and death in severe cases. Elevated levels of dietary selenium may also increase superoxide radicals. Consumption of high intake of selenium should be avoided because the safe range of dietary intake for selenium is narrow.

#### 1.5.3.2.3 Zinc

Zinc deficiency can increase oxidative damage, decrease the activity of various zinc-containing enzymes, weaken the immune function of the body, and affect the growth and development of the fetus. There is an evidence that zinc toxicity that impairs immune response and turns down the levels of high-density lipoprotein (HDL) cholesterol is caused by long-term high doses intakes (about 6–20 times the recommended nutrient intake (RNI)). Zinc toxicity also induced copper deficiency. Symptoms of acute zinc toxicity include vomiting, epigastric pain, fatigue, and lethargy. Zinc supplementation decreases the accumulation of lead in tissues. Zinc protects against methylmercury damage. On the other hand, lead inhibits zinc for intestinal absorption and replaces zinc on hem enzyme; cadmium also inhibits zinc for absorption and replaces zinc on metallothionein.

#### 1.5.3.2.4 Other Minerals

Long-term excessive intake of sodium chloride may have influences on the amount of cytochrome P450 and glutathione peroxidase activity. After a 360 days' feeding of an 8% sodium chloride diet, a phenomenon that a decrease of cytochrome P450 content and an increase of glutathione peroxidase activity may lead to influence the capacity of activating benzo[*a*]pyrene to Ames-positive mutagens was found, suggesting that the detoxification ability of the body was affected under the condition of long-term excessive intake of sodium.

Magnesium poisoning is not easy to occur under normal circumstances, but in patients with renal insufficiency or treatment with magnesium, magnesium poisoning is often caused by excess magnesium in the body. Excessive intake of magnesium can lead to diarrhea, nausea, gastrointestinal spasm, and other gastrointestinal reactions; serious patients may appear to have drowsiness, muscle weakness, weak knee tendon reflex, muscle paralysis, and other clinical symptoms. It was found that the production of glucuronic acid in isolated hepatocytes was reduced by about 50% after 10 days of feeding the rats with magnesium-deficient feed.

Excessive copper intakes over the binding capacity of the liver may cause toxicity. Early signs of copper toxicity include weakness, listlessness, and anorexia, followed by hepatic necrosis, vascular collapse, coma, and even death. Manganese poisoning mainly damages the central nervous system and causes reproductive secretion dysfunction.

### 1.5.4 Phytochemicals

With the discovery of phytochemicals in cruciferous vegetables, more and more evidences show that these secondary metabolites can improve certain biological functions, and their content in food is low.

The major sulfur-containing constituent in the garlic, diallyl sulfide (DAS), is proved to have the function of inhibiting the process of cytochrome P450-catalyzed oxidations. Under the catalysis of cytochrome enzyme, DAS is oxidized to DASO and finally to DASO2, competitively inhibiting cytochrome P450 2E1. On the other hand, garlic oil has the hypolipidemic function, which is attributed to the activation of lipase and the deprivation of NADPH. In addition, as an inhibitor and inactivator of P450 2E1, the sulfur-containing compound disulfiram can inhibit aminopyrine demethylation in vivo and in vitro.

Naringin is the active form of a flavonoid from citrus fruits. Naringenin, the aglycone of naringin, is readily formed in human bodies after consuming grapefruit juice. This compound can inhibit the oxidative metabolism of drugs by inhibiting cytochrome P450 3A4. The function of the active ingredients in the grapefruit juice has been reported to promote drug metabolism. Felodipine, the dihydropyridine calcium channel antagonist, increased over fivefold greater in the plasma under the action of grapefruit juice compared to the water. One study reported that the cytochrome P450 3A4-mediated 7-hydroxylation of coumarin in the human bodies is not inhibited by naringin when given in water instead of grapefruit juice, as measured by the urinary excretion of 7-hydroxcoumarin, while various amounts of naringin added to grapefruit juice show a dose-dependent inhibition. Thus, it can be deduced that a possible synergism between grapefruit juice (or one of its components) and naringin and the inhibitory potency of grapefruit juice can be amplified by naringin.

Prenylated flavonoids extracted from hops are found to be very effective in inhibiting human P450 enzymes in vitro. Hops contain a large number of flavonoids, of which xanthohumol has the strongest inhibitory effect on the 7-ethoxyresorcinol deethylase activity expressed by human P450CYP1B1 cDNA expressed at low molar concentrations. This suppression is inherently competitive. However, whether the intake of hop-related inhibitors is sufficient to be effective in the human body remains to be verified. Since isoxanthohumol and 8-pentenylnaringenin can also inhibit the in vitro metabolic activation of carcinogen AFB1, the abovementioned phytochemicals may prevent the carcinogenic effects of AFB1.

## 1.6 Other Interactions of Nutrients/Nutrition and Toxicants

Nutrition and toxicants are connected in three main ways (Fig. 1.3). First, food is the medium of exposure to poison and enhance the individual exposure chances. Second, when a toxicant enters into the human body, it might also have an interaction with an individual's nutrition status and have an effect on the quantity of

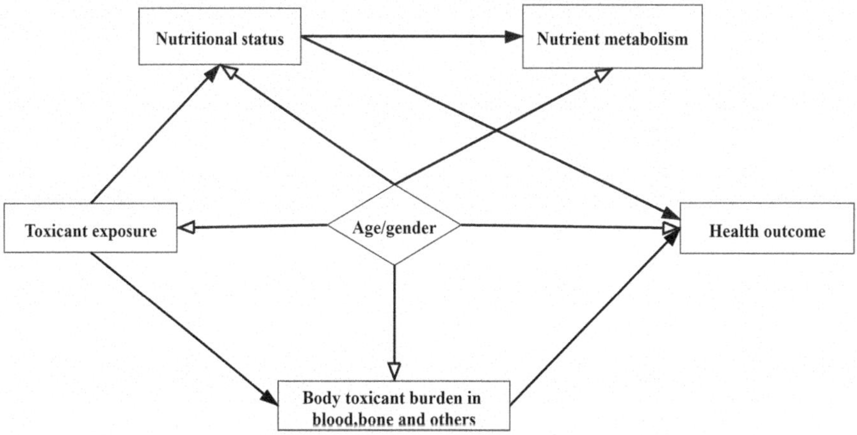


**Fig. 1.3** A model of interaction between nutrients and toxicants

toxicant retainment and bioavailability. Toxicants may affect nutrient absorption and stores as well. Third, nutrients and their metabolism may also interact with the toxicant via affecting a specific health outcome in the body. Other factors should be considered at the same time, since they influence both nutritional status and toxicant exposure, such as age and gender.

### *1.6.1 Nutritional Deficiencies May Influence the Exposure and Toxicity of Toxicants*

Lead, as a heavy metal used in chemical industry, can be found everywhere in our lives. However, if consumed in excess, it can lead to lead poisoning, especially in children and pregnant women, who are at high risk of lead exposure. Moreover, lead exposure causes irreversible damage to the growth and development, psychological and behavioral development, intellectual development, and potential development of fetus and infant [39]. The interactions between lead and micronutrients occur in intestinal absorption, brain neurochemistry, and cognitive function. The divalent metal transporter 1 is a common intestinal transporter for iron and lead. So iron deficiency may increase the absorption of lead.

A number of evidences prove that higher doses intake of calcium may inhibit lead absorption in blood. Placental transfer of lead is lower in pregnant women who consume diets rich in iron and diets with higher hemoglobin levels. Nutritional interventions may contribute to the prevention against lead exposure in the condition of extensive exposure or the source beyond the control. Iron and calcium supplementation shows some positive effects.

Cadmium exposure is associated with renal tubular toxicity and bone metabolism disorder. Health effects associated with cadmium are more common in women than men. Cadmium is not an essential element in the body, but is absorbed from the environment after birth, mainly through foods such as shellfish, leafy vegetables, rice, grains, and beans, which may contain relatively high levels of cadmium, water, and air. Tobacco smoke is an important source of environmental cadmium. The absorption mechanism of cadmium is basically the same as that of iron, calcium, and zinc. Dietary protein, vitamin D, calcium, zinc, and other elements can affect the absorption of cadmium. Vitamin D can promote the absorption of cadmium, and cadmium can, in turn, interfere with the absorption of dietary iron. Low protein, low calcium, and low level of iron contribute to the absorption of cadmium. Low iron stores and intake may result in higher cadmium burdens in the body [40]. Both cadmium and iron are ingested into the colon by means of the divalent metal transporter 1, and after that, cadmium (and lead) is likely rearranged into the circulation system through calcium transporters and ferroportin. It is possible that the absorption of cadmium may increase at very early stages of iron deficiency.

### *1.6.2 Nutrients Affect the Disposition of Toxicants*

Besides the influence of absorption, nutrients can also affect the toxicity of xenobiotics by sequestration into body depots, such as the deposition of chlorinated hydrocarbons in adipose tissue or lead in the bone. Then, these compounds have relatively less chances for adverse effects on metabolism than the case that they would be deposited in the liver or kidney. For example, chlorinated hydrocarbons are mainly stored in adipose tissue of mammals. Under the condition of starvation for short periods, fat is mobilized from the depots for energy provision, and simultaneously, the excretion of dieldrin obviously elevates in blood of rats [41]. Moreover, the duration of exposure to dieldrin needed to produce overt poisoning is connected with the size of adipose tissue in dogs, while force-feedings of dogs may delay or prevent dieldrin poisoning, possibly by sequestering the dieldrin in adipose tissue. On the other hand, when given low-calcium diets, bone formation is lower, or the bone is resorbed to provide calcium, and lead deposition is dramatically increased in the kidney and blood (Fig. 1.4).

### *1.6.3 Nutrient Deficiencies and Toxicants Affect Similar Outcomes*

Arsenic toxicity is caused by excessive consumption of arsenic from food and water contaminated with arsenic, as well as long-term exposure to arsenic at work. Arsenic is linked to lung, bladder, and skin cancers and is related to an increased risk of diabetes mellitus and hypertension, particularly in populations with high exposures.

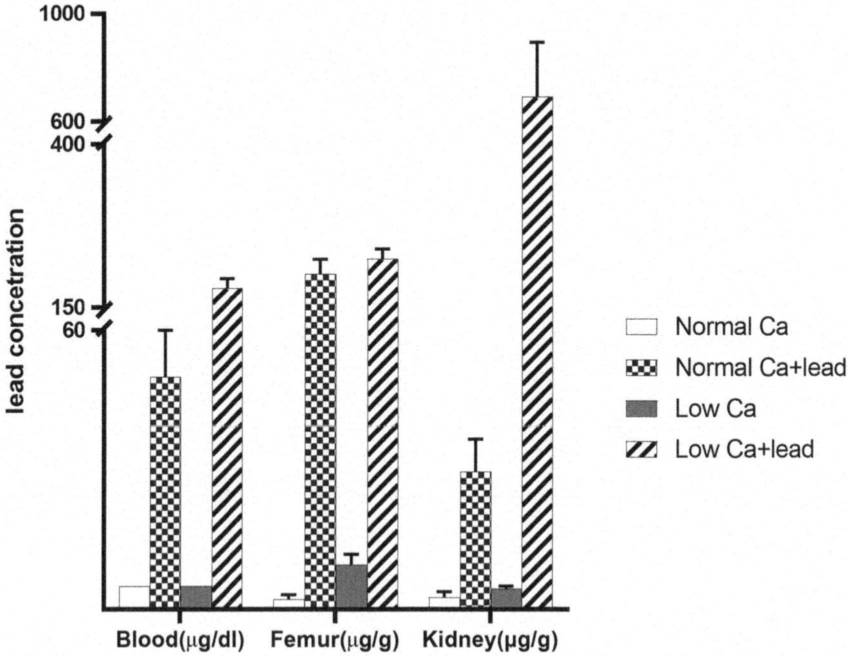


**Fig. 1.4** Effects of dietary calcium deficiency on lead disposition

An interaction may occur between arsenic and nutrient status and metabolism (folate) on health outcomes [42].

Malnourished individuals may be more susceptible to adverse health effects of toxic substances. Because of the physiological changes in women of reproductive age and the growth and development of characteristics of young children, they are more likely to suffer from malnutrition and toxicant exposure. Chemical exposures may cause neurodevelopmental disorders.

Fetal growth requires a lot of nutrition, so women of reproductive age are more likely to be undernourished during pregnancy. In addition, the environmental exposure of women of reproductive age is particularly dangerous because of their connection to fetuses and infants through placental exchange and breast milk. Finally, exposure to toxicant may accelerate the development and course of chronic diseases in adulthood, particularly neurodegenerative diseases. Exposures to toxicants during early life may program later disease and adverse outcomes through epigenetic processes [43].

The fraction of environmental exposure–related disease may be small, but the importance of nutrition in affecting health cannot be neglected. The interaction between chemical exposures and poor nutrition may result in chronic diseases with high costs to health.

Nutrients and toxicants may also regulate the function of the same enzymes and organ systems. For example, the activity of the enzyme 8-aminolevulinic acid dehydratase (ALAD) requires zinc but is inhibited by lead. Zinc can antagonize ALAD inhibition generated by lead in vivo and in vitro [44]. Lead can inhibit the synthesis of protein, and a portion of this inhibition can be overcome by higher concentrations of Fe in vitro.

Similarly, nutrients and toxicants often influence the same organ systems or parameters of toxicity, although affecting different enzymes or different steps in biochemical synthesis. For example, in the hematopoietic system, an anemia occurs with deficiencies of cobalt, copper, iron, and fluorine as well as with toxicities of cadmium, lead, manganese, molybdenum, selenium, and zinc. The development of poisoning-triggered anemia can be dependent on the levels of the nutrients [45]. The anemia of manganese toxicity is associated with low iron levels in the serum, liver, kidney, and spleen and may be improved by increasing dietary iron. Anemia due to cadmium toxicity is associated with low levels of iron in plasma and the liver. The increase in dietary intake of ascorbic acid may improve the anemia through increasing liver iron contents.

### *1.6.4 Nutrient Deficiencies May Produce More Complicated Alterations of Toxicity*

Nutritional deficiency and toxic substances can also arouse more complicated alterations of toxicity, such as nonspecific behavioral changes including growth inhibition or animal anorexia. In certain situations, the decrease in the dietary levels of the toxic compound may improve the behavioral defect. For example, iron deficiency and lead poisoning in children can lead to clinical behavioral defects, such as hyperkinesia. Iron deficiency, which is common in children, is often accompanied by high exposure to lead, so lead-related behavioral changes may be due in part to iron deficiency [46].

The interactions between nutrients and toxicants are complex. Simply feeding high levels of nutrients does not certainly protect against toxic compounds. Dietary intake of a nutrient from deficient to adequate levels may provide protective effects against toxicity. However, much higher levels of the nutrient may lead to increased risk of toxicity itself, but also, the protection from the toxicants may be reversed or not needed. In some cases, ingesting high levels of nutrients may simply speed up their transformation into more active forms of metabolism.

### 1.6.5 Some Nutrients in Food Can Block Nitroso Reaction

*N*-nitroso compounds exhibit strong carcinogenic properties with specific targeting for the liver. Vitamins C and E, phenolic acids, and flavonoids have strong effect to block the nitroso reaction. Some foods, such as tea, kiwi, and sea buckthorn fruit juice can prevent against the toxicity of nitrosamines. Garlic may inhibit the activities of nitrate-reducing bacteria in the stomach, therefore significantly decreasing nitrite contents in the stomach. As precursors of *n*-nitroso compounds, nitrates, nitrites, and amines exist widely in the living environment. Due to the presence of pH value, catalysts, inhibitors, and other factors, the absorption rate of the precursor will be affected, thus affecting the reaction rate and yield of the synthesis of *n*-nitroso compounds. Until now, it has been impossible to estimate the number of nitrosamines formed in the stomach from ingested precursors. As a result, their overall effectiveness cannot be assessed.

### 1.6.6 Some Nutrients in Food Can Influence the Metabolism of Toxicant

In addition, the levels of various key enzymes necessary for metabolism, transformation, degradation, and excretion of toxins in the body can be disturbed by the availability of nutrients, and these interactions may alter toxicity. For example, cytochrome P450, a member of a heme-thiolate protein superfamily, is involved in the metabolism of endogenous and exogenous substances, including drugs and environmental compounds. It has been shown to act a pivotal part in the metabolism of POPs and *N*-nitrosamines, and it and flavoproteins are the major enzymes involved in the reduction of exogenous substances in vivo. There is higher oxidoreductase activity in intestinal microflora, which acts a pivotal part in the reduction of xenobiotics [47, 48]. In all, many dietary nutrients including dietary protein, fat, sugar, vitamin C, vitamin E, folate, betaine, iodine, and selenium may alter Cyp450 enzyme levels via various pathways.

## References

1. Kirchmair J, Howlett A, Peironcely JE, Murrell DS, Williamson MJ, et al. How do metabolites differ from their parent molecules and how are they excreted? J Chem Inf Model. 2013;53:354–67.
2. Goodman BE. Insights into digestion and absorption of major nutrients in humans. Adv Physiol Educ. 2010;34:44–53.
3. Björck I, Granfeldt Y, Liljeberg H, Tovar J, Asp NG. Food properties affecting the digestion and absorption of carbohydrates. Am J Clin Nutr. 1994;59:699S–705S.

4. Dick HM, Opperhuizen A. Biotransformation rates of xenobiotic compounds in relation to enzymes activities: a critical review. Toxicol Environ Chem. 1989;23:181–90.
5. Atashgahi S, Shetty SA, Smidt H, Willem M. Flux, impact, and fate of halogenated xenobiotic compounds in the gut. Front Physiol. 2018;10:1–9.
6. Gibaldi M, Boyes RN, Feldman S. Influence of first-pass effect on availability of drugs on oral administration. J Pharm Sci. 1971;60:1338–40.
7. Varma VS, Obach R, Rotter C, Howard R, George C, et al. Physicochemical space for optimum oral bioavailability: contribution of human intestinal absorption and first-pass elimination. J Med Chem. 2010;53:1098–108.
8. Kuramoto N, Baba K, Gion K, Sugiyama C, Taniura H, Yoneda Y. Xenobiotic response element binding enriched in both nuclear and microsomal fractions of rat cerebellum. J Neurochem. 2003;85:264–73.
9. Almazroo OA, Miah MK, Venkataramanan R. Drug metabolism in the liver. Clin Liver Dis. 2017;21:1–20.
10. Iyanagi T. Molecular mechanism of phase I and phase II drug-metabolizing enzymes: implications for detoxification. Int Rev Cytol. 2007;260:35–112.
11. Ioannides C. Up-regulation of cytochrome P450 and phase II enzymes by xenobiotics in precision-cut tissue slices. Xenobiotica. 2013;43(1):15–28.
12. Stavropoulou E, Pircalabioru GG, Bezirtzoglou E. The role of cytochromes P450 in infection. Front Immunol. 2018;9:89.
13. Kolrep F, Rein K, Lampen A, Hessel-Pras S. Metabolism of okadaic acid by NADPH-dependent enzymes present in human or rat liver S9 fractions results in different toxic effects. Toxicol In Vitro. 2017;42:161–70.
14. Jancova P, Anzenbacher P, Anzenbacherova E. Phase II drug metabolizing enzymes. Biomed Pap Med Fac Univ Palacky Olomouc Czech Repub. 2010;154:103–16.
15. Kedderis GL. Biotransformation of toxicants. Comprehensive. Toxicology. 2010;1:137–51.
16. Planning Committee for a Workshop on Potential Health Hazards Associated with Consumption of Caffeine in Food and Dietary Supplements, Food and Nutrition Board, Board on Health Sciences Policy, Institute of Medicine. Caffeine in food and dietary supplements: examining safety: workshop summary. Washington, DC: National Academies Press; 2014.
17. Sadek K, Saleh E. Fasting ameliorates metabolism, immunity, and oxidative stress in carbon tetrachloride-intoxicated rats. Hum Exp Toxicol. 2014;33(12):1277–83.
18. Rasmussen MK, Bertholdt L, Gudiksen A, Pilegaard H, Knudsen JG. Impact of fasting followed by short-term exposure to interleukin-6 on cytochrome P450 mRNA in mice. Toxicol Lett. 2018;282:93–9.
19. de Vries EM, Lammers LA, Achterbergh R, Klümpen HJ, Mathot RA, Boelen A, Romijn JA. Fasting-induced changes in hepatic P450 mediated drug metabolism are largely independent of the constitutive androstane receptor CAR. PLoS One. 2016;11(7):e0159552.
20. Geissler C, Powers H. Human nutrition. 13th ed. Oxford: Oxford University Press; 2017.
21. Walter-Sack I, Klotz U. Influence of diet and nutritional status on drug metabolism. Clin Pharmacokinet. 1996;31(1):47–64.
22. Ronis MJ, Rowlands JC, Hakkak R, Badger TM. Inducibility of hepatic CYP1A enzymes by 3-methylcholanthrene and isosafrole differs in male rats fed diets containing casein, soy protein isolate or whey from conception to adulthood. J Nutr. 2001;131(4):1180–8.
23. Amelizad Z, Narbonne JF, Daubeze M, Bonnamour D, Oesch F. Monooxygenase activity of systems reconstituted with fractions from rats fed standard and low protein diets. Biochem Pharmacol. 1986;35(18):3169–71.
24. Weatherholtz WM, Campbell TC, Webb RE. Effect of dietary protein levels on the toxicity and metabolism of heptachlor. J Nutr. 1969;98(1):90–4.
25. Clinton SK, Imrey PB, Alster JM, Simon J, Truex CR, Visek WJ. The combined effects of dietary protein and fat on 7,12-dimethylbenz(a)anthracene-induced breast cancer in rats. J Nutr. 1984;114(7):1213–23.

26. Tatsuta M, Iishi H, Baba M, Uehara H, Nakaizumi A, Iseki K. Reduction in NaCl-enhanced gastric carcinogenesis in rats fed a high-protein diet. Cancer Lett. 1997;116(2):247–52.
27. Yamada A, Shimizu N, Hikima T, Takata M, Kobayashi T, Takahashi H. Effect of cholesterol on the interaction of cytochrome P450 substrate drug chlorzoxazone with the phosphatidylcholine bilayer. Biochemistry. 2016;55(28):3888–98.
28. La Merrill M, Emond C, Kim MJ, Antignac JP, Le Bizec B, Clément K, Birnbaum LS, Barouki R. Toxicological function of adipose tissue: focus on persistent organic pollutants. Environ Health Perspect. 2013;121(2):162–9.
29. Teel RW, Strother A. Glucose alters rat liver S9-mediated mutagenesis, metabolism and DNA-binding of aflatoxin B1. Cancer Lett. 1990;54(3):163–9.
30. Mehta RS, Harris SR, Gunnett CA, Bunce OR, Hartle DK. The effects of patterned calorie-restricted diets on mammary tumor incidence and plasma endothelin levels in DMBA-treated rats. Carcinogenesis. 1993;14(8):1693–6.
31. Manjgaladze M, Chen S, Frame LT, Seng JE, Duffy PH, Feuers RJ, Hart RW, Leakey JE. Effects of caloric restriction on rodent drug and carcinogen metabolizing enzymes: implications for mutagenesis and cancer. Mutat Res. 1993;295(4–6):201–122.
32. Schwalfenberg GK, Genuis SJ. Vitamin D, essential minerals, and toxic elements: exploring interactions between nutrients and toxicants in clinical medicine. Sci World J. 2015;2015: 318595.
33. Podszun M, Frank J. Vitamin E-drug interactions: molecular basis and clinical relevance. Nutr Res Rev. 2014;27(2):215–31.
34. Yoo JS, Park HS, Ning SM, Lee MJ, Yang CS. Effects of thiamine deficiency on hepatic cytochromes P450 and drug-metabolizing enzyme activities. Biochem Pharmacol. 1990;39(3): 519–25.
35. McKay RH, Higuchi DA, Winder WW, Fell RD, Brown EB. Tissue effects of iron deficiency in the rat. Biochim Biophys Acta. 1983;757(3):352–8.
36. Stål P, Johansson I, Ingelman-Sundberg M, Hagen K, Hultcrantz R. Hepatotoxicity induced by iron overload and alcohol. Studies on the role of chelatable iron, cytochrome P450 2E1 and lipid peroxidation. J Hepatol. 1996;25(4):538–46.
37. Shrimali RK, Irons RD, Carlson BA, Sano Y, Gladyshev VN, Park JM, Hatfield DL. Selenoproteins mediate T cell immunity through an antioxidant mechanism. J Biol Chem. 2008;283(29):20181–5.
38. Luo H, Yang Y, Duan J, Wu P, Jiang Q, Xu C. PTEN-regulated AKT/FoxO3a/Bim signaling contributes to reactive oxygen species-mediated apoptosis in selenite-treated colorectal cancer cells. Cell Death Dis. 2013;4:e481.
39. Anticona C, San SM. Anemia and malnutrition in indigenous children and adolescents of the Peruvian Amazon in a context of lead exposure: a cross-sectional study. Glob Health Action. 2014;7:22888.
40. Olsson IM, Bensryd I, Lundh T, Ottosson H, Skerfving S, Oskarsson A. Cadmium in blood and urine—impact of sex, age, dietary intake, iron status, and former smoking—association of renal effects. Environ Health Perspect. 2002;110(12):1185–90.
41. Zabik ME, Schemmel R. Dieldrin storage of obese, normal, and semistarved rats. Arch Environ Health. 1973;27(1):25–30.
42. Spratlen MJ, Gamble MV, Grau-Perez M, Kuo CC, Best LG, Yracheta J, Francesconi K, Goessler W, Mossavar-Rahmani Y, Hall M, Umans JG, Fretts A, Navas-Acien A. Arsenic metabolism and one-carbon metabolism at low-moderate arsenic exposure: evidence from the strong heart study. Food Chem Toxicol. 2017;105:387–97.
43. Goodrich JM, Dolinoy DC, Sánchez BN, Zhang Z, Meeker JD, Mercado-Garcia A, Solano-González M, Hu H, Téllez-Rojo MM, Peterson KE. Adolescent epigenetic profiles and environmental exposures from early life through peri-adolescence. Environ Epigenet. 2016;2(3): dvw018.
44. Tian L, Zheng G, Sommar JN, Liang Y, Lundh T, Broberg K, Lei L, Guo W, Li Y, Tan M, Skerfving S, Jin T, Bergdahl IA. Lead concentration in plasma as a biomarker of exposure and

risk, and modification of toxicity by δ-aminolevulinic acid dehydratase gene polymorphism. Toxicol Lett. 2013;221(2):102–9.
45. Horiguchi H, Oguma E, Kayama F. Cadmium induces anemia through interdependent progress of hemolysis, body iron accumulation, and insufficient erythropoietin production in rats. Toxicol Sci. 2011;122(1):198–210.
46. Kordas K, Stoltzfus RJ, López P, Rico JA, Rosado JL. Iron and zinc supplementation does not improve parent or teacher ratings of behavior in first grade Mexican children exposed to lead. J Pediatr. 2005;147(5):632–9.
47. Yamamoto FY, Diamante GD, Santana MS, Santos DR, Bombardeli R, Martins CC, Oliveira Ribeiro CA, Schlenk D. Alterations of cytochrome P450 and the occurrence of persistent organic pollutants in tilapia caged in the reservoirs of the Iguaçu River. Environ Pollut. 2018;240:670–82.
48. Sasaki S, Sata F, Katoh S, Saijo Y, Nakajima S, Washino N, Konishi K, Ban S, Ishizuka M, Kishi R. Adverse birth outcomes associated with maternal smoking and polymorphisms in the N-nitrosamine-metabolizing enzyme genes NQO1 and CYP2E1. Am J Epidemiol. 2008;167(6):719–26.

# Chapter 2
# Effects of Xenobiotics on Nutrition

**Peiyu Xu**

**Abstract** Nutrients and toxicants can mutually affect each other from the steps of digestion, absorption, metabolism, and excretion. Here, we first introduce the way toxicants affect nutrients by modifying the nutrient contents of foods and altering the nutrient metabolism. Then, the effects of nutrients on toxicants were discussed, including nutrient's effect on toxicant absorption and storage organs and/or tissue of toxicants and cellular enzymes, like 8-aminolevulinic acid dehydratase (ALAD). The relationship between nutrients and toxicants is complex and close, requiring more scientific experiments to verify. The working definition of a drug-nutrient interaction (DNI) is that which results from a physical, chemical, physiological, or pathophysiologic relationship between a drug and a nutrient, multiple nutrients, or food in general. Drugs affect the intake and the absorption of nutrients. Then, there are some special interactions between drugs and foods, such as tyrosine reaction and disulfiram reaction. The interaction is considered significant from a clinical perspective if therapeutic response is altered or nutritional status is compromised.

**Keywords** Xenobiotics · Nutrients · ADME · Drug-food interaction (DIN)

## 2.1 Effects of Toxicants on Nutrients/Nutrition

Nutritional toxicology studies the relation between toxicants and nutrition. Such studies are important due to the increasing realization that nutrients and toxicants can mutually affect each other. In general, the interaction between toxicants and nutrients occurs in a variety of steps, including digestion, absorption, metabolism, and excretion of nutrients [1, 2].

P. Xu (✉)
Department of Nutrition, Food Safety and Toxicology, West China School of Public Health and West China Fourth Hospital, Sichuan University, Chengdu, People's Republic of China



L. Zhang (ed.), *Nutritional Toxicology*,
https://doi.org/10.1007/978-981-19-0872-9_2

### 2.1.1 Effects of Toxicants on Nutrition

Some toxicants exert considerable influence on certain nutrition dependent on the location and their chemical properties. The influence of toxicants on nutrients takes place at different stages, including absorption, distribution, metabolism, and others.

The earliest effect that toxicants affect nutrition is to modify the nutrient contents of foods. This phenomenon is observed in the interaction between ascorbic acid and nitrite. The reaction of ascorbic acid with nitrite not only detoxifies the nitrite but also converts the ascorbic acid to dehydroascorbic or its oxidative products, resulting in decreased concentration of available ascorbic acid. In the next step, some toxicants inhibit the digestion of nutrients. Ethanol and other general gastrointestinal tract inflammatory agents, for example, decrease the efficiency of stomach acidity, resulting in decreased ability in denaturing proteins. In the step of absorption, some toxicants have inhibitory effects on nutrient absorption. For example, alcohol inhibits the absorption of thiamine and folate. Mineral oil solubilizes the fat-soluble vitamins and thereby prevents their absorption. Metabolic inhibitors such as cyanide and dinitrophenol inhibit many nutrient absorption [2].

A more broad effect of toxicants on nutrients is the alternation of nutrient metabolism. Metabolism of nutrients includes modification, activation, and synthesis into the ultimately active forms of the nutrients and metabolic processing and degradation into forms suitable for excretion. Some toxicants may inhibit both the synthesis of the ultimately active form of the nutrient and its function after synthesis. For example, some toxicants can induce the activity of microsomal cytochrome P450-dependent mixed-function oxidase system. Such induction can enhance both activation and inactivation reactions for vitamin D, with the net effect of decreased vitamin D function [3]. Usually, nutrients involved in metabolic conjugation of toxicants are needed in increased amounts when the organism is exposed to those toxicants.

Competition between nutrients and toxicants is another process that toxicants interact with nutrients. Such interaction is based on their chemical similarity. The function of essential trace element could be influenced by some heavy metals by such competitive inhibitory mechanism. For example, cadmium is chemically similar to zinc and has increased toxicity in zinc deficiency.

Heavy metals, such as lead (Pb), cadmium, and mercury (Hg), are absorbed in the intestines from contaminated food sources and water, and they can also be inhaled from cigarette smoke and polluted air [4, 5]. The absorption of divalent cations such as calcium and zinc is affected by heavy metals, and metallothionein assists in their absorption and delivery into the circulation [6, 7]. For example, low iron intake can lead to increased absorption of lead and cadmium, and iron-deficiency anemia elevates levels of divalent metals, including some heavy metals [8]. Additionally, dietary fiber inhibits uptake of divalent cations, including heavy metals, from food sources and bile containing heavy metals in the body. The study suggests blood lead and mercury levels are negatively associated with body fat percentage only in men, and blood lead, cadmium, and mercury levels are positively related to hemoglobin levels in both genders. Selenium (Se) is an essential trace element for humans. It is found in the enzyme glutathione peroxidase. This enzyme protects the organism against certain types of damage. Some data suggest that Se plays a role in the body's metabolism of mercury (Hg). Selenium and mercury bind in the body to each other, and Se could reduce the toxicity of Hg [9]. It is not totally clear what impact the amount of Se has in the human body on the metabolism and toxicity of prolonged Hg exposure.

### 2.1.2 The Effects of Nutrients on Toxicants

Conversely, nutrients also have effects on toxicants, and such nutritional effects on toxicants take place at different stages of intoxication, including the stages of absorption, distribution, metabolism, and others [10–13].

Absorption is the earliest stage of intoxication. Many toxicants enter the body through the gastrointestinal tract. The nutritional influence on toxicant absorption via the gastrointestinal tract has great effects on their toxicity. For example, lead absorption via the gastrointestinal tract is known to be altered by several dietary metals, including calcium, phosphorus, and iron. Interestingly, lead absorption via the gastrointestinal tract is also affected by fat and vitamin D [12].

Certain toxicants are usually stored in specific organs and/or tissues [14]. Such organ–/tissue-specific storage usually confines the toxicity. Change of the storage organs and/or tissues could result in increased toxicological effects. Some nutrients can cause the change of organ and/or tissue distribution of toxicants. For example, lead is normally stored in the bone. Low calcium diets cause the bone to be resorbed in order to provide calcium, resulting in dramatic increase of lead in the kidney and blood. In another example, chlorinated hydrocarbon insecticides are mainly stored in adipose tissue, but starvation increases its excretion from adipose tissue, resulting in increased poisoning effects.

Some nutrients and toxicants may target the same cellular enzymes. For example, lead and zinc target the same enzyme 8-aminolevulinic acid dehydratase (ALAD) but have opposite effects. Lead inhibits ALAD, whereas zinc activates its activity. Thus, zinc has an antagonistic effect on ALAD inhibition produced by Pb. In vitro, lead inhibits ALAD activity over a range of lead concentrations. Addition of zinc to incubation media containing lead activated the enzyme; the degree of activation was proportional to zinc concentration above a threshold zinc concentration. Iron is important for protein synthesis, but lead inhibits this process. Therefore, inhibition by lead can be overcome by adding higher concentrations of iron [8, 15].

A few nutrients react directly with toxicants and thereby limit their toxicity. For example, selenium forms a complex with mercury, and ascorbic acid reacts with nitrite. Those are the examples of nonenzymatic reactions between nutrient and toxicant.

In a more complex circumstance, nutrients could alter the metabolism of toxicants, which have fundamental influence on toxicity. Nutrients achieve such effects mainly through the alterations in the concentrations of key enzymes involved in detoxification reactions. For example, protein-deficient diets reduce the cytochrome P450 concentration in liver microsomes. Deprivation of ascorbic acid also causes the decrease of cytochrome P450 in several organs, including the liver, kidney, and adrenals. Iron-deficient diets reduce the concentration of cytochrome C and myoglobin. Reduced concentration of such important enzymes has broad effects on detoxification of many toxicants.

## 2.2 Effects of Drugs on Nutrients/Nutrition

The working definition of a drug-nutrient interaction (DNI) is that which results from a physical, chemical, physiological, or pathophysiologic relationship between a drug and a nutrient, multiple nutrients, or food in general. The interaction is considered significant from a clinical perspective if therapeutic response is altered or nutritional status is compromised [3]. The potential number of interactions and permutations seems infinite. But it remains unclear what proportion of these has actually been identified and, more to the point, what number of the identified subset may be considered clinically significant.

Firstly, drugs affect the intake of nutrients by suppressing appetite, changing the taste, producing gastrointestinal reactions, inhibiting the function of the CNS, and increasing appetite. Secondly, drugs affect the absorption of nutrients through intestinal factors, the intestinal-mucosal factors, the metabolism, and the excretion. Thirdly, nutrients/drug-food interaction has been discussed from many aspects. Then, there are some special interactions between drugs and foods, such as tyrosine reaction, flushing reaction, disulfiram reaction, hypoglycemia reaction, and milk-alkali syndrome. Finally, we introduced the factors that affect the interaction between food and drugs; body factors, drug and food factors, the physical and chemical properties of the drug, and bioavailability (BA) are important parameters of drug-food interaction. The drug-food interaction has been widely concerned in clinical medication; the more researchful efforts would increase in the new drug research and development [3].

DNI could be described or examined based on five general categories:

1. The impact of nutritional status on drug disposition and effect.
2. The impact of food on drug disposition and effect.
3. The impact of specific nutrients on drug disposition and effect.
4. The impact of drugs on nutritional status.
5. The impact of drugs on the disposition and effect of specific nutrients.

Keep in mind that the precipitating factor could be multifaceted, including the interplay of disease and genotype.

A great amount of medical studies has shown that there are interactions between food and drugs. They will reduce the bioavailability of drugs if we have improper diet during the period of taking medicine, Further cause deficiency of nutrients and metabolic disorder even produces a series of diseases. Therefore, the drug-food interaction has been widely concerned in clinical medication. It has been listed as an important content to guide the rational use of drugs in clinic, and the researchful efforts have been increased in the new drug research and development as well. FDI includes the interactions of both pharmacodynamics and pharmacokinetics. The interaction of pharmacodynamics mainly refers to the effect of food on drug therapy, and there are few reports about it currently.

### 2.2.1 Drugs Affect the Intake of Nutrients

1. Suppression of appetite: Most drugs have a side effect of decreasing appetite. Among them, amphetamine is the most obvious. Such drugs can stimulate the center of satiety in the CNS; they are used to reduce body weight and treat obesity in clinic as the suppressants of appetite. Because of its sedative effect, it can also be used to treat ADHD in children. The growth-retarded effect of L-amphetamine sulfate is stronger than that of methylphenidate. In addition, the drug swelling in the stomach can also inhibit appetite, such as taking volumetric pectin and carboxymethyl cellulose; stomach absorbs large amounts of water, and this expansion will lead to satiety, thereby inhibiting appetite [14].
2. Changing the taste: Drugs can make abnormal taste, amblygeustia or discomfortable aftertaste. D-Penicillamine, clofibrate and 5-fluorouracil, and other drugs can affect appetite by reducing or changing the taste.
3. Gastrointestinal reactions: After taking the drug, gastrointestinal side effects including nausea, vomiting, and others will affect the feeding. Some drugs can damage the digestive mucosa causing similar symptoms, such as long-term taking of digitalis and anticancer chemotherapy drugs.
4. Inhibition of the function of CNS: Larger doses of sedatives can reduce the level of human's consciousness and result to loss of appetite, while low doses of sedatives can eliminate people's anxiety and increase appetite.
5. Increasing appetite: Drugs that can promote appetite including insulin, steroid hormones, sulfonylurea, and so on have been used for the rehabilitation of patients in poor nutritional status and physical weakness. Sedative drugs can also significantly increase appetite and increase body weight, such as stability, chlorpromazine, amitriptyline, monoamine oxidase (MAO) inhibitors, and antidepressant drugs.

### 2.2.2 Drugs Affect the Absorption of Nutrients

Factors affecting the absorption of nutrients including intestinal anatomy and structure of the intestinal mucosa are shown below [15].

1. Intestinal factors: Drugs can change the transit time of food in the intestine and reduce the absorption of nutrients. For example, cathartic agents can shorten the time of intestinal transit and nutrients and minerals lost with the feces easily; laxatives like citrate, used commonly, can reduce the absorption of glucose, like tincture and phenolphthalein. Drugs can change the intracavitary environment to prevent the absorption of certain nutrients. Antacids can change the pH in the stomach; it can reduce the absorption of iron in the diet when used frequently. In the acidic environment, high-priced iron can be converted into ferrous iron, which is absorbable, while it will be blocked in alkaline environment. Fat-soluble vitamins are dissolved in mineral oil with the feces, leading the absorption to be greatly reduced.

   Drugs can affect the activity of cholic acid: The effect of drugs on the activity of cholic acid affects the absorption of body fat, cholesterol, vitamins, and other, such as tranquilize and cholestyramine for hypoglycemia; neomycin for the inhibition of intestinal flora can also inhibit the digestion and absorption of lipids and fat-soluble vitamins in the intestine by inhaling or inactivating bile salts.

   Drugs can combine with nutrients: drugs can interfere with the absorption of nutrients. The complex of neomycin and bile salt can inhibit the role of lipase that the absorption of fat and fat-soluble vitamins is affected. The complex combines with tetracycline, calcium, iron, or magnesium ions that can interfere with the absorption of these minerals. Drugs inhibit the absorption of nutrients directly: long-term use of phenytoin can inhibit the absorption of folic acid and vitamin C. Cholesterol can reduce the absorption of cholesterol but also affect the absorption of fat-soluble vitamins, iron, folic acid, and vitamin B12.

   Drugs can kill the normal intestinal flora, which is an important source of B vitamins, and many antibiotics can kill the normal intestinal flora that can synthetize vitamins B, resulting in synthesis of the vitamin reduced. Antibiotics can cause the biosynthesis disorder of vitamin K in the intestine, like sulfonamides [16].
2. The intestinal-mucosal factors of drugs affect the nutrient absorption: Drug can damage small intestine mucosa: long-term abuse of laxatives can damage the structure of intestinal villi that results to microvilli obstruction, followed by affecting the absorption of nutrients by inhibiting the brush border enzymes and transit system of the small intestine, causing malnutrition or anemia if things go on like this. It has been reported that 6 h after taking neomycin, there were some histopathological changes to the intestinal mucosa that the absorption of sucrose and xylose decreased and the excretion of protein, sodium, potassium, and calcium increased. Neomycin also can decrease nutrition absorption by promoting the precipitation of bile salts that can inhibit two mechanisms of pancreatic lipase; among them, mucosal injury is the most important factor. Colchicine can affect the cell division of the intestinal mucosa and inhibit the growth of epithelial cells. Methotrexate also has this damaging effect. These medications can cause diarrhea, reduce nutrients, and cause loss.

   Drug may effect the transport mechanism of the small intestinal mucosa. Colchicine that cures gout can affect the transport system of the intestinal mucosa with the excretion of cholic acid, fat, protein, sodium, and potassium in the feces of patients increased; serum cholesterol decreased; and vitamin B12 absorption blocked. Biguanides that treat diabetes can block the enzyme system of the intestinal mucosa and prevent the sugar into the blood by intestinal mucosal transportation, which can finally lead to malabsorption of carbohydrate. Sulfasalazine that treats gastric ulcers and colitis inhibits the enzyme system in the jejunum, reduces absorption of folic acid and transportation in the small intestine.

3. Drugs affect the metabolism of nutrients: Drugs can interfere with the physiological functions of active vitamins by inhibiting the transformation of vitamins into the corresponding coenzyme or inhibiting enzyme system that the vitamin is involved in. Some drugs can also compete with vitamin receptors and cause symptoms of vitamin deficiency. In addition, some drugs can activate the activity of liver enzymes, promoting the catabolism of certain vitamins and leading to a decrease in body storage, such as the use of phenobarbital during pregnancy, which can cause vitamin K deficiency in infants. Similarly, anticonvulsant drugs can accelerate the metabolism of vitamin D and folic acid.
4. Drugs affect the excretion of nutrients: Many drugs can increase the excretion of nutrients, resulting in nutrient deficiency or lack thereof, such as isoniazid that can increase the excretion of vitamin B6. Thiazide diuretics can cause potassium to be lost in large amounts from urine. Calcium in antacids can interfere with the absorption of phosphorus. In order to maintain the balance of calcium and phosphorus in the body, the kidney will increase the excretion of calcium accordingly, so it may cause softening after using antacids 3 weeks later.

### 2.2.3 *Nutrients/Drug-Food Interaction*

The functional interaction in the gastrointestinal tract has special significance because it can alter the digestion and absorption of drugs and some nutrients. The absorption rate will decrease if transit time decreases, and many drugs can affect gastrointestinal dynamics. Different food ingredients affect gastrointestinal motility as well. Dietary fiber not only promotes gastrointestinal motility but also reduces the contact area between drugs or nutrients and the gastrointestinal tract to reduce its absorption.

Drugs could change gastrointestinal secretion. The reduction in chloride production and the large increase in inhibitory drugs of gastric acid will alter the balance among the ionic therapies. The reduction in concentration of bile acid may be caused by an increase in binding, an increase in excretion, and a decrease in secretion. For example, antibiotic neomycin can bind to bile acids that increase its excretion in the feces, reducing the blood concentration of the drug, the absorption of fat-soluble vitamins accordingly, and the cholesterol levels by reducing the reabsorption. Changes in the flora can affect the biological function of drugs because many drugs are sensitive to bacterial metabolism. Some drugs are cleared through the intestinal microorganisms, which is very necessary for full play of the efficacy. Drugs can also inhibit the absorption of nutrients by affecting the protein synthesis of intestinal mucosal cells directly. This inhibition can result in a decrease in nutrient uptake because most transport systems require the synthesis and transport of protein actively, and some drugs require primary metabolism in intestinal mucosal cells before entering the bloodstream, so alterations in protein synthesis of intestinal mucosa and damage to transport function of intestinal epithelium both can affect the efficacy of drugs [15–17].

The biological function between drugs and nutrition will interact or antagonize after a certain period of time and cause a series of histological changes correspondingly. The common mechanisms of action are the following: Drugs in the blood exist free or with other components, usually in the form of protein binding. The binding portion is more stable in physiological conditions relatively, but its stability varies with electrolyte balance and the presence of competitive molecules. The major transporter in plasma is albumin; the number of drugs transported by the protein can be affected by the concentration and other compounds that bind to albumin. Some drugs can activate the detoxification system; for example, the activation of cytochrome P450 pathway will make metabolism be enhanced of some nutrients. There are some drugs on the metabolism of nutrients that have a direct impact, like anticonvulsants that can promote the decomposition of vitamin D in the liver. Antagonistic effects occur when the physiological functions of drugs and nutrients are opposite, such as the antagonistic effects of vitamin K and salicylic acid drugs during coagulation. Many drugs will promote the excretion of nutrients in the kidneys directly or indirectly, such as antibiotics that can cause increased excretion of electrolytes in the kidney and barbital drugs that can cause increased excretion of ascorbic acid in the kidney.

Long-term use of minocycline or doxycycline can inhibit the normal bacteria in the intestine, affecting the biosynthesis of vitamin K and resulting to deficiency. Prednisone, dexamethasone, and other corticosteroid drugs are taken for a long time, which can cause protein synthesis reduction in the body and further promote protein into glycogen, reducing the use and absorption of glucose in renal tubule. Oral contraceptives are not conducive to the body's storage of glucose and also can affect the synthesis of nicotinic acid and protein in vivo.

Nutrition and nutritional status can affect the function and distribution of drugs. The most significant factor affecting drug distribution is the change in protein synthesis. This change is usually caused by inadequate protein intake or serious illnesses that can affect the absorption, transport, and metabolism of all of these protein-dependent processes. Protein-energy malnutrition will affect the role of albumin. Malnutrition also brings other effects. The blood concentration of the drug increases due to the damage of the scavenging mechanism. The concentration of plasma amino acid can affect the entry of drugs into the CNS. A large number of neutral amino acids that transport into the brain will certainly compete for transporting to binding sites. Dietary composition can change the clinical efficacy significantly of certain drugs by affecting the spectrum of amino acid in blood. Drugs for the treatment of Parkinson's disease, such as L-dopamine, are produced by decarboxylation of peripheral nerve tissue, causing severe side effects. Because the pyridoxal phosphate is a cofactor of amino acid decarboxylase, increased intake of pyridoxine increases the activity of the enzyme and enhances the side effects of L-dopa. A decarboxylase inhibitor (methyldopa and hydralazine) can be used as an adjunct to reduce the side effects of L-dopa in the surrounding tissue after decarboxylation. Chemical actions will occur between some of the nutrients in food and molecules of drug, such as chelating reaction, neutralization reaction, metathesis reaction, and redox reaction. Finally, it will become precipitation or differences in pH caused by nutrient damage or drug effects.

The earliest study found that tetracyclines can be combined with divalent metal ions in food to produce insoluble calcium and magnesium salts, such as calcium ions and magnesium ions. This is mainly due to two structures of tetracycline drug (phenolic hydroxyl and enol group) so that two tetracycline molecules can be chelated with divalent metal. Nutritionists have suggested that some food like edible milk, dairy products and gypsum, and magnesium halide salt produced by tofu cannot take tetracycline drugs at the same time; otherwise, it will interfere with the absorption of drugs and nutrients. Although tetracycline is not used in clinics, the structural transformation of drugs still retains the phenolic hydroxyl- and enol-based structure, which the parent nucleus has, like doxycycline and dimethylamine tetracycline. In addition, quinolone antibiotics such as ciprofloxacin can also be chelated with metal ions to form a precipitate. The precipitate can be formed through combining isoniazid with calcium or magnesium ions. For isoniazid, the antituberculosis effect is achieved by binding to metal ions in vivo. The combination of isoniazid with these components in the gastrointestinal tract can affect absorption and finally affect efficacy. In addition, the difference of pH between the drugs and nutrients will cause interactions that make both drugs and nutrients to be destroyed or reduced at the same time. Acidic vitamin C combines with alkaline drugs like sodium bicarbonate. Not only vitamin C is destroyed, but also the acid-reducing capacity of sodium bicarbonate is also reduced. Taking aminophylline with fruits full of vitamin C at same time, it will be destroyed. Similarly, when taking acidic drugs, do not drink alkaline drinks, such as coffee.

Food will affect the pH of the gastrointestinal tract. The pH of gastric juice will decrease after eating. Erythromycin is a drug that is unstable in the stomach and stimulates the stomach. Therefore, pharmaceutical preparations such as cellulose acetate and cellulose resin are insoluble in acidic medium and aqueous solution but can be quickly dissolved in alkaline aqueous solution and organic solvent. Therefore, this type of enteric preparations should be fasting. On the one hand, stomach acidity is high when fasting and has no effect on the case. On the other hand, fasting drugs can be quickly discharged from the stomach into the intestine; the drug can quickly be released and absorbed. Eating alkaline foods, beverages, and wines when taking these medications may cause the enteric coating to be damaged by the pH and/or type of media. Nutrients in food contain a variety of antioxidant ingredients, such as vitamin C and vitamin E. In addition, taurine, flavonoids, sulfur compounds, etc. may be associated with certain antibiotics that cause redox reaction, such as doxycycline. Taking lactic acid bacteria and other probiotics products with drugs will lead to probiotics inactivation; it is best not to use erythromycin and sulfonamides at the same time.

### 2.2.4 Special Interactions Between Drugs and Foods

Special interactions existing between nutrients and drugs are worthy to be noticed. Due to the special composition of food that there will be a special interaction between the drug and food, drugs can also interfere with the metabolism of food-induced adverse reactions. Typical examples including tyramine reaction, flushing reaction, disulfiram reaction, and hypoglycemia are discussed below [15, 17, 18].

1. Tyrosine reaction: Tyramine, which is rich in cheese, wine, lentils, and other foods, can stimulate the sympathetic that elevates blood pressure. Patients who take monoamine oxidase (MAO) inhibitors also suffer from acute hypertension or high adrenergic crises when they are exposed to tyramine or other drugs containing quasi-sympathetic amine; the mechanism is that the excretion of monoamine oxidase in the metabolism of monoamine oxidase cannot be metabolized; then, the accumulation of cheese in the body caused by the central nervous system to thyroid and catecholamine levels increased rapidly, resulting in headache, hallucinations, vomiting, flushing, and other symptoms. Severe hypertensive emergencies can occur and even cerebrovascular rupture and death that is caused by tyrosine reaction. Common monoamine oxidase inhibitors that cause tyramine reaction are linezolidone, furazolidone, isoniazid, selegiline, perphenazine, and moclobemide, the severity of the symptoms is proportional to the dose of the drug and the amount of tyrosine ingested. Therefore, you must pay attention to the diet when using the monoamine oxidase inhibitors in patients, and try not to eat or eat less with food content of high heparin. Smoked or fermented fish and poultry meat products, broad beans (also containing ingredients with levodopa), aged cheese, malate dextrose yeast, sauerkraut, soy products, and other foods containing more tyrosine that are use with MAOI-AB and MAOI-A simultaneously should be prohibited. Tyramine is less in fresh fish and poultry meat, filling and bottled beer, yogurt, and other foods; this food may be considered in a small amount with drugs. The tyramine reaction has been used to improve the emotional state of patients with severe depression. This drug can improve the adrenaline and serotonin of the CNS. Only one of these substances can enhance cardiovascular effects, such as tyramine.
2. Flushing reaction: If patients are drinking after taking some of the central inhibitor, their looks will show flushing, accompanied with difficulty in breathing, headache, and other symptoms; the reason is that the sedative effect of alcohol and drugs makes the central inhibitor excessively inhibited. Commonly induced substances are sedative hypnotics, analgesics, narcotics, and antihistamines.

3. Disulfiram reaction: Disulfiram absorbed by the body can cause facial flushing, palpitations, difficulty in breathing, and other symptoms. Especially after drinking, because of its inhibition of acetaldehyde dehydrogenase in the liver, the metabolism of acetaldehyde (an intermediate product of ethanol) is blocked and accumulated in vivo, causing a variety of serious discomfort, such as flushing, headache, palpitations, nausea and vomiting, and chest and abdominal pains. So we can use this drug for abstinence. Other drugs, such as metronidazole and hibernation, have similar effects. The analysis of 320 cases of drug-induced disulfiram reaction showed that the drugs can cause the disulfiram reaction including cefuroxime, metronidazole, patchouli water, hydrocortisone, cefazolin, glibenclamide, and furazolidone. It is reported that disulfiram reaction occurred when the patients were sterilized with alcohol cotton balls during using cefazolin. Therefore, the doctor should remind patients to not drink or use alcoholic drugs for 1 week during using these drugs, especially in elderly patients and cardiovascular disease patients.
4. Hypoglycemia reaction: In diabetic patients taking hypoglycemic drugs after drinking or eating some sweets after fasting, drinking may lead to hypoglycemia; the symptoms are feeling weak, unconsciousness, excitement, etc. This is because ethanol stimulates insulin secretion in the body, especially when released rapidly, and ethanol and hypoglycemic agents both can inhibit gluconeogenesis; both of them can cause severe hypoglycemia and neurological disorders.
5. Milk-alkali syndrome: In 1949, the US doctors found that taking baking soda and milk several times can lead to a large amount of calcium and alkali absorption being induced in a short period of time during the treatment process of gastric ulcer, which will make patients with hypercalcemia, if not treated, have metastatic calcification and renal failure, thus the name milk-alkaline syndrome (MAS). In recent years, with the wide use of calcium carbonate preparations, the number of MAS adverse reactions reported increased.

   Lithium is used for the treatment of manic depression. Sodium intake is known to have an effect on lithium excretion. Limiting sodium intake can cause accumulation of lithium and poisoning in patients, while increasing sodium intake can increase excretion of urinary lithium. According to a report, caffeine can increase the excretion of lithium in the kidneys; if the patient stops drinking coffee, it can lead to a decrease in the excretion of lithium and further aggravate it.

### 2.2.5 Factors that Affect the Interaction Between Food and Drugs

The important factors that influence the interaction between food and medicine are body, drug, and food factors. Different individuals or the same individuals under different physiological conditions have different effects on the use of drugs and food intake, and the interactions are different too. It can be affected by age, sex, body shape, body composition, heredity, lifestyle, potential diseases, and other factors.

Because of the decline in physiological function, the body composition of the elderly will change, fat and protein components will increase, and water will decrease; the quantity and opportunities of using drugs more than other people will increase too. So the incidence of drug-food interactions also increased. Studies have shown that both malnutrition and obesity can amplify the effects of drugs and nutrients. At the same time, genetic polymorphisms between the drug and methylene tetrahydrofolate reductase may affect the requirements of vitamins B6, B12, and B2 and folate [18]. In some diseases, polymorphisms may protect patients from drug-food interactions significantly.

The physical and chemical properties of the drug are the most important factors for the interaction of the rest of the food. Different doses of the same drug or different products of the same drug may have different physical and chemical properties, resulting in different interactions. The interactions of drug and food may also depend on the dosage, time, way, and amount of food intake and composition and the interval time. Many interactions are avoided by altering the route of administration or changing the interval time between drugs and food.

An important parameter of drug-food interaction is the change in bioavailability (BA) [16, 18]. BA depends on absorption and the effect of first-pass elimination. The most important pharmacokinetic interaction is the absorption changes of drugs caused by a chemical reaction (such as chelation) or a physiological response (gastric acidity, bile secretion, and gastrointestinal motility) caused by eating. Interactions that only affect the absorption of drugs are common but have little clinical significance. However, changes of absorption in some concentration-dependent drugs and time-dependent drugs may affect efficacy. Interactions that affect drug distribution, metabolism, and exclusion are rare, with grapefruit juice as exception. Grapefruit juice is known to be a potent inhibitor of cytochrome P4503A3 system, so it can increase the BA, which can be metabolized by CYP3A4 enzymes. Pharmacodynamics is usually manifested in synergistic or antagonistic effects of drugs and nutrients at the receptor level, resulting in increased or decreased efficacy.

## References


1. Mahaffey KR, Vanderveent JE. Nutrient-toxicant interaction: susceptible population. Environ Health Perspect. 1979;29:81–7.
2. Hathcook JN, editor. Nutritional toxicology, vol. 1. New York: Academic Press; 1982.
3. Walter-Sack I, Klotz U. Influence of diet and nutritional status on drug metabolism. Clin Pharmacokinet. 1996;31:47–64.
4. Madeddu R, Solinas G, Forte G, et al. Diet and nutrients are contributing factors that influence blood cadmium levels. Nutr Res. 2011;31:691–7.
5. Akesson A, Berglund M, Schutz A, Bjellerup P, Bremme K, Vahter M. Cadmium exposure in pregnancy and lactation in relation to iron status. Am J Public Health. 2002;92:284–7.
6. Swiergose-Kowalewska R. Cadmium distribution and toxicity in tissues of small rodents. Microsc Res Technol. 2001;55:208–22.
7. Qin YY, Leung CKM, Leung AOW, Wu SC, Zheng JS, Wong MH. Persistent organic pollutants and heavy metals in adipose tissues of patients with uterine leiomyomas and the

association of these pollutants with seafood diet, BMI, and age. Environ Sci Pollut Res. 2010;17:229–40.

8. Sunmin P, Byung-Kook L. Body fat percentage and hemoglobin levels are related to blood lead, cadmium, and mercury concentration in a Korean adult population (KNHANES 2008–2010). Biol Trace Elem Res. 2013;151:315–23.
9. Geir B. Selenium as an antidote in the treatment of mercury intoxication. Biometals. 2015;28: 605–14.
10. Fleisher D, Li C, Zhou Y, Pao L-H, Karim A. Drug, meal and formulation interactions influencing drug absorption after oral administration: clinical implications. Clin Pharm. 1999;36:233–54.
11. Kirk JK. Significant drug-nutrient interactions. Am Fam Physician. 1995;51:1175–82.
12. Thomas JA, Tschanz C. Nutrient-drug interactions. In: Kotsonis FN, Mackey M, Hjelle JJ, editors. Nutritional toxicology, target organ toxicity series. New York: Raven Press; 1994. p. 139–48.
13. Thomas JA, Stargel WW, Tschanz C. Interactions between drugs and diet. In: Loannides C, editor. Nutrition and chemical toxicity. New York: Wiley; 1998. p. 161–82.
14. Tschanz C, Stargel WW, Thomas JA. Interactions between drugs and nutrients. Adv Pharmacol. 1996;35:1–26.
15. Meltzer HM, Brantsaeter AL, Borch-Iohnhsen B, et al. Low iron stores are related to higher blood concentrations of manganese, cobalt and cadmium in non-smoking, Norwegian women in the HUNT 2 study. Environ Res. 2010;110:497–504.
16. Yamreudeewong W, Henann NE, Fazio A, et al. Drug-food interactions in clinical practice. J Fam Pract. 1995;40:376–84.
17. Chengyu H, Yun L, Guo Z, et al. Medical nutrition. Beijing: People Health Press; 2003. p. 356–63.
18. Boullata JI, Armenti VT. Handbook of drug-nutrient interactions. Berlin: Springer Science +Business; 2004. p. 5–8.

# Chapter 3
# Effects of Antinutritional Factors on the Digestibility, Bioavailability, and Quality of Protein

**Yujuan Shan and Sicong Tian**

**Abstract** A number of antinutritional factors are naturally existed in foods, especially the plant-derived products. These antinutritional factors largely influence the digestibility, bioavailability, and even their bioactivities. In this chapter, we mainly focused their effects on protein, one of the most important macronutrients in foods. In this case, the leading antinutritional factors include natural sources such as trypsin inhibitor, tannin, and phytic acid. Moreover, we also paid a close attention on the products that is formed during the processing of foods, such as Maillard reaction products (MRP), protein-bound D-amino acids, and lysinoalanine. Finally, we tried to analyze the interaction of nutrients and antinutritional factors.

**Keywords** Antinutritional factors · Protein digestibility · Toxins and nutrients interact

## 3.1 Introduction

As we know, some naturally existing antinutritional factors such as trypsin inhibitors in beans, tannins in beans and grains, and/or phytates in grains may potentially lower the digestibility values of protein and amino acid (up to 50%) and protein quality (up to 100%) in animal models [1]. Moreover, during heat and/or alkaline processing of proteins such as Maillard reaction compounds, D-amino acids, and/or lysinoalanine (LAL), quite a few of antinutritional factors highly formed. These antinutritional factors also decreased protein digestibility by up to 28% in vivo and remarkably reduced protein quality (up to 100%) by rat growth methods [2].

A number of basic measures were used to determine the quality of a protein sources, including protein digestibility, bioavailability of amino acids, the amount of indispensable amino acids (IAA) and nonessential nitrogen as well as their dietary proportions. Based on regional and national balance data, dietary surveys, and

Y. Shan (✉) · S. Tian
Wenzhou Medical University, Wenzhou, People's Republic of China
e-mail: yujuanshan@wmu.edu.cn



L. Zhang (ed.), *Nutritional Toxicology*,
https://doi.org/10.1007/978-981-19-0872-9_3

evidence of protein/amino acid composition, significant differences in the availability of protein and dietary essential amino acid were obtained between the developed and the developing countries. Protein digestibility and quality of mixed diets in developing economies were also obviously lower than of those in developed countries.

Diets in developed countries were mainly derived from highly digestible proteins of animal and vegetable origin. However, diets of developing countries were predominantly based on nondigestible proteins, such as less refined grains and beans, which contain large amounts of nondigestible protein components, insoluble fiber, and antinutritional factors. Some antinutritional factors may be endogenic, and others may be formed during thermal/alkaline processing of proteins. For instance, the following antinutritional factors such as trypsin inhibitors and hemagglutinins in beans, tannins in beans and cereals, phytates in cereals and oilseeds, glucosinolates in mustard and canola protein products, and yeast protein products and antinutritional factors such as urinary acid nucleobases naturally occurred [1]. And the main antinutritional factors formed during thermal/alkaline processing included Maillard reaction products (MRP), oxidized forms of sulfur-containing amino acids, D-amino acids, and lysinoalanine (LAL, an unnatural nephrotoxic amino acid derivative) [2].

## 3.2 Antinutritional Effects of Trypsin Inhibitors

Soybeans are the main source of trypsin inhibitors in common food and feed products [3]. Concentrations of trypsin inhibitors (mainly the Kunitz inhibitors) in soybeans were varied at 8.6–48.2 mg/g sample or around 20.3–122.6 mg/g protein. This obvious change in trypsin inhibitors may be due to differences in cultivars and using of different measurements. Once the soybeans were properly processed, the activity of trypsin inhibitor inside would be damaged and largely decreased. Autoclaved soybeans, for example, contained trypsin inhibitors at a concentration of 3.7–8.1 mg/g sample or about 15.9–21.5 mg/g protein. Otherwise, the concentration of trypsin inhibitor was also lowered in peas and other grain legumes, compared with raw soybeans [3].

Trypsin inhibitors were characteristic of protein, which could be inactivated by heat treatment such as boiling, autoclaving, steaming treatment, or tableting or could be removed by fractionation [4]. The leading causes of heat inactivation of trypsin inhibitors included temperature, heating time, particle size, humidity and initial endogenous level, and perhaps crop species and cultivar.

In some animal experiment, intake of raw soybean protein preparations or inhibitors extracted from soybean could cause pancreatic enlargement. Similarly, soya protein extract feeding significantly increased pancreatic weight of male and female rats, while it reduced the spleen weight. Soybean trypsin inhibitors could induce the synthesis and secretion of proteases (such as trypsin, chymotrypsin, and elastase) as well as the hypertrophy and hyperplasia of the pancreas in vivo. The increased secretion of proteases suggested that an endogenous loss of amino acids in the form of enzymes secreted by the overactive pancreas may lead to the growth depression. Trypsin and chymotrypsin were particularly abundant in sulfur-containing amino acids [5]. As a result, the effect of an overactive pancreas may transfer these amino acids existing in the synthesis pool of tissue proteins to the synthesis of enzymes and then lost in the feces.

According to reports, protein quality and protein and amino acid digestibility were negatively influenced due to high levels of trypsin inhibitors in diets and other antinutritional factors in soybeans, kidney beans, and other grain legumes in vivo. The impacts of a food-grade defatted soya flour (Nutrisoy) and autoclaved Nutrisoy on the digestibility of protein and amino acid in growing pigs were explored. There was significant difference in the trypsin inhibitor activities between the Nutrisoy and autoclaved Nutrisoy diets [6]. Compared with the autoclaved Nutrisoy-based diet, the unheated Nutrisoy-based diet is 50% lower in ileal protein digestibility and about 27–51% lower in some amino acids' digestibility like arginine, histidine, isoleucine, and leucine [6].

The effect of heat treatment on the digestibility of amino acids in red kidney beans (*Phaseolus vulgaris* L.) had been studied [7]. In a diet containing 10% protein, kidney beans were selected as the only source of dietary protein, and true fecal digestibility was determined in rats. And the raw soybean diet containing the highest antinutritional factors inversely exhibited the lowest amino acid digestibility. The digestibility for the first limiting amino acid (methionine + cystine) and valine in the raw soybean diet were negative, with −19% and −8% digestibility rate, respectively, while the digestibility values for other IAA, as arginine and histidine, were 4–32%. Moreover, the digestibility of amino acids was obviously improved by home processing and commercial canning. In particular, the beneficial effect of home processing was more significant (68–88% in digestibility) than commercial canned food (40–80% in digestibility). The amino acid digestibility values of the canned beans were much lower when compared to the home-processed beans, which indicated that the intense heating during canning process had an adverse effect on amino acid digestibility, especially the first limiting amino acid (methionine + cystine), which presented 40% digestibility rate by canned but 68% via home-cooked [8].

In lentils and legumes, the true fecal digestibility of restrictive amino acids such as tryptophan, methionine, cystine, and threonine was lower than that of protein. For example, the value of true digestibility in autoclaved pinto bean was 80%; however, the maximum value in restrictive amino was only 68%. This may imply that protein digestibility was actually not an ideal approximation of the bioavailability of amino acids in grain legumes.

Trypsin inhibitors in beans not only had adverse effects on the digestibility of protein and amino acids but also had negative effects on protein quality [9]. Three soybean breeding lines, TG-923-2, TG-1019-2, and TG-1497-1, for instance, developed in the form of raw materials in Nigeria, contained 20.3–51.1 mg/g trypsin inhibitors, had true fecal protein digestibility of 47–58%, and had negative protein efficiency ratio (PER) (20.46–0.88) compared to casein control (2.47). Compared to autoclaved TG-1019-2, boiling was more conductive to reduce levels of trypsin inhibitors at 41.3 mg/g protein and increases true fecal protein digestibility by 35% and PER by 1. A previous investigation in human found that the nitrogen retention of ingested raw soybean meal was reduced by 20% than heated soybean meal [10].

## 3.3 Antinutritional Effects of Tannin

Tannins, naturally water-soluble polyphenolic compounds, were able to complex and precipitate proteins in aqueous solutions and widely present in a variety of plant species such as grains and legume seeds [11]. Tannins were famous for its potential protein precipitants, and they had been reported to reduce protein and amino acid digestibility of animals fed cereals and grain legumes containing tannins [12]. In addition, a casein diet containing a rising dose of tannin-rich fava bean hulls caused a linear decrease in amino acids digestibility [13]. The digestibility of some dispensable amino acids, particularly glycine, glutamic acid, and proline, was largely influenced, while most IAA was hardly affected. It was postulated that a significant reduction in digestibility of the dispensable amino acids was mainly due to the interactions of tannins with the proline-abundant proteins [14].

Increasing the quantities of dietary tannin-rich fava bean husks led to a linear increase in the number of proline-rich proteins in the glands and the parotid glands [15]. Similarly, intake of high tannin sorghum or purified concentrated tannins could increase the size of the parotid gland, as well as the synthesis and secretion of proline-rich proteins in rats [16]. These findings implied that proline-rich proteins were produced in the saliva and combine with dietary tannins in the mouth to protect dietary protein. The combination of tannins with diets and endogenous proteins had been postulated as the main mechanism for reducing digestibility of protein and amino acid in tannin-containing diets [17].

## 3.4 Antinutritional Effects of Phytic Acid

Phytic acid (myo-inositol hexaphosphoric acid), collectively known as phytate, was a naturally occurring substance in the plants. It was mainly found in grains, seeds, and nuts and serves as a source of minerals and inositol during germination [18]. Phytate, rich in negatively charged phosphate groups, was well known to chelate with several essential minerals in the gastrointestinal tract of humans and animals, causing their low bioavailability. Phytate disturbed homeostasis of zinc and also blocked the bioavailability of other nutrients including proteins. In addition, phytate could weaken the activity of digestive enzymes, such as carboxypeptidases and aminopeptidases, through chelating with mineral cofactors or interacting with the protein (either enzyme or substrate) [19].

Phytic acid both in vegetables and animal proteins interfered with the proteolysis of pepsin, possibly via the formation of phytate-protein complexes at lower pH value. The effects of phytic acid on protein and amino acid digestibility and protein quality were reported in feeding poultry and swine with phytase supplementation to production rations [20]. In these studies, microbial phytase was provided to animal rations to degrade phytate, which consequently increased bioavailability of phosphorus and reduced the environmental impact of high phosphate wastewater from intensive pig and poultry units [21].

The roles of adding phytase in ameliorating protein and amino acid digestibility and protein utilization, especially in pigs and poultry, had been generally investigated [18]. While the effect of adding phytase on the ileal digestibility of amino acids was less reported, a few literatures showed that digestibility of some amino acids was improved by an increase of 3–10% in the ileal, which may be related to the increase in growth rates and protein retention.

In 5-week-old broilers, phytase supplementation to a variety of feedstuffs including corn, sorghum, and wheat as well as four oilseed meals and two grain by-products (wheat middling and rice polishing) significantly improved the digestibility of protein and amino acids fed. In the latter study, ileal protein digestibility was obviously increased by phytase supplementation, with average improvements of 4–8% in the various feedstuffs for individual IAA. Particularly, the greatest improvement was observed in threonine and valine. Dietary phytate concentration was negatively correlated with the inherent protein digestibility and amino acid digestibility of the feedstuffs.

## 3.5 Important Antinutritional Factors Formed During the Processing of Foods

During thermal processing and the food storage, proteins undergo chemical changes causing a possible loss of their nutritional value. For example, Maillard reaction was well known as a nonenzymatic browning process in which reducing sugars and lysine in various products such as dairy foods, eggs, and grains react with each other.

Although antinutritional effects of MRP had been mainly reviewed in rodents, only a few related data on protein digestibility and utilization from human were reported. Adolescent males aged 11–14 years easily and usually consume large quantities of fast foods and snacks, leading to high intake of MRP [22]. In a two-period crossover trial, healthy adolescent males were randomly assigned to two different diets: the brown diet (rich in MRP) and the white diet (low in MRP). The results showed that a diet rich in MRP negatively associated with protein digestibility. While the possible long-term influence of an excessive intake of MRP during adolescence deserves further investigation.

*Protein-bound* D*-amino acids and* LAL (an unnatural amino acid derivative) were formed when proteins were exposed to heat and/or an alkaline solution [23]. During this process, two mainly chemical changes would occur, including racemization of amino acids to D-enantiomers and simultaneous formation of LAL. Formation of D-amino acids in proteins depended on pH, time, and temperature. The presence of D-amino acids in proteins would impair digestibility and nutritional quality of protein and amino acid, which led to the different nutritional utilization in animals and humans [24].

Protein-bound D-amino acids might adversely affect protein digestibility and safety of processed foods [25]. Upon being absorbed, D-amino acids might be utilized by racemases, epimerases, or D-amino acid oxidases. The amino acid oxidase system would be saturated when foods rich in D-amino acids were consumed. Proteins containing D-amino acids could be hydrolyzed at peptide bonds containing L-amino acids. However, the hydrolysis rates might be slower than that of the corresponding unprocessed proteins. These changes adversely affect the nutritional quality and safety of foods used to produce nonnutritious forms of amino acids. In addition, these racemic proteins might compete with other proteins for the active site of digestive proteinases in the intestine and thereby limited the effective utilization of these un-racemized proteins. The slower absorption of free and peptide-bound D-amino acids may decrease the responding protein's digestibility [25, 26]. It was not confirmed at present whether D-amino acid containing oligopeptides could change the composition of microflora in the digestive tract, while free D-methionine was well used by mice and rats but was rarely utilized by humans, which indicated the possible involvement of gut microflora.

The formation of D-amino acids and LAL, as well as their effect on protein digestibility, had been investigated. Three sources of protein (casein, β-lactoglobulin, and wheat protein) were heated and alkaline treated (heating at 65 °C for 6 or 24 h, pH 10.5–11.5) [27]. For instance, about 11–15% of L-asparagine and aspartic acid, the most sensitive amino acids, were racemized in above protein sources after 24-h heating. Moreover, the alkaline/heat exposure increased the quantities of LAL, and in these protein sources, approximately 10–12% of the total lysine was converted to LAL [24]. In contrast, true ileal protein digestibility of the treated casein, β-lactoglobulin, and wheat protein was reduced by 13%, 14%, and 17%, respectively. The heat/alkaline treatment of casein decreased the digestibility of aspartic acid, serine, and glycine. However, digestibility of other amino acids was not affected. This study demonstrated that even small amounts of D-amino acids and LAL in protein can do harm to protein digestibility.

LAL, a strong chelator, might exert its toxic effect by conjugating with minerals including calcium, iron, copper, and zinc in renal tubule cells [24]. Compared with several animals, the human kidney was more susceptible to LAL damage. The altered mineral (iron and copper) status due to intake of LAL was also reported in rats [28]. Iron levels in the liver and kidney were largely decreased. As we know, iron content of the elderly and infants is generally reduced, and they may eat food from a single source. The human susceptibility to the nephrotoxic effect of LAL remains unclear [24]. Since long-term feeding of alkali-treated soy protein to baboons, no nephrotoxic effects had been observed, so it might be safe for human to consume low levels of dietary LAL. However, further studies were required to evaluate the actual amount of LAL ingested [25]. In addition, the effect of long-term consumption of alkaline-treated foods with high LAL content on the balance of copper and other minerals should also be checked.

Protein digestibility and quality of some enteral products were based on casein, and soy protein isolate (SPI) were not as good as casein. Intestinal products contained higher levels of LAL than casein (998–2333 vs. 0 mg/g protein). The formation of LAL was produced in a variety of proteins under non-alkaline conditions. Therefore, the lower protein digestibility of the intestinal products was reasonably related with the presence of LAL in these products [29]. Similarly, heat processing at pH 12.2 caused an obvious decrease in protein digestibility and net protein utilization (NPU) of casein and soybean [24].

For the protein nutritional values of European infant formula, the formation of LAL was identified as the most sensitive predictors to assess protein damage caused by heat treatment [30]. The results show that liquid milk formula contains up to ten times more LAL than powder forms, which suggested more rigid heat processing during the producing of liquid milk formulas was used, while there were many unresolved questions including the detailed chemical composition of amino acids responsible for lower protein digestibility.

The low levels of nutritive value in a racemized, heat, and alkali-treated dietary protein were closely associated with human nutrition. A reduction in the quantity of L-amino acid and in the overall digestibility contributed to the lower nutritional value. In contrast, many IAA amino acids were not racemized into D-isomers and would be absorbed after protein breakdown, while adjacent D-amino acids might disturb their absorption [31]. The impact of adjacent D-amino acids on the bioavailability of methionine in some tripeptides existed in casein and soya protein has been reported [32].

## 3.6 Interaction of Nutrients and Antinutritional Factors

Nutrients and toxins interacted with each other in various manners, including ingestion, digestion, and absorption and postabsorptive effects. As for intake, an individual current nutritional state played powerful role in its propensity to ingest toxins [33]. For instance, sea urchins (*Arbacia punctulata*) escaped from foods containing the diterpenoid pachydictoyl A when fed ad libitum but consumed these foods in the first 2 days of food deprivation [34]. Therefore, they once again refused the changed foods, due to the either changes in nutritional status or aversion learning.

In addition to recent feeding history, the nutrients coexisting with toxins in food could also affect the quantities of the toxins ingested. For instance, a compensatory response to the dilution of the overall nutrient in the diets caused the velvetbean caterpillar (*Anticarsia gemmatalis*) to consume toxic levels of caffeine, while other findings had confirmed a specific effect of macronutrient balance on the digest of toxins [34]. The water amounts of food were also reported to influence the detergency of toxins directly, rather than via through its role as a dietary nutrient diluent.

Some evidence indicated that non-nutrient components of plants affected the digestion and absorption of foods through interacting with nutrients. The most extensive attention was played on polyphenols, which combined with dietary proteins and other macromolecules in the intestinal tract of consumers and thereby reducing the digestibility and availability of these nutrients [35]. The quantities of nutrient in foods and the individual nutrition state both similarly affected the post-processing of toxin absorption. Cytochrome P450 enzymes, specifically metabolic xenobiotic, were sensitive to dietary levels of various nutrients. Nutritional deficiencies slow down the metabolism of xenobiotics mediated by cytochrome P450. In contrast, thiamine deficiency might even increase the activity of these enzymes, which indicated the further investigation required [36].

The interactions between nutritional and nonnutritive components in the diets were universal and complex. In the locust experiment, the dietary protein/digestible-carbohydrate macronutrient balance and the dietary levels of tannic acid were changed [37]. As a result, the mortality was increased in diets with tannic acid and increasing macronutrient imbalance, compared to the diet supplemented with a balance of protein and carbohydrate. Under the condition of excess carbohydrate relative to protein, tannic acid mainly acted on the blocking feeding. In reverse, when foods contained excess protein, tannic acid did not affect intake but worked after digestibly.

## References

1. Liener IE. Implications of antinutritional components in soybean foods. Crit Rev Food Sci Nutr. 1994;34:31–67.
2. Anderson RL, Wolf WJ. Compositional changes in trypsin inhibitors, phytic acid, saponins and isoflavones related to soybean processing. J Nutr. 1995;125:581S–5S.
3. Miyagi M, Shinjo S, Nishida R, et al. Trypsin inhibitor activity in commercial soybean products in Japan. J Nutr Sci Vitaminol. 1997;43:575–80.
4. Giami SY. Chemical composition and nutritional attributes of selected newly developed lines of soybean (*Glycine max* (L) Merr). J Sci Food Agric. 2002;82:1735–9.
5. Radha C, Kumar PR, Prakash V. Enzymatic modification as a tool to improve the functional properties of heat-processed soy flour. J Sci Food Agric. 2008;88:336–43.
6. Peace RW, Sarwar G, Touchburn SP. Trypsin inhibitor levels in soy-based infant formulas and commercial soy protein isolates and concentrates. Food Res Int. 1992;25:137–41.
7. Gatel F. Protein quality of legume seeds for non-ruminant animals: a literature review. Anim Feed Sci Technol. 1994;45:317–48.
8. El-Adaway TA. Nutritional composition and antinutritional factors of chick peas (*Cicer arietinum* L.) undergoing different cooking methods and germination. Plant Foods Hum Nutr. 2002;57:83–97.
9. Vadivel V, Janardhanan K. Nutritional and antinutritional characteristics of seven South Indian wild legumes. Plant Food Hum Nutr. 2005;60:69–75.

10. Shimelis EA, Rakshit SK. Effect of processing on anti- nutrients and in vitro protein digestibility of kidney bean (*Phaseolus vulgaris* L.) varieties grown in East Africa. Food Chem. 2007;103:161–72.
11. Betancur-Ancona D, Gallegos-Tintoré S, Delgado-Herrera A, et al. Some physiochemical and antinutritional properties of raw flours and protein isolates from *Mucuna pruriens* (velvet bean) and *Canavalia ensiformis* (jack bean). Int J Food Sci Technol. 2008;43:816–23.
12. Li S, Sauer WC, Huang S, et al. Response of pancreatic secretions to feeding diets with low and high levels of soybean trypsin inhibitors in growing pigs. J Sci Food Agric. 1998;76:347–56.
13. Wu W, Woodie P, Williams M, et al. Amino acid availability and availability-corrected amino acid score of red kidney beans (*Phaseolus vulgaris* L.). J Agric Food Chem. 1996;44:1296–301.
14. Sarwar G, Peace RW. Comparisons between true digestibility of total nitrogen and limiting amino acids in vegetable proteins fed to rats. J Nutr. 1986;116:1172–84.
15. Jansman AFJ, Frohlich AA, Marquardt RR. Production of proline-rich proteins by the parotid glands of rats is enhanced by feeding diets containing tannins from faba beans (*Vicia faba* L.). J Nutr. 1994;124:249–58.
16. Duodu KG, Taylor JRN, Belton PS, et al. Factors affecting sorghum protein digestibility. J Cereal Sci. 2003;38:117–31.
17. Elkin RG, Freed MB, Hamaker BR, et al. Condensed tannins are only partially responsible for variations in nutrient digestibilities of sorghum grain cultivars. J Agric Food Chem. 1996;44: 848–53.
18. Selle PH, Ravindran V, Caldwell RA, et al. Phytate and phytase; consequences for protein utilisation. Nutr Res Rev. 2000;13:255–78.
19. Li Z, Alli I, Kermasha S. In vitro alpha-amylase inhibitor activity-phytate relationships in proteins from Phaseolus beans. Food Res. 1993;26:195–201.
20. Vaintraub IA, Bulmaga VP. Effect of phytate on the in vitro activity of digestive proteinases. J Agric Food Chem. 1991;39:859–61.
21. Antony U, Chnadra TS. Enzymatic treatment and use of starters for the nutrient enhancement in fermented flour of red and white varieties of finger millet (Eleusine coracana). J Agric Food Chem. 1999;47:2016–9.
22. Seiquer I, Díaz-Alguacil J, Delgado-Andrade C, et al. Diets rich in Maillard reaction products affect protein digestibility in adolescent males aged 11-14 y. Am J Nutr. 2006;83:1082–8.
23. Friedman M. Chemistry, nutrition, and microbiology of D-amino acids. J Agric Food Chem. 1999;47:3457–79.
24. Friedman M. Chemistry, biochemistry, nutrition and microbiology of lysinoalanine, lanthionine, and histidinoalanine in food and other proteins. J Agric Food Chem. 1999;47: 1295–319.
25. International Life Science Institute. Mechanism of toxicology of lysinoalanine. Nutr Rev. 1989;47:362–4.
26. Sternberg M, Kim CY, Schwende FJ. Lysinoalanine: presence in foods and food ingredients. Science. 1975;190:992–4.
27. de Vrese M, Frik R, Roos N, et al. Protein-bound D-amino acids, and to a lesser extent lysinoalanine, decrease true ileal protein digestibility in minipigs as determined with 15N-labeling. J Nutr. 2000;130:2026–203.
28. Sarwar G, L'Abbé MR, Trick K, et al. Influence of feeding alkaline/heat processed proteins on growth and protein and mineral status of rats. Adv Exp Med Biol. 1999;459:161–77.
29. Sarwar G, Peace RW. The protein quality of some enteral products is inferior to that of casein as assessed by rat growth methods and digestibility-corrected amino acid scores. J Nutr. 1994;124: 2223–32.
30. Pompei C, Rossi M, Mare F. Protein quality in commercial milk-based infant formulas. J Food Qual. 1987;10:375–91.
31. Paquet A, Thresher WC, Swaisgood HE, et al. Synthesis and digestibility determination of some epimeric tripeptides occurring in dietary proteins. Nutr Res. 1985;5:891–901.

32. Sarwar G, Paquet A, Peace RW. Bioavailability of methionine in some tripeptides occurring in dietary proteins as determined by rat growth. Nutr Res. 1985;5:903–9.
33. Slansky F. Allelochemical-nutrient interactions in herbivore nutritional ecology. In: Rosenthal GA, Berenbaum MR, editors. Herbivores: their interaction with secondary plant metabolites. New York: Academic Press; 1992. p. 135–74.
34. Slansky F, Wheeler S. Caterpillars' compensatory feeding response to diluted nutrients leads to toxic allelochemical dose. Entomol Exp Appl. 1992;65:171–86.
35. Bennick A. Interaction of plant polyphenols with salivary proteins. Crit Rev Oral Biol Med. 2002;13:184–96.
36. Yang CS, Brady JF, Hon J-Y. Dietary effects of cytochromes P450, xenobiotic metabolism, and toxicity. FASEB J. 1992;6:737–44.
37. Simpson SJ, Raubenheimer D. The geometric analysis of nutrient-allelochemical interactions: a case study using locusts. Ecology. 2001;82:422–39.

# Chapter 4
# Natural Toxicants Originating from Food/Diet

**Yan Yu, Kingsley Katleho Mokoena, and Crystal Ethan**

**Abstract** Even though many of the toxic chemicals found in our environment that concern the general public are man-made, there are also hundreds of naturally occurring poisons from animal, plant, fungal, and microbial sources. In fact, the most toxic chemicals known to man are natural poisons, and it would be unreasonable to suggest, as some of the advertising claims do, that natural constituents of food occur just like they do with synthetic additives.

It has been known for centuries that some natural toxic substances are harmful and are sometimes used for murder, suicide, or even misguided medical treatment. Although natural poisons are relatively rare among accidental poisoning cases compared with poisoning by drug overdose, accidental poisoning by natural substances still happens occasionally. Toxins of natural origin have diverse structures and mechanisms of action, making it impossible to discuss and cover all of them in this chapter individually because of their many categories. In light of this, we will not examine every toxic substance derived from animals, plants, fungi, or microorganisms, but rather just a few interesting and important examples.

**Keywords** Food safety · Nutrition · Food-borne diseases · Microorganisms · Additives · Toxins

## 4.1 Bacterial Toxins

Food-borne pathogens cause a great number of diseases with significant effects on both human health and economies globally. There are numerous cause of food-borne diseases, and bacteria are by far the common prevalent of the causes. With microbial illnesses noted to be on the increase, the Center for Disease Control and Prevention (CDC) reported that approximately half of all incidences of diarrhea in the United

Y. Yu (✉) · K. K. Mokoena · C. Ethan
Department of Nutrition and Food Safety, School of Public Health, Xi'an Jiaotong University, Health Science Center, Xi'an, People's Republic of China
e-mail: yuyan@mail.xjtu.edu.cn



L. Zhang (ed.), *Nutritional Toxicology*,
https://doi.org/10.1007/978-981-19-0872-9_4

States were of food-borne origin [1]. The following are the causes behind the growth in food-borne illnesses:

1. There are more enhanced epidemiological procedures, such as better case-reporting protocols and faster detection and identification of food-borne diseases.
2. Significant changes in people's lifestyles and food consumption patterns during the previous few decades. Over the previous few decades, people's lifestyles and food consumption patterns have changed dramatically. The younger generations want to eat out regularly and travel more frequently, sampling a wide range of international cuisines. As a result of globalization, vegetables and fruits now come from a variety of nations, each with its own cleanliness standards and microorganism strains [2]. As a work-oriented nation, we have moved further away from food preparation due to urbanization, infrastructure development, and technological advancements; as a result, learning about food preparation has become less important. Currently, food preparation standards are seldom included in college curricula, i.e., home economics courses or lessons. Most people are eating more in restaurants these days, where food passes through multiple hands, raising the risk of food contamination due to improper handling [3]. Furthermore, consumer's demand for natural or organic foods has caused a spike in unpasteurized food production.
3. Food distribution has become far more global, which benefits a wider potential population; however, this has increased the risk of food-related problems spreading further [4]. On a broad scale, a single mistake in the food preparation chain can affect a huge number of people, frequently with disastrous results. Cross-contamination is also a risk in major food manufacturing operations.
4. The progression of microorganisms' resistance to antibiotics, a phenomenon where microorganisms like fungi and bacteria develop the ability to thrive even when antibiotics are administered. Hence, antibiotic overuse has multiplied the incidence and reduced the capability to treat microbial illnesses, an eventuality that is becoming a major international public health threat. As human beings, we are also to blame for often failing to use proper food handling and safety methods in our homes. Our improper food handling practices, such as poor hygiene (or lack of good cleaning techniques) in the kitchen, reckless hand-washing, and failure to observe appropriate food temperatures, have placed us at an increased danger to food-borne diseases [5]. Quite a number of people are of the belief that maintaining a cold chain is critical to prevent food spoiling; thus, food should be allowed to get to room temperature before being refrigerated. This is based on the observation of food deterioration, which occurs when food is refrigerated at high temperatures, causing the cooling system to lose its coolness, permitting rotting to take place.

Most food-borne diseases are microbial-related and attributed to either infections or intoxication [6]. The difference between these two is that diseases caused by pathogens are known as infections while those that involve the pathogen's toxins and toxic metabolites as causative agents are referred to as intoxication (poisoning).

Additionally, food-borne illnesses occurring from other natural environment sources such as soil, air, and water are also referred to as intoxication.

## *4.1.1 Intoxication*

### 4.1.1.1 *Bacillus* Species

*Bacillus* is a gram-positive, rod-shaped bacterium genus that produces endospores either aerobically or facultatively anaerobically [7]. *Bacillus* bacteria can remain in a dormant state for years in the environment and thus are considered as endospores. Endospores are a novel type of bacterial vegetative cell that differs from the parent cell in terms of enzyme content, chemical structural properties, and corporeal functions. Moreover, the spores are resistant to different weather conditions, heat, dehydration, radiation, cold, and disinfectants. These new types of cells carry all the properties necessary for the formation of another vegetative cell through germination, once conditions are favorable. *Bacillus* species have a wide range of physiological characteristics with the ability to produce a range of enzymes, antibiotics, and different metabolites. As such, *Bacillus* species are used in various industries, including agricultural, medical, and pharmaceutical processes [8]. In the *Bacillus* group, only two species, *B. anthracis* and *B. cereus*, are acknowledged as opportunistic pathogens in human and animals, while the large majority of *Bacillus* species are considered harmless saprophytes.

*Bacillus cereus* is a nonencapsulated, motile, and hemolytic bacteria. This bacterial group is extensively distributed in the environment and can reside in a broad range of ecological environs, including soil and water. In 1949, Hauge conducted a classical study in Oslo, Norway, that implicated *B. cereus* in food poisoning [9]. *Bacillus cereus* infected 600 patients in 3 health centers and a seniors' institution, with the source identified as a vanilla sauce prepared the day before and left at room temperature in a jar. Hauge detected a thick infestation of big gram-positive rods by direct microscopic inspection of the vanilla sauce. It was in 1969 when the first case of *B. cereus* diarrheal food poisoning in North Carolina with 209 incidents was linked to food [10]. The emetic syndrome of *B. cereus* has been associated with comestibles ranging from vegetables and salads to meat and casserole meals. Rice meals have also been recurrently associated with the emetic effects of *B. cereus*; however, the rationale supporting this association remains unknown. The bacteria typically cause a sort of self-limiting food poisoning, and in rare situations, it is deadly for immuno-compromised individuals.

*Bacillus anthracis* on the other hand has its habitat in the soil and is responsible for the deaths of many animals due to anthrax poisoning. The bacteria are also found in the gastrointestinal (GI) tracts of animals that survived a previous anthrax poisoning. Anthrax is generally regarded as an occupational disease, given that veterinarians, butchers, ranchers, and individuals working mostly and in close proximity with animals are the most affected. *Bacillus anthracis* gains entry into the body using the cutaneous route via abrasions or scratches on the skin. Inhalation has been established as another means by which *B. anthracis* enters in the body; however, this means is not the most common route of entry. Inhalation of *B. anthracis* causes the woolsorter's disease. *Bacillus anthracis* infections are highly fatal (>80%) especially the inhalation form, except if treatment with antibiotics is administered. Consumption of infected food is another potential route of entry and causes gastrointestinal anthrax. This is the most frequent means of exposure to *B. anthracis*, and fatality is between 25% and 60%. Anthrax is not an airborne disease; therefore, the infectious rate is low; nonetheless, there is an exception with cutaneous anthrax, since drains from open sores spread the bacteria easily. In 1958, there were approximately 100,000 cases of anthrax globally; however, the exact figures for incidents in African countries do not exist because of continued challenges in reporting. Moreover, anthrax remains endemic in African and Asian countries even though there are vaccination programs in place. In the aftermath of the September 11, 2001, anthrax became notorious as a suspected bioterrorism weapon [11]. Many individuals were infected with the fatal inhalation of anthrax and many more with the cutaneous anthrax after exposure to a powdery substance dispatched through mails. Consequently, the American postal service has to employ irradiation as a safety measure and has been searching for other means to ensure the safety of its employees.

##### 4.1.1.1.1 Mode of Action

*Bacillus cereus* food intoxication had been linked with two major enterotoxins, and each is supposedly associated with the two major symptoms of the condition [12]. The diarrheal condition is caused by a single enterotoxin; however, antibodies are already in existence to detect the presence of enterotoxin protein. The second enterotoxin that is referred to as the emetic toxin is associated with the emetic symptom and has been challenging to distinguish from others, owing to the deficit of a practical model system.

Cell division of *Bacillus anthracis* cells occurs locally, leading to an acute inflammatory reaction. The capsule then inhibits phagocytosis of polymorph nuclear leukocytes causing the bacteria to thrive and further multiply. The bacteria's exotoxin is discharged topically, and the bacteria spread out rapidly, encroaching on neighboring tissues. This is succeeded by necrotic black lesions, serosanguinous fluid, septicemia, and extensive invasion of tissue. The woolsorter's disease likewise has toxemic and invasive symptoms [13]. The bacterium swiftly transforms into vegetative cells in the respiratory system, producing exotoxin and causing pulmonary necrosis, septicemia, and meningitis in about 24 h. Bloody vomiting, abdominal discomfort, nausea, and diarrhea are all symptoms of gastrointestinal anthrax. To combat the bacteria, it is necessary to administer antibiotics in the initial phases of the infection. Cipro (Bayer Pharmaceuticals) is the main medication used for treatment of inhaled anthrax.

4.1.1.1.2 Clinical Symptoms

Two prominent kinds of illnesses are related with *B. cereus* food poisoning: the diarrheal and emetic illness. The diarrheal illness exhibits abdominal pain and diarrheal within 4–16 h of incubation. However, symptoms often lessen within 12–24 h. The emetic illness is characterized by nausea and vomiting, with a shorter incubation period of 1–5 h.

Anthrax is a disease that is both totemic and invasive, as shown by symptoms outlined earlier. Usually, malaise, fatigue, and at times dry cough ensue from inhalation of anthrax. The infected person often experiences melioration, followed by dwindle in health, which is primarily characterized by breathing difficulty, sweating, and bluish skin discoloration. Approximately 24–36 h after these severe symptoms commence, the infected individual experiences shock and eventually death. In a case of cutaneous anthrax, mild fever, skin lesions, and swelling of the skin occur.

#### 4.1.1.2 *Clostridium botulinum*

*Clostridium botulinum* bacteria are gram-positive, mobile, cylindrical-shaped, and anaerobic spore formers, which produce the botulism toxins. Their usual habitat is the soil, and as such, their spores are commonly found on fresh produce. The spore form is an essential contaminating component that can withstand different states of the environment. *Clostridium botulinum* encompasses a diverse group of pathogenic bacteria: the eight types, which include A, B, C1, C2, D, E, F, and G, all of which have the capability to produce neurotoxins [14]. Each letter represents an anti-genetically distinct type. Types A, B, E, and F are deadly to human beings, while types B, C, and D are deadly to livestock, and types C and E are dangerous to birds.

*Clostridium botulinum* requires oxygen-free environment for survival; hence, packaging of vegetables (such as mushrooms) with plastic wraps for prolonged time should be evaded. The most fatal *C. botulinum* is type A; 25% of its outbreaks result in death, while only about 8% of type B outbreaks are fatal. Although irradiation is useful in food sterilization process, the spores of type A and B can withstand the rays, making the method less effective. Also, radiation kills off some useful bacteria at even low doses, making the use of irradiation in food safety a little complex. *Clostridium botulinum* is restricted by a low water activity and an acidic pH (5). The vapor pressure of the water in food is divided by the vapor pressure of pure water to compute the water activity (aw) in food. The maturation of bacteria is inhibited by a value of 0.95 or more than 30% sucrose or more than 10% salt.

Botulinum (botulism) is coined from the *Latin* word "botulus" meaning sausage, given that this was the initial food related to botulism cases. There are different classes of botulism, depending on the means through which the toxin was acquired. For the category arising from food poisoning, individuals typically take in the toxin with contaminated food. Another category is the rare wound botulism wherein the botulinum toxin is formed in an infected tissue. The infant botulism is the third category, in which *C. botulinum* colonizes the lumen of newborns' intestines.

Food poisoning caused by botulinum toxin is infrequent in the United States [15]. On few occasions, outbreaks occur from homemade meals, especially fruits and veggies. Commercially canned foods have the tendency to initiate large outbreaks; nonetheless, the food industry has developed cautionary steps to impede such occurrences.

In regard to infant botulism, American-produced honey is the only infant food with consistent presence of *C. botulinum* spores. Spores can be inhaled from soil conveyed on children clothes during outdoor activities. Inhaled spores have more effects on children than adults due to the fact that the gut flora of adults consists of competitors of *C. botulinum* while that of infants do not. Spores taken in with contaminated food germinate and produce the botulinum toxin in the intestine.

This spore type is the most significant contaminant since they can endure many environmental circumstances. Soil environs offer the types A and B great possibilities to develop. Growth depends on presence or absence of oxygen; thus, the latter favors the growth of *C. botulinum*. Contamination of uncooked food items generally happens during the growing and harvesting periods of the crops.

Since spore growth and development are hindered at pH 5, acidic pH is usually used in the food industry as a food safety mechanism. Hence, high sucrose (>30%) and salt (>10%) contents are efficient means to limit bacterial proliferation. Hardening meat and meat products with nitrites also prevent growth of the bacteria.

#### 4.1.1.2.1 Mode of Action

Botulism is typically credited to food made with deficient heat (improper sterilization). Botulinum toxin in foods can be deactivated by heating to 80 °C for at least 30 min. Decontaminated cell extracts are noted to have low-level toxicities. However, it seems that intact protein becomes toxic in an event of exposure to endogenous protease. Most of the time, these toxin-protein combinations are found with components of another meal that protect the toxin from advanced degradation by the GI tract. The neurotoxin is absorbed by the upper intestine and carried by the blood to the peripheral neurons. The toxins bind to receptor sites and prevent acetylcholine from completing the reaction, resulting in paralysis.

#### 4.1.1.2.2 Clinical Symptoms

Typical symptoms of botulism include nausea, vomiting, headache, fatigue, and muscular paralysis. These are manifested between 12 and 72 h after consumption of contaminated food. The first indications are general body and muscle weakness, followed by effects on the eyes such as drooping eyelids, delayed pupil response to light, and impaired (blurry) vision. Infected people also experience a dried mouth as well as difficulties speaking and swallowing. These symptoms persist for about

1–10 days, with high death rate. The progression of paralysis of the muscles is enervated by the cranial nerves. Non-administration of treatment in the early stage results in the paralysis and dysfunction of the respiratory organs and eventually death. Treatments include the usage of gastric washouts, emetics, and mechanical ventilation. Horse serum antitoxin is also an alternative treatment. Recuperation from the effects of the bacteria could take several months.

Wound botulism is basically tetanus-like and entails an infection at the injury site, while infant botulism infects toddler $\leq 1$ year old. The symptoms involved in infant's botulism differs from that of adults' food poisoning; at the initial stage, an infected child experiences constipation, succeeded by an evident paralytic reaction. In the subsequent hours or days after the initial symptoms manifest, breathing difficulty ensues, and if not treated appropriately, death occurs. In this instance, mortality is commonly categorized as sudden infant death syndrome. Ordinarily, symptoms are reverted by the use of mechanized ventilating procedures and treatment.

#### 4.1.1.3 Staphylococci

Staphylococci are gram-positive, nonmotile, facultative anaerobic cocci. There are 23 species making up the genus *Staphylococcus*; of these, *Staphylococcus aureus* is liable for majority of staphylococcal food poisoning occurrence. The main feature of *S. aureus* is its possession of coagulase and thermonuclease (TNase). Nevertheless, several intermediary species have both coagulase and TNase and might be implicated in food poisoning. Coagulases are soluble enzymes that cause coagulation of plasma elements, while TNases are heat-stable phosphodiesterases that could split DNA or RNA to 3′-monophosphate nucleosides. *Staphylococcus aureus* is pervasive in nature, owing to the fact that it is a component of the natural flora of the skin, nose, throat, and GI tract of different animals and humans. The bacteria thrive in both aerobic and anaerobic environments and at low water activity ($a_w$ 0.83). However, it does not develop properly in the presence of other bacteria.

The record of *Staphylococcus* and food poisoning links to M. A. Barber recounting memories of food poisoning from unrefrigerated cow milk in the Philippines in 1914 [16]. He succeeded in tracking the implicating substance to cream left at room temperature with a cow with staphylococcal mastitis. Using himself as a test subject, he discovered it was possible to inoculate sterile milk with contaminated milk, inducing an emetic symptom and diarrhea within 2 h. In 1930, Dack et al. deduced that a toxin referred to as enterotoxin was responsible for the milk-related poisoning [17]. The term was used to depict the toxin's effects on the intestine, i.e., its action on the viscera, which results in a nauseating feeling, stomach cramp, diarrhea, and vomiting, typically lasting for a couple of hours. Enterotoxins are primary amino acid molecules that are resistant to proteolytic catalysts and stable when heated up to 100 °C for about 30 min. The enterotoxins A (SEA), B (SEB), $C_1$ ($SEC_1$), $C_2$ ($SEC_2$), $C_3$ ($SEC_3$), D (SED), and E (SEE) are causative agents of *Staphylococcus*-induced illnesses. They impact the intestinal tract and are detected through their response to certain antibodies. About 0.1–1 μg of enterotoxin is needed to induce an illness.

Although staphylococcal food poisoning occurs throughout the world, it is rarely reported due to its comparatively temperate symptoms. Majority of *S. aureus*-related food poisoning events affect a small amount of people, like family members; however, some larger outbreaks have been reported, like the outbreak affecting 1300 persons attending a picnic in Indiana [18]. *Staphylococcus aureus* in the United States has been linked to approximately 241,000 cases of food-borne illness between 2000 and 2008, while in European countries, it was linked to approximately 8% of all food-borne illnesses in 2016 [19, 20]. In 2013, 12.5% of all food-borne bacterial outbreaks in China were linked to *S. aureus* [21]. Staphylococcal poisoning is generally not fatal and can be treated without complications. Fermented sausages, salads, custards, and cream-filled confectioneries are frequent origins of outbreaks.

#### 4.1.1.3.1 Mode of Action

Enterotoxins A and D are responsible for staphylococcal food poisoning; they cause diarrhea by inhibiting water absorption from the lumen. They also have actions on the emetic receptor sites, which induces vomiting. On the other hand, enterotoxin B causes colitis by damaging the epithelium of the intestine. Several strains of *S. aureus* produce hyaluronidase, a hydrolyzing enzyme that acts on the hyaluronic acid of connective tissues and eases spreading of the bacteria in tissues. Staphylokinase (fibrinolysin) is also a by-product of many *S. aureus* strains; its key action is the dissolution of fibrin clots. Generally, *S. aureus* contains lipase, which causes disruption of cell membrane and leucocidin, which causes destruction of leukocytes.

There are three modes of action, namely, the intestinal tract (or emetic response); circulatory system impacts, low blood pressure, decreased urination, excess fluid in the lungs, and blood pooling in vascular plains that lead to shock and death; and others, such as allergic impacts, that occur especially in health facility employees operating with biological samples.

#### 4.1.1.3.2 Clinical Symptoms

The damages caused in the infected person is owing to the consumption of food carrying enterotoxins A and D, which acts on the emetic receptors, resulting in vomiting, and inhibits absorption of water from the intestine, resulting in diarrhea. However, the illness does not observe fever (increase in body temperature). Following oral treatment of broad-spectrum antibiotics, the actions of enterotoxin B can be elicited in people with normal bowel movements. Antibiotics of this sort could exclusively permit antibiotic-resistant, enterotoxin-producing *S. aureus* strains to overgrow.

## 4.1.2 *Infections*

### 4.1.2.1 *Salmonella*

Nearly all recognized species of *Salmonella* (Enterobacteriaceae family) are pathogenic to humans and have their reservoirs in a multitude of livestock. *Salmonella* virulence in humans depends on the strain and how susceptible an individual is; those in a bad state of health are susceptible to *Salmonella* diseases. Approximately 1% of the recognized serotypes of *Salmonella* are linked with a single species, called the adapted host. For instance, in poultry birds, *S. pullorum* and *S. gallinarum* cause diarrhea and fowl typhoid, respectively, although they are seldom seen in other species.

Salmonellosis is an illness that is associated with the *Salmonella* bacteria and sometimes includes septicaemia, typhoid fever, and enteric sickness. Of all salmonella-related illnesses, only the food-borne gastroenteritis type is mostly referred to as salmonellosis. Non-typhoidal salmonellosis caused approximately 90,300 deaths globally in 2015, while approximately 178,000 deaths were associated with typhoidal salmonellosis [22]. It is important to note that this estimation may be greatly underreported. Salmonellae inhabit the gastrointestinal passage of several animals, like poultry, rodents, birds, and others, in which they are either innocuous or morbific. In recent years, commercial poultry and its related habits such as overcrowding and poor hygiene have been the leading source of human infection. Hence, poorly cooked poultry meat could function as sources of food-borne infection. It is established that eggs get infected in the hen as reproduction is initiated; thus, consuming uncooked eggs is quite unsafe. *Salmonella* carriers are uncommon (approximately 1%); in few cases where food dealers and babysitters are suspected to be carriers, these individuals are banned by domestic health agencies from functioning at workplaces until their feces samples reveal negative results.

Most occurrences of food poisoning may be traced back to the ingestion of previously infected food that was poorly prepared. Meat and dairy items, as well as undercooked eggs, cheese, salads, and cold sandwiches, are common causes. Pet turtles were discovered to be a source of salmonellosis in the 1950s and 1960s, with the bacteria being given to newborns who came into touch with the turtles and transferred the infection to their mouths [23]. The infectious dosage of salmonellosis is approximately $10^5$, yet in some instances, fewer organisms cause infection. Salmonellae are destroyed at high temperature of $\geq$60 °C for about 15–20 min or by antiseptic and chlorine.

*Salmonella* bacteria are gram-negative bacilli that do not form spore; however, species such as *S. typhimurium* has a capsule. They are facultative anaerobes and thrive on just about any kind of medium. They also have the capacity to withstand the acidity of the gut and colonize the ileum and colon. *Salmonella* invades the epithelium of the intestine and are phagocytized by macrophages, where they multiply within the cells. While in the macrophage, they have resistance to antibiotics, necessitating several interventions to deter a recurrence. Endotoxins are released during the response of polymorphonuclear leukocytes to the intrusion of the bacteria, leading to the destruction of the bacteria.

##### 4.1.2.1.1 Clinical Symptoms

The bacteria incubated between 8 and 48 h in an infected individual, yielding the following symptoms: fever, diarrhea, nausea, headache, and chills. Self-recovery occurs in roughly 2 days; however, immuno-compromised individuals are more susceptible. Very rarely, enteric fever or blood poisoning develops when the bacteria are generated within the gut or transported through the blood. In these instances, the bacteria infiltrate the gastrointestinal passage, followed by the circulatory system through the lymphoid system, resulting in a systemic disease. Occasionally, complexities involving secondary pneumonia and osteomyelitis may arise and might be fatal (70% mortality rate) if proper treatment is not administered. Antibiotics is used to treat enteric fever; however, for very frequent food poisoning conditions, the use of antibiotics is not encouraged in order to avoid the evolution of resistant strains.

The most severe type of *Salmonella* illness is the typhoid fever. Globally, an approximated 11–17.8 million typhoid fever incidences occurred annually in the past 5 years [24]. These cases are often traced to individuals who undertook recent overseas trips. Humans are adapted host for *S. typhi* (3%), and carriers can sustain the bacteria. *Salmonella typhi* is spread in water and food through contamination with infected feces and sometimes from the hands of carriers (Typhoid Mary). Although Chloramphenicol is the preferred medication for treatment, approximately 15% of strains have resilience to this drug.

#### 4.1.2.2 *Campylobacter jejuni*

*Campylobacter jejuni* and *Salmonella* have been established as the two most common causative agents of food-borne illnesses. *Campylobacter jejuni* is a gram-negative, slender, curved, and motile rod bacterium. It is a microaerophile, meaning it requires low levels of oxygen to survive; oxygen and carbon dioxide of about 3–5% and 2–10%, respectively, are required for optimum propagation. The bacteria are also fairly frail and vulnerable to environmental stressors such as oxygen level of ≥21%, desiccation, high temperatures, disinfectants, and acidic conditions. Based on different surveys, campylobacter infections are found widely in the environment, and similarly to shigellosis, *C. jejuni* has been established to be the foremost cause of acute bacterial diarrhea globally and most likely is responsible for more deaths than *Shigella* and *Salmonella* combined [25]. The bacteria are usually found in healthy cattle, chicken, birds, and flies; however, humans are not usual carriers. It is sometimes found in unchlorinated water and water bodies like streams and ponds. *Campylobacter jejuni* is responsible for campylobacteriosis, which is otherwise referred to as *C. enteritis* or gastroenteritis.

#### 4.1.2.2.1 Clinical Symptoms

*Campylobacter jejuni* infection results in a diarrheal condition, characterized by runny or gluey stool frequently containing invisible blood and fecal leukocytes [6]. Other clinical indications usually associated with campylobacteriosis include fever, abdominal pain and nausea, headache, and muscle pain. Onset of the illness takes place 2–5 days after ingesting tainted food or drinking tainted water and typically lasts between 7 and 10 days; however, relapses occur in about 25% of incidences. In majority of the cases, no treatment with antibiotics is required, as self-healing occurs. *Campylobacter jejuni* typically has a low infective dose; approximately 400–500 bacterial cells are capable to induce sickness in some persons, while a higher count is required in others. The mechanisms via which the bacteria act are not fully comprehended. Asides being an intrusive microbe, it also yields a heat-liablc toxin capable of causing diarrhea. Complications are uncommon, although infections have been linked with labile arthritis, hemolytic uremic syndrome, and septicaemia. An estimation of 0.1% has been established as the case-fatality rate for *C. jejuni* infections, which reveals that one death occurs for every thousand incidences. Fatality is uncommon in healthy persons and most times occurs in cancer patients and other immuno-compromised individuals. Just 20 incidences of bacterial-induced abortion evoked by *C. jejuni* have been recorded by various researchers. Other complications (uncommon ones) associated with *C. jejuni* infections are meningitis, recurrent colitis, acute cholecystitis, and Guillain-Barré syndrome. *Campylobacter jejuni* infections are common in <5-year-old children and adults between 15 and 29 years of age; nonetheless, people of all ages are susceptible to the pathogen. Reactive arthritis, which is an uncommon complication of the infection, is correlated with individuals possessing the human lymphocyte antigen B27 (HLA-B27). Outbreaks typically involve a few individuals (<50), although a large outbreak affecting 2000 people was reported in Bennington, VT, during a period when the town used a unchlorinated water source as its water supply [26]. Various small-scale outbreaks have been documented among children who consumed raw milk while on class excursions to dairy farms. One outbreak has also been linked with eating uncooked clams. According to a survey, roughly 50% of campylobacteriosis is linked with the consumption of improperly prepared or recontaminated chicken or handled chicken. It's the major cause of intermittent bacterial diarrhea disease in America.

### 4.1.2.3 *Clostridium perfringens*

*Clostridium perfringens*, formerly *Clostridium welchii*, are a spore-forming, anaerobic bacteria that are distributed in the soil and are a regular occupant of the intestinal tract of humans and animals [7]. It is differentiated from other clostridia species by its immobility and ability to obtain nitrite from nitrate reduction and ferment lactose in milk. *Clostridium perfringens* is prevalent in the United States and Europe; they are encountered in prepared meals since they possess heat-resistant endospores

[27]. This bacterium develops quickly in moderately hot food as well as the human intestine; in these environs, an enterotoxin is given off.

For nearly all cases, the ill use of temperature on prepared food has been identified as the precise reason of *C. perfringens* food poisoning. The small numbers of the bacteria are usually in meals after preparation and multiply to large numbers (capable of causing food poisoning) during the cooling off and storing of cooked meals. Meats, meat products, and gravy are the commonly involved foods. *Clostridium perfringens* poisoning is one of the major frequently reported food-borne illnesses in the United States, and normally, dozens to hundreds of individuals are affected. There is a possibility that several incidences are not reported, owing to the fact that the contaminated food or patient's feces are not regularly tested for the bacteria or its toxin. Based on CDC reports, an estimate of approximately one million incidences occurs per annum in the United States [28].

##### 4.1.2.3.1 Clinical Symptoms

Incubation of the bacteria occurs between 8 and 22 h subsequent to the ingestion of food contaminated with high counts of the bacteria that have the capacity to produce the food poisoning toxin. The most frequently occurring type of *C. perfringens* infection is characterized by intense abdominal cramps and diarrhea. The illness normally resolves within a day, although some individuals may continue to experience less critical symptoms for about 7–14 days. Fatality is very rare and usually results from dehydration and other complications; hence, only a small number of deaths have been documented.

Necrotic enteritis originating from *C. perfringens*' infection is usually deadly. This condition also originates from the consumption of an enormous amount of the causal agent in contaminated food. Necrotic enteritis or pig-bel syndrome–related mortality is due to the intestines being infected and occurrence of cells death in the intestines as well as the consequent septicaemia. Clinical indications of the disease originate from the intake of very high counts ($>10^8$) of vegetative cells. The generation of toxin by *C. perfringens* in the digestive tract is connected to sporulation.

#### 4.1.2.4 *Shigella*

Kiyoshi Shiga is the Japanese scientist who first identified and named *Shigella* in 1898 during an outbreak in Japan. *Shigella* is a gram-negative, rod shaped, nonmotile bacteria; it has similarity with *E. coli*, particularly its DNA [29]. The genus has following species: *dysenteriae*, *flexneri*, *boydii*, and *sonnei*. *Shigella* is easily destroyed by heat, precisely at 63 °C held for 5 min, and survives only within the pH range of 5–8. It is the causal agent for bacterial dysentery also known as shigellosis. Incidences of shigellosis are less frequent than salmonellosis, and it accounts for just 2% of all documented food poisoning outbreaks. Transmission of the bacteria is normally via direct contact (person to person) or by the fecal-oral route (through contamination of food or water). House flies also play an important role in the distribution of the bacteria on food (especially already cooked meals). Contamination of food and water relies greatly on personal and environmental hygiene. The temperature at which food is cooked plays a vital role in eliminating the bacteria; food prepared and held at low temperature is certain of not destroying the bacteria. Shigellosis is a challenging food-related health issue in places with unhygienic conditions, especially in developing countries. In impoverished African and Asian countries, shigellosis is ubiquitous. Globally, antibiotic-resistant strains of various *Shigella* species and serotypes have also emerged. In addition, overpopulated environments especially small units such as day-care or boarding schools are challenged with frequent outbreaks of the illness [30]. Salads (particularly ones with shrimp, tuna, or chicken), raw oysters, watermelon, pasta, beans, apple cider, cream puffs, and hamburger are the usual food-borne sources.

The bacteria produce a toxin, generally referred to as Shiga toxin. The Shiga toxin is a protein possessing enterotoxin, neurotoxic, and cytotoxic actions. The following harmful health effects have been associated with Shiga toxin:

1. Attack on the nervous system, leading to paralysis and destruction of neurons
2. Attack on the gastrointestinal system, leading to fluid accumulation and dysentery
3. Inhibition of protein synthesis
4. General effects on different cells, usually resulting in cell death

##### 4.1.2.4.1 Clinical Symptoms

Symptoms differ depending on the severity of the infection. Some cases are asymptomatic, while others are characterized with moderate diarrhea or dysentery. Critical incidences might exhibit the following: bloody feces containing with mucus and pus, dehydration, fever, chills, toxemia, and vomiting. The emergence of the illness is between 1 and 7 days; symptoms may continue for some days to weeks. The ileum and colon are the main sites of infection. *Shigella* gets into the epithelium, where it multiplies within the epithelial cells. The bacteria move to neighboring cells causing death of these cells, which consequently lead to ulcers that exude blood. Shigella is highly infectious; about $10^2$–$10^4$ cells can induce an infection. In several cases, the use of antibiotics is not required for treatment, since the illness is self-limiting and resolves within a week to 2 weeks. However, shigellosis is rather devastating in children and aged and immuno-compromised individuals. Multiple occurrences of *Shigella*-induced diarrhea are supposedly a cause for chronic rheumatoid arthritis.

#### 4.1.2.5 *Escherichia coli*

*Escherichia coli* is a gram-negative, lactose-fermenting, non-spore-forming rod bacteria that usually remain harmlessly confined to the intestinal lumen

[31]. Majority have flagella for motility and give off gas from the breakdown of glucose, while a few forms have capsules. It exists as a normal flora of the intestinal tract of almost all mammals. They are passed in feces and easily contaminate water and food. Almost all strains of *E. coli* are innoxious; however, a few are pathogenic to humans. From as far back as 1980, *E. coli* O157:H7 strain has garnered a lot of attention as an important food- and waterborne pathogen and the major causal agent of Shiga toxin *E. coli*-induced human disease such as diarrhea, hemorrhagic colitis, and hemolytic-uremic syndrome (HUS) [32]. Nonetheless, prevailing evidence suggests there is a necessity to gain more knowledge about the O157:H7 strain as well as other *E. coli* serotypes. The six known groups of enterovirulent *E. coli* (EEC group) that are causative agents for human gastroenteritis include the following:

1. Enteropathogenic *E. coli* (EPEC): these are causative agents for infant diarrhea, and usually, transmission occurs through contaminated and unchlorinated well water.
2. Enteroinvasive *E. coli* (EIEC): these are causative for symptoms linked to shigellosis or a type of bacillary dysentery transmissible by water, food, and direct person contact.
3. Enterotoxigenic *E. coli* (ETEC): notorious traveler's diarrhea.
4. Enterohemorrhagic *E. coli* (EHEC, also known as the Shiga toxin-producing *E. coli* (STEC) and previously known as verotoxin-producing *E. coli* (VTEC)): toxin producing *E. coli* O157:H7.
5. Enteroaggregative *E. coli* (EAggEC): this is an increasingly recognized enteric pathogen that causes acute or persistent diarrhea in adults and children in both developed and developing countries.
6. Attaching and effacing *E. coli* (A/EEC): this causes attaching-effacing lesions in intestinal epithelial cells.

#### 4.1.2.5.1 Enteropathogenic *Escherichia coli* (EPEC)

There are several controversies surrounding the sources and prevalence of EPEC due to the fact that food-borne outbreaks are unpredictable. Uncooked beef and chicken are the usual food involved in EPEC outbreaks; however, food exposed to fecal contamination had the tendency to evoke the illness. The disease most frequently linked with EPEC infections is infantile diarrhea [33]. This is based on the fact that infants are the most affected, particularly bottle-fed infants living in third-world countries. Therefore, it can be deduced that contaminated water is usually used in rehydrating infant food formulas. Infantile diarrhea may sometimes be prolonged, resulting in dehydration, electrolyte imbalance, and death. In developing countries, death rates of 50% have been documented [33].

*Clinical Symptoms*

Enteropathogenic *Escherichia coli* infection results in watery or bloody diarrhea linked with the adherence of the bacteria to the intestine and physical transformation of the integrity of the intestine. On the other hand, bloody diarrhea is linked with adherence of the bacteria to the intestine and an acute tissue destruction process, which is possibly brought about by toxin (verotoxin) analogous to the one observed in *Shigella dysentery* [29]. The infection dose for EPEC is supposedly quite low in infants; hence, it is extremely infectious in infants as compared to adults; the dose in adults is assumed to be analogous to that for other pathogens ($>10^6$ total dose).

#### 4.1.2.5.2 Enteroinvasive *Escherichia coli* (EIEC)

Enteroinvasive *Escherichia coli* (EIEC) is capable of producing bacillary dysentery. Based on current knowledge, the foods that harbor this form of *E. coli* is not well-known; however, food or water contaminated with fecal matter from an infected person can lead to the illness in others. A couple of events have been linked to hamburger meat and unpasteurized milk. Enteroinvasive *Escherichia coli* is the enteropathogen that is most frequently isolated from children above the age of 2 years. Although EIEC is considered to be an enteropathogen of concern in terms of diarrheal outbreaks, epidemiological studies estimating the actual burden of disease globally still lack.

*Clinical Symptoms*

The intake of EIEC leads to the microorganisms permeating the epithelial cells in the intestine, resulting in mild dysentery, which is frequently misdiagnosed as dysentery from *Shigella* species [34]. Enteroinvasive *E. coli* is defined by blood and mucus in infected people's excretions. As few as ten organisms of EIEC are believed to be infectious, similarly to *Shigella.* The incubation period for EIEC is usually 12–72 h after ingestion of contaminated food. Symptoms of the disease include pinches in the abdomen, puking, pyrexia, chills, diarrhea, and an overall discomfort. Generally, the mild bacillary dysentery is self-limiting without any known complications. However, in pediatric cases, hemolytic uremic syndrome (HUS) is common.

#### 4.1.2.5.3 Enterotoxigenic *Escherichia coli* (ETEC)

ETEC class consists of a comparatively small portion of *E. coli* species that has been linked worldwide with diarrheal illnesses within the various age groups. ETEC often times cause diarrhea in individuals living in industrial hubs and toddlers in under-developed regions. It's been almost two decades since the pathogenesis of this cholera-like disease has been understood, and ETEC is not regarded as a severe food-borne illness in modernized countries due to their high hygienic standards and practice. In less modernized regions of the world, contamination of water with human fecal matter could possibly result in the contamination of food [34]. ETEC are not frequently found in dairy products like semisoft cheese, and in most instances, infected food handlers are possible sources of contamination. A comparatively large dose ($10^8$–$10^{10}$) of ETEC is likely required in order to overtake the small intestine, where the bacteria multiply, and produces toxins that bring about secretion of fluids. Depending on the infectiousness of the dose, the incubation period of the bacteria may differ; high infectious dose normally induces diarrhea within a day post consuming polluted/contaminated food. For infants, a lesser count of bacteria will give rise to the illness. Moreover, ETEC has been noted to result in 30–60% of traveler's diarrhea [34].

*Clinical Symptoms*

The usual clinical indication of ETEC infection includes pinches in the abdomen, puking, obnoxiousness, low pyrexia, chills, runny diarrhea, and an overall discomfort. All these are normally self-resolved within a short period of time. However, it is essential for infants or weakened aged people to obtain proper electrolyte replenishment therapy.

#### 4.1.2.5.4 *Escherichia coli* O157:H7 (Enterohemorrhagic *E. coli* or EHEC)

Enterohemorrhagic *E. coli* (EHEC) has been noted to have similarity with *Shigella* in regard to the kind of toxin produced by both organisms [29]. Shiga toxin given off in *Shigella dysentery* type 1 and the toxin produced by *E. coli* O157:H7 instigated researchers to label the *E. coli* O157:H7 toxin as Shiga-like toxin or Shiga-toxin-producing *E. coli* (STEC). It was also termed verotoxin by other researchers since it could destroy Vero cells in identified tissue cultures of the African green monkey's kidneys [35]. The relationship between STEC serotypes and other adverse health conditions has been established and considered a critical concern for human well-being and food safety. Shiga-toxin-producing *E. coli* is known to be associated with HUS in 5–10% of patients with STEC infection; also, an association with kidney failure is observed in 5% of patients. Approximately 50% of STEC patients suffer from prolonged renal dysfunction, with close to 5% death. Infection can cause thrombotic thrombocytopenic purpurea (HUS with neurological consequences), which is potentially fatal [36]. Shiga toxins have been shown to penetrate the gut's epithelial cell barriers and target microvascular endothelial cells in the intestine, kidney, and central nervous system. The development of a functional inhibitor of Bcl-2, which plays a key role in cell death by subduing cells and allowing them to succumb to apoptosis pathways, might be the cause of cell death. STEC infection may affect anybody, but it is more dangerous for newborns, the elderly, and the ill. The average infective dose ranges from 50 to 100 bacteria; however, in few instances, as little as 10 bacteria can establish the illness. In addition to the leading sources of exposure (contaminated raw meat, raw milk, and ruminant products), cross-contamination of other foods like bean sprouts or apple juice is becoming a growing concern [37]. *E. coli* O157 may survive for long periods of time in water at extremely low temperatures and can adapt to acidity. Frequent routes of transmission include hand-to-hand contact and flies perching.

The damaging effects of *E. coli* O157:H7 on a host have been attributed to three sets of genes identified in the bacteria. These genes have the capability to yield Shiga-like toxins, which binds to the lining of epithelial cells of the intestines and has a plasmid. These two Shiga toxins, Stx1 (VT1) and Stx2 (VT2), are encoded on a bacteriophage and may be easily transferred from one creature to another by the virus. The use of antimicrobial drugs to stimulate O157 to produce additional toxins may be justified because of the bacteriophage-encoded Shiga toxins. 4-Fluoroquinolones are examples of complexes that are capable of inducing bacteriophages. Treatment with antibiotics has been linked with the development of HUS in children; these occurrences suggest a clear scenario of contraindication.

The binding gene in *E. coli* may transfer from one *E. coli* bacterium to another, and genes on the plasmid can also conjugate to other plasmids. Therefore, the risk of genes' movement worsens the severity of STEC infections. In several countries including the United States, different researchers have identified non-O157 serotypes that are capable of producing toxins; these serotypes have also been linked with severe human illnesses. The number of known non-O157 serotypes is considered to be moderate, given that only a fraction of the small number of laboratories investigating new sources of Shiga toxin is interested in identifying non-O157 serotypes. The serotypes O111, O26, O145, and O103 have been identified in the United States.

In the quest of discovering *E. coli*-related food-borne pathogens, the search for O157 serotype only may not be sufficient to provide food manufacturers and health practitioners with adequate information to ensure food safety and good health. However, it is not expected that all STEC must be discovered, since not all forms might have critical health concerns and the expenditure involved in these researches is quite enormous. Moreover, research aiming to further investigate and provide in-depth knowledge on gene transfer and its effect on human health is a dire need in terms of food safety and health discovery.

Given that exposure to STEC results in destructive outcomes, the possibility of more STEC-related non-O157 serotypes is very critical and depicts a major threat to vulnerable populations. Thus, there is a necessity to improve diagnosis, provide efficient means to contain distribution of the organism, and also enlighten consumers and health practitioners about the organism. Raw or poorly cooked beef has been involved in several reported outbreaks; sprouts, unpasteurized fruit juices, dry-cured salami, lettuce, game meat, and cheese curds have also been involved in *E. coli* O157:H7.

*Clinical Symptoms*

The clinical indication of this illness includes severe cramping, occasional vomiting, low-level fever, and diarrhea (which is watery at first and turns bloody over time). In some cases, fever might be completely absent, and some individuals may pass out only watery diarrhea. These symptoms persist for about 8 days, and self-healing normally occurs without the need of serious health therapy. In infants, clots of blood may block coiled tubes in the kidney, causing a disruption of normal kidney functions. This condition may be irreversible and might necessitate dialysis treatment. In severe situations, a comatose state occurs, and eventually, the child dies. Hemorrhagic colitis is the most grave disease caused by *E. coli* O157:H7. It is not a frequently occurring infection; however, its documented incidence is not a true reflection of its frequency. In some regions like the Pacific Northwest, *E. coli* O157:H7 is considered the second after *Salmonella*, as a cause of bacterial diarrhea. Due to the glaring evidence of profuse and visible blood in critical cases, infected individuals quickly seek medical help; in comparison, there are fewer severe cases and numerous regular ones.

#### 4.1.2.6 *Listeria monocytogenes*

*Listeria monocytogenes* is a gram-positive, nonencapsulated short rod that was first described in 1923 under the name *Bacillus hepatis*. It is also a facultative intracellular parasite. *Listeria monocytogenes* is sometimes mistakenly identified as streptococci due to its coccobacillary form, β-hemolytic feature on blood agar, and a proclivity toward growing in short chains. Water, soil, and sewage have been established as the usual habitat for *L. monocytogenes*, although inappropriately pasteurized food products have also been noted to be linked with the organism [38]. *Listeria monocytogenes* has been found in the gastrointestinal passage, vaginal tract, and throat of humans and animals. It multiplies and survives in these environments for long period of time; however, it's killed by antiseptic chemicals and heat treatment. The disease listeriosis is found mainly in animals, although some severe illnesses can also be experienced in human fetus, newborns, pregnant women, and ill individuals [38, 39]. Exogenous infection occurs via an infected female genital organ, in the uterus, or during the process of child birth, resulting to disease in the fetus or newborn.

##### 4.1.2.6.1 Clinical Symptoms

The organism gains entry to the circulatory system, where it is taken up by macrophages, in which cell multiplication occurs, eventually leading to septicemia. As the bacteria cell multiplies, hemolysin is produced and causes lysis of the macrophages by destroying their cell membrane. The bacteria span from one cell to another and sometimes are conveyed to the meninges, resulting in meningitis. Additionally, the bacteria enter the liver and spleen, causing several boils and granulomas. Listeriosis infection in expectant women may end in loss of the pregnancy. Treatment is usually inefficacious; however, early treatment has a possibility of success.

#### 4.1.2.7 *Vibrio*

*Vibrio* is a genus made up of gram-negative, arched, and encapsulated rod bacteria possessing a singular polar flagellum for motility and ubiquitous in aquatic environments [40]. They are also facultative anaerobes, and depending on somatic polysaccharide antigens, they are classified into six subcategories. One of the most pathogenic species of the genus is *Vibrio cholerae*; it is the causative agent of the dreadful cholera. Cholera is endemic in regions that have poor water and sanitation, as well as poor hygiene infrastructures [41]. The illness is associated with a toxin produced by the organism, known as cholera toxin or choleragen. *Vibrio cholera* is found in plankton and shellfish and can survive in salty water for a lengthy period of time. Cholera is transmitted through ingestion of contaminated water and/or food particularly in vulnerable communities such as those affected by famine, poor water and sanitation infrastructure, war, and natural disasters [41]. Transmission is largely through fecal-oral route via the ingestion of contaminated water, uncooked or undercooked seafood, and/or consumption of raw fruits and vegetables that were fertilized with contaminated feces. About $10^{11}$ bacterial count has been established as the dose at which the illness manifests. *Vibrio cholera* is inactivated by boiling food or water for at least 10 min.

##### 4.1.2.7.1 Clinical Symptoms

Infected individuals undergo a sudden onset of runny diarrhea, puking, and abdominal cramps within some hours to days after intake of the bacteria. There is a heavy release of body fluids, about 20–30 times more than the usual amount of body fluid loss. The organism invades the mucosal surface and quickly multiplies on it. The mucosa of the intestines gets shredded due to impairments of the intestine, leading to the release of bits of mucus from intestinal cells; these mucus bits have resemblance with rice grains, resulting in the derivation of the term rice-water stool. Death is less than 1% if fluids are replaced promptly; however, if the illness remains unattended to, 60% of infected individuals goes into a coma and eventually dies. Around a core polypeptide A component, the choleragen is made up of five polypeptide B subunits. The B subunit binds to ganglioside receptors, whereas the A subunit crosses the cell membrane and activates the adenyl cyclase system, resulting in c-AMP production. Enhanced intracellular c-AMP induces the entry of fluids and electrolytes into the intestinal lumen, consequently resulting in the hyper-secretion of fluids and chloride, the inhibition of sodium absorption, and diarrhea. The toxin also adheres to ganglioside receptors of epithelial cells. Other toxins with less well-understood functions are also produced.

#### 4.1.2.8 *Yersinia enterocolitica*

*Yersinia enterocolitica* is a member of the genus *Yersinia* in the family of Enterobacteriaceae and a pleomorphic, gram-negative organism (sometimes ovoid and sometimes rod shaped). At 37 °C, the organism is nonmotile, but when cultivated at 30 °C, it produces peritrichous flagella, which allows it to move. It is one of only four pathogenic species in the *Yersinia* genus and the causative agent of diarrhea and is commonly found in domestic animals. Human outbreaks have been linked to tainted milk, water, and a variety of foods, with pigs serving as the primary source of the bacterium [42].

*Yersinia enterocolitica* has an invasive outer-membrane protein with a high molecular weight that facilitates cell adhesion and penetration. Lipoprotein factors that impede phagocytosis are also produced. Several strains produce a heat-stable enterotoxin that is related to *E. coli*'s Shiga toxin.

##### 4.1.2.8.1 Clinical Symptoms

Fever, stomach discomfort, headache, and vomiting, as well as peripheral neutrophilia, are all symptoms that are comparable to those seen in acute appendicitis. Acute watery diarrhea is commonly associated with the illness. Incubation period is between 24 and 36 h and lasts for 1–3 days.

## 4.2 Mycotoxins

The term "mycotoxin" was created and first used near London, England, in 1962, when close to 100,000 turkey poults died following an unusual veterinary crisis [43]. Mycotoxins are secondary metabolites emanating from fungi and the potential to be detrimental to vertebrates when consumed or when they come into contact with the human skin. Molds and humans share a love-hate relationship. Certain molds have been used as a source of food for humans (such as the ripening of cheese) and in medicines [44]. Mold development has been known to cause physical alterations for quite some time. Extracellular proteolytic and lipolytic enzymes are found in molds, and they catalyze the processes that cause food to soften. Mold development in food alters the taste, odor, and appearance of the product. Molds, particularly filamentous fungus, generate compounds (mycotoxins, *myo* meaning fungal) that can cause cancer, estrogenic effects, mutagenicity, and teratogenicity in humans and animals.

The Food and Agriculture Organization (FAO) of the United Nations estimated that approximately 25% of the cereals produced in the world are contaminated by mycotoxins; other nourishments, like nuts, condiments, produce, and their by-products, can too be sullied by these contagious fungal metabolites [45]. Various stages of the food chain are involved in the production of agricultural mycotoxins, including during pre-harvest, harvesting, and drying, as well as during the storage period. Mycotoxin production from fungal growth can be enhanced by poorly managed agricultural and harvesting practices, improper handling, packaging, and storage, as well as poor logistical conditions. When food is processed, food commodities are held under conditions that prevent fungal contamination and growth as well as possible mycotoxin production. This is especially true when water content is low enough to prevent mold growth and mycotoxin production. To have mycotoxin-free products, this is crucial. In addition, when the water activity of processed and stored food products increases to levels allowing fungal growth, mycotoxin production will also develop.

Mycotoxins can be decreased in meals by the following methods: (1) oil refining, (2) grain milling, (3) mold prevention (sodium bisulfite, sorbate, propionate, nitrate), and (4) ammonia and ozone treatment of cereals. Many mycotoxins are heat and regular cooking stable. Some aflatoxins are destroyed by moist heat, such as when peanuts are roasted. In the United States, refining techniques are efficient in eliminating aflatoxins; nevertheless, in the Orient, the flavor of crude peanut oil is favored because of its taste, despite the presence of substantial mycotoxins. Mold development is prevented by using forced air-drying methods.

### *4.2.1 Aflatoxin*

Initially, the words mycotoxin and aflatoxin were frequently interchanged, owing to aflatoxin's status as the most well-known and significant mycotoxin in food. Nonetheless, in recent times, mycotoxin has become a generic term used to define any toxin of fungal origin. Additionally, these mycotoxins of superior agro-economic importance referred to as aflatoxin are generated by *Aspergillus flavus* [43]. Corn, cottonseeds, nuts, and figs are all greatly affected by the mold. Aflatoxicosis affects the majority of animals, including poultry and people. The mouse is also totally immune to the carcinogenic effects of aflatoxin. A clear link exists between dietary aflatoxin and liver cancer in underdeveloped nations, with human males being more vulnerable [45]. In India and Africa, where people have been forced to consume moldy grains to survive, the number of cases has been particularly high. Within 3 weeks of consumption, acute toxicity might develop. According to epidemiological statistics, individuals must be exposed to aflatoxin B1 (AFB1) and hepatitis to have a substantial risk of cancer [46]. Hepatitis B and liver cancer rates in the United States are both low.

Aflatoxins are coumarins that have been heavily modified. B1, B2, G1, and G2 are the four primary kinds of aflatoxins; each nomenclature is dependent on whether they glow blue or green under UV light. A cyclopentanone ring is found in the B-group, whereas a lactone ring is found in the G-group. The most powerful carcinogen is AFB1 ($LD_{50} = 0.3$–$9.0$ mg/kg), followed by G1, B2, and G2 in that order. The potent carcinogens are among the metabolites triggered by cytochrome P450 to 2,3 epoxides of aflatoxin, not the original parent molecules. These chemicals are strong electrophiles that have the potential to attach to proteins and DNA covalently. Aflatoxin M1, a 4-hydroxyl precursor of AFB1 present in cows' food, can be detected in dairy and may pose a health risk. In some animal species, AFB1 is a strong carcinogenic substance. In humans, epidemiological research on various regions of the globe has found a link between aflatoxin levels in the diet and the risk of primary hepatocellular carcinoma. Aflatoxin pollution may be a key pathogenic component resulting in synergistic effects with hepatitis B virus exposure in many of the same places across the world where the incidence of hepatitis B virus is also linked with liver cancer. In AF samples, AFB1 is sometimes found to be the highest component in the mixture. As per a report from the European Food Safety Authority (2007) that pooled 34,326 samples from EU members (including export samples, import samples, company samples, and a marker control sample), Brazil nuts (pistachios) and spices (Limit of Detection (LOD) was 0.1–0.2 μg/kg for AFB1 and 0.2–0.4 μg/kg for total AFs) had the highest percentage of positive samples [47]. There was a significant increase in the mean and maximum values of Brazil nuts and pistachios compared to all other food groups. In addition, figs, peanuts, spices, hazelnuts, and almonds consistently scored 97.5th percentile values above 2 μg/kg for AFB1 and more than 4 μg/kg for total AFs. In the majority of food categories, other than baby foods and maize, the maximums values were fairly high. The carcinogenic potency of AFM1 is about half that of AFB1.

Aflatoxins are heat-stable and easily transformed to toxic products. Treatment with ammonia reduces and inactivates aflatoxins. Lactic fermentation at pH <4.0 results in the conversion of AFB1 to AFB2a, which is less toxic. Other environmental conditions, such as the presence of organic acid, also irreversibly convert AFB1 to aflatoxicol B, which is 18 times less toxic than AFB1.

The acute toxicity of aflatoxins, as well as their propensity to cause cancer, is two major concerns. Aflatoxin has a lethal dose of 0.5 mg/kg of body weight, and death occurs within 72 h. Hepatocellular injury and peritoneal cavity hemorrhage are the causes of death. Animals become more resistant as they get older. Aflatoxin is very poisonous when compared to a chemical like lead, which has a toxicity of around 500 mg/kg of body weight. When a sublethal dose of aflatoxin is consumed over several days to weeks, it causes moderate to severe liver damage. Biliary hyperplasia, or increased cell proliferation in the bile duct area of the liver, is one type of hepatic injury. Fat buildup and a shift in the color of the liver from purple-red to yellow-red are also common.

### *4.2.2 Ochratoxin A*

The microflora *Aspergillus ochraceus* produces ochratoxin A (OTA), which is present in cereals and is promoted by moist circumstances and moderate temperature. In European Union nations like Bulgaria and Romania, as well as the former Republic of Yugoslavia, the mycotoxin has been linked to Balkan endemic nephropathy (BEN), notably in meals produced from wheat collected after heavy rain. It may be found in cereals, soybeans, peanuts, and cheese and is thus regarded as a significant mycotoxin in terms of food safety [45]. The mycotoxin was discovered in Danish pigs a few years ago, and it spread to ham tissues sold to other nations. In birds, fishes, and mammals, the OTA compounds ($LD_{50}$ of 20–50 mg/kg) induces nephrotoxicity, as well as teratogenic effects in chicken and rats. The central nervous system is also targeted by ochratoxin. The structures of OTA compounds comprise a dihydroisocoumarin derivative connected to phenylalanine via an amide bond, as well as a chloride atom in certain cases [48].

The Scientific Co-operation on Questions relating to Food (SCoOP) report (2002) stated more than 50% of positive samples for cacao and derived products (81%), win (59%), dried fruits (73%), cereals (54%), and spices (52%). The percentage of positive samples depended largely on LOD levels [49]. On the other hand, dried green coffee, fruits, and spices had the highest mean levels of contamination. Even though no data were provided regarding OTA contamination on cocoa beans, results of the processed products revealed that cocoa beans were a raw material prone to OTA contamination. Among an assortment of condiments, coriander, nutmeg, pepper powder, and paprika were the most common and highly contaminated ingredients. Dry fruit statistics, in particular products derived from vine fruits, are reportedly highly susceptible to contamination with OTA as a result of the drying process. It has been confirmed that green coffee is a raw material that can be contaminated with OTA.

## *4.2.3* Fusarium *Toxins*

### 4.2.3.1 Fumonisin

Food commodities contaminated the most with fumonisin B are wheat and maize during the milling process and the resulting milling fractions. As part of the food industry, grits (mean 347 g/kg) and flour (mean 408 g/kg) are often used to further modify ingredients before they are consumed in the final product [50]. In contrast, sweet corn is less likely to be contaminated by *Fusarium* than other maize varieties. It has been proven that chemical processes, such as extrusion, can reduce fumonisin B levels to some extent. In general, cornflakes, for example, have low levels of fumonisin B.

### 4.2.3.2 Zearalenone

Zearalenone can be discovered in moldy grains (such maize and wheat) in periods of high humidity and moderate temperatures [51]. EFSA reports that there is a significantly higher frequency of ZEN in maize (33%) with a mean content of 15 μg/kg among grains intended for human consumption [51]. Grain mill products followed the same trend, even though wheat bran (33 μg/kg) had extremely high levels.

Consequently, ZEN levels and prevalence in the analyzed cereal products for human consumption were relatively low, with the exception of oils derived from vegetables (specifically maize germ oil) with approximately 86% positive samples and a mean level of around 72 μg/kg [51].

Generally, ZEN is being redistributed between different fractions of milling. By-products (such as dust, hulls, and others) from the process of cleaning the raw cereal grains were then characterized according 3-to-30-fold higher ZEN concentrations than the cleaned cereal grains, while bran contained up to twofold higher concentration levels. ZEN is generally not affected by the cooking process. We observed more than 40% reduction only when alkaline conditions were used or when extrusion cooking was performed.

#### 4.2.3.3 Trichothecenes

In 1913, Russians (near Siberia) were compelled to consume millet, wheat, and barley that had been stored outside throughout the winter due to a severe famine. The melting snow increased the moisture or dampness content, which aided mold development and resulted in a massive mycotoxicosis outbreak. The illness, known as alimentary toxic aleukia (ATA), spread like wildfire and was connected to *Fusarium* species infection of cereals [50]. Alimentary toxic aleukia causes bone marrow shrinkage, agranulocytosis, necrosis, ischemic chest pain, septicemia, and eventually death. The illness manifests itself in three phases: (1) swelling of the mouth and throat, diarrhea, and puking; (2) an asymptomatic second phase when immunodepression sets in; and (3) a deadly stage with pinpoint skin bleeding and necrotic ulcers in various regions of the body. The immediate symptoms of poisoning are often neurological, while the chronic effects of toxicity are characterized by massive gastrointestinal tract necrosis, which causes swelling, bleeding, and low count of the white blood cells. Trichothecenes are a collection of over 80 sesquiterpenes that are by-products of 12,13-epoxytrichothecene and are the major mycotoxins generated by *Fusarium*. T-2 toxin, vomitoxin (deoxynivalenol), HT-2 toxin, neosolaniol, and diacetoxyscirpenol are the most common trichothecenes. These chemicals can be deacetylated, hydroxylated, and glucuronidated in the liver and kidney in vivo, potentially lowering the danger of eating trichothecenes. The acute toxicity ($LD_{50}$) ranges from 50 to 70 mg/kg. The potency is due to the presence of an epoxide ring in their formation. Trichothecenes cause toxicity by inhibiting protein synthesis at the initiation, elongation, and termination stages [52].

In barley, oats, sorghum, rye, and safflower seeds, the contamination concentrations of vomitoxin range from 1 to 20 ppm. In both animals and humans, vomitoxin induces loss of appetite and emesis. Human poisoning incidents caused by vomitoxin contamination of grains are a serious issue for several nations. Vomitoxin has a fatal dosage of 50–70 mg/kg.

T-2 toxin has neurobehavioral impacts, which are cytotoxic, and causes skin necrosis, bleeding, and edema. Meal refusal, lack of appetite, and depression are examples of neurological dysfunctions. The hematological system is the most severely affected, with fast drops in counts of white blood cells and platelets of 10–75% and significant cellular destruction in the bone marrow, spleen, and lymph nodes. T-2 toxin is an immunosuppressant that causes injuries in the lymph nodes, spleen, thymus, and bursa of Fabricius. T-2 toxin may be present in barley, corn, oats, and wheat; however, it is less common than vomitoxin. T-2 toxin, on the other hand, is more dangerous ($LD_{50}$ in mice is 2–4 mg/kg).

*Fusarium* produces fumonisins, fusarochromanones, zearalenone, fusarins, moniliformin, and wortmannin, which are non-trichothecene mycotoxins. Animal toxicoses, such as equine leukoencephalomalacia (ELEM) and swine pulmonary edema, are linked to fumonisins in both animals and humans. Corn is the most common dietary source for mycotoxin; in a 1990–1991 investigation, fumonisin B1 and B2 were found in substantial quantities in 124 commercial specimens of corn-based meals from the United States and Africa [53]. Corn exposure has been linked to a significant risk of human esophageal cancer; however, the liver is the primary source of damage, culminating in hepatocellular carcinoma and nephritis.

#### 4.2.3.4 Emerging *Fusarium* Mycotoxins

In-depth surveys reveal the far-ultraviolet spectrometer (FUS) and ENN contamination of raw and processed food products remain scarce and largely limited to the Mediterranean and northern Europe. There has been a report of beauvericin (BEA) or ENNs in the past several decades in a number of European countries including Croatia, Italy, Finland, Norway, Spain, Slovakia, and Switzerland. Various literature investigated grain cereal samples such as barley, maize, rice, oats, and wheat, with few including cereal product samples such as bread as well as baby food products. Moreover, several surveys have determined that ENNs are highly prevalent. Across southern, central, and northern Europe, the sums of BEA and ENNS have recently been found in high concentrations (up to 10–500 mg/kg) in wheat, barley, and corn. There have been reports of high concentrations of ENNs in Spain and northern Africa (maximum concentration of enniatin A 814 mg/kg in rice in the Spanish market). Northern Europe had significantly lower concentrations (maximal enniatin B in wheat from Finland was 18.3 mg/kg) [54]. Enniatin B was the most commonly found mycotoxin in pastas and baby foods, and its concentrations ranged from <LOQ to 106 and from <LOQ to 1100 μg/kg, respectively. It was found that multigrain cereals and baby food contained high levels of these mycotoxins (70,3%). In regions of cooler climates such as southern Europe and Morocco, BEA is not significant in grains occurring at concentrations of tens of mg/kg [54]. On the contrary, a maximum BEA level of 844 μg/kg in a Tunisian sample of wheat pasta was observed in a study of cereal-based products from the Mediterranean country. There have been fewer investigations into FUS. It appeared as high as 19.6 mg/kg occasionally in rice from Morocco, while its natural occurrence did not seem to be widespread in cooler climates. Studies available on the presence on these mycotoxins in grains and foods suggest a continuous low level of exposure to these toxic metabolites may exist. Thus, the toxicological impact of mycotoxin

mixtures as well as other fusarium mycotoxins should be further investigated, including butenolide, chlorine compounds, chlamydosporol, ENNs, fusarochromanones, fusarin C, and wortmannin.

### 4.2.4 *Penicillia Mycotoxins*

Derived from the *Penicillium* mold, *penicillin* is still one of the first and the most widely used antibiotic agents. *Penicillium notatum* is the bacterium that produces the widely used antibiotic penicillium, which efficiently inhibits bacterial cell wall production.

#### 4.2.4.1 Patulin

Patulin is a secondary metabolite produced by several spoilage fungi species including and [55]. It is a mycotoxin produced by *Penicillium expansum*. The mold and mycotoxin are produced when the fruit deteriorates. Patulin is toxic to gram-positive and gram-negative bacteria, and at some point, it was considered for use in the medical fraternity as an antibiotic, but its role in causing disease in animals and human remains unclear. However, patulin has raised some public health concern as it has been reported to possess some potential carcinogenic properties. Patulin has been shown to induce tumors and thus considered to be very toxic ($LD_{50}$ of 15–35 mg/kg). In alkaline conditions, the mycotoxin is unstable but stable under acidic conditions, which can be inactivated by adding ascorbic acid.

#### 4.2.4.2 Rubratoxins

*Penicillium rubrum* is a mold that produces rubratoxin, a mycotoxin that is commonly linked to cattle illnesses. Rubratoxin is divided into two types: A and B. An illness in cattle and swine was discovered to have been caused by eating moldy maize wherein death of the swine takes place approximately 24 h after consuming roughly 0.5 pound of moldy maize ($LD_{50} = 6.6$ mg/kg of body weight, which causes liver and kidney damage) [56].

#### 4.2.4.3 Yellow Rice Toxins

Yellow rice toxins (such as citrinin and citreoviridin) are mycotoxins generated by penicillia, a fungus that grows on rice, especially rice that has already been milled. Citrinin has an $LD_{50}$ of 50 mg/kg, and it affects the kidneys, inducing tubular injury. Citreoviridin, meanwhile, has an $LD_{50}$ of 20 mg/kg and leads to respiratory disorders and paralysis of hind limbs.

### 4.2.5 *Ergot Alkaloids and Other Mycotoxins*

Mycotoxicoses have historically been known. Ergotism, induced by *Claviceps purpurea*, first appeared in Europe during the Middle Ages, in the 1400s. The intake of ergot (fungus called *Claviceps purpurea*) that grows on rye causes drunkenness as well as hallucinations, disorientation, and convulsions, ultimately leading to arteriolar spasms and gangrene [57]. The gangrenous results are attributed to alkaloids, which are partial-adrenergic agonists and promote constriction of blood vessels, and the hallucinating effects are due to derivatives of the hallucinogen lysergic acid, the most significant of which are ergonovine (ergometrine) and ergotamine [57]. Each of these chemicals is pharmacologically functional. Ergotamine causes blood vessels to constrict, and it has historically been used in medicine to alleviate headaches. Ergotamine, on the other hand, can induce gangrene if used excessively. Ergonovine strongly effectuates uterine constriction. Ergot alkaloids were first connected to ergotism, often described as St. Anthony's fire or the holy fire, in the 1850s. In 1951, an incident in Pont St. Esprit, France, afflicted a large section of the populace after consuming homemade bread in which ergot-contaminated flour was used.

*Claviceps purpurea* is a prevalent pre-harvest grain mold that develops on grasses and cereals' ears. Between 10 and 30 °C and high relative humidity, ergot production is encouraged. Sclerotia are produced during the hibernating period and finally turn out in the wheat grain after harvest. Ergotism has virtually been eradicated as a modern-day issue; nevertheless, a handful of instances were documented in Ethiopia from 1977 to 1978, and the ergot alkaloid may still be present at low levels in rye [58]. Ergot is not a modern health concern to humans or animals because of quality assurance mechanisms in place.

Ergot alkaloids cause peripheral arterial constriction and neurological problems by affecting smooth muscles. The early symptoms include severe prinking in the limbs, as well as heat and cold feelings. The gangrenous impacts of α-adrenergic ergot agonists are caused by lengthy and severe peripheral vasoconstriction. Itching, muscular cramping, psychiatric problems, puking, migraine, numbness, spasms, and convulsion are all indications of the central nervous system attacks [59].

## 4.3 Plant Toxicants

Roughly 700 types of around 30,000 types of North American plants are viewed as toxic. Toxic species are found all through the kingdom Plantae, in ferns, algae, angiosperms, and gymnosperms. Several well-known plant toxins exist, ranging from the irritant formic acid (commonly found in nettles and ants) to more toxic compounds like atropine found in deadly nightshade berries (*Atropa belladonna*), cytisine in laburnum, and coniine in hemlock.

In certain circumstances, groups of genera inside a single family display comparable toxicity. Conversely, the toxic nature of a few plants of the same genus varies greatly. For example, *Solanum nigrum* (the nightshade) is recognized to be very toxic, while *Solanum intrusum* is regarded as nontoxic. *Solanum intrusum* is known as nursery huckleberry (or wonder berry), and it is often prescribed to home gardeners for its palatable organic products. In another example, *Euphorbia pulcherrima* (also known as poinsettia) was reported as the cause of human deaths in the early 1900s in Hawaii. Manipulation by agriculturist and horticulturist for differently colored and longer-lasting poinsettia could be the reason for their loss in toxicity, as indicated by lack of toxic effects in rodents that were given huge amounts of the red bracts.

The idea that possibly harmful chemicals may be present in ordinary meals is hard for the average individual to comprehend, even for some learned persons. From a sentimental point, food is considered as something that supports life; therefore, it ought to be wholesome and unadulterated and occasionally have a spiritual ambiance. As a result, most people are surprised to learn that plants and certain animals used for food may create a variety of potentially hazardous compounds [59]. The large array of toxic chemicals produced by plants (phytotoxins), usually referred to as secondary plant compounds, are often held to have evolved as defense mechanisms against herbivorous animals, particularly insects and mammals [59]. These compounds may be repellent but not particularly toxic, or they may be acutely toxic to a wide range of organisms. They include sulfur compounds, lipids, phenols, alkaloids, glycosides, and many other types of chemicals. Many of the common drugs of abuse such as cocaine, caffeine, nicotine, morphine, and cannabinoids are plant toxins.

Toxic dietary elements of plant source are difficult to characterize. Several plants create metabolic products (waste chemicals) or secondary compounds that, if ingested in large enough amounts, might have negative consequences. A few are natural low-molecular-weight poisons, whereas others are secondary metabolic products. With species having varying characteristics (like flavors, pigments, and/or protective compounds), secondary metabolism is also species-specific. Secondary metabolic products can be classified into a number of categories, namely, carcinogens, growth inhibitors, mutagens, neurotoxins, and teratogens [59]. Testing for these secondary metabolic products is assumed to be costly, and without any mandatory government regulations, most are not tested for. In the plant kingdom, many phytochemicals are produced as secondary metabolites, e.g., metabolic by-products of metabolism, excretion, and elimination. Through evolution, some of these secondary metabolites have become important defense chemicals used by the plant against insects and other organisms [59]. Biochemically, the plant's weaponry is highly advanced. Chemicals called primary metabolites perform significant functions in physiological plant activities as photosynthesis, lipid energy, and nucleic acid metabolism, as well as biosynthesis [59]. Secondary metabolites apparently meliorated in reaction to and interactions with animal and plant kingdom creatures, as well as herbivores and diseases. The by-products of basic plant metabolism include macro- and micronutrients. Latest advancements in genetically engineered crops have taken use of this understanding to create plants that can protect themselves from diseases and predators in improved ways.

Although evolved, this evolution was not specifically for human consumption as only a few species of plants are used for human consumption, with the majority toxic for human consumption. Sinclair and co-authors theorized that humans were initially carnivores until they gained knowledge on how to use fire to cook plants to remove or deactivate their toxicity [60]. Many chemicals that have been shown to be toxic are constituents of plants that form part of the human diet. For example, the carcinogen safrole and related compounds are found in black pepper. The nightshade family includes potatoes and tomatoes, which carry likely harmful compounds (alkaloids). Solanine is a chemical present in the eyes and skins of potatoes, and if burnt (green beneath the skin) or blighted, solanine levels can rise sevenfold, enough to kill a young kid. Cooked potatoes containing high levels of solanine contain a harsh flavor and might provoke throat irritation. In animals, solanine has been found to have teratogenic consequences. The beetle and leafhopper are thought to use solanine as a natural insecticide. Tomatine, another toxic molecule present in tomatoes, is likewise an alkaloid and may act as a natural plant pesticide. Psoralen is a substance generated by a stressed plant and can be detected in parsnip, carrots, and celery. Psoralen is a skin irritant that causes rashes and other skin issues. Quinines and phenols are widespread in food. Livestock poisoning by plants is still an important veterinary problem in some areas. Consumers were advised to refrain from purchasing and consuming potatoes that are green with signs of germination, physical damage, or decomposing since glycoalkaloids are not broken down by cooking. Cooking finely cut and chopped cyanogenic plants in boiling water assists in the release of toxic hydrogen cyanide prior to consumption. Consumption of dry-cooked or low-moisture-cooked cyanogenic plants should be limited to small quantities.

### *4.3.1 Goitrogens*

In some parts of the world, human goiter, which occurs due to lack of iodine, is serious and a critical health problem. In certain parts, cruciferous plants are said to be a contributing factor leading to human goiter being endemic. Goiter develops in animals in the event of consuming seeds from several *Brassica* plants. Thyroid growth may be caused by excessive eating of *Brassica* species such as broccoli, cabbage, mustard greens, rutabaga, and turnip, but not when consumed as portion of a balanced diet [61]. Cruciferae's glucosinolate produces goitrin, which is a goitrogenic chemical. Glucosinolates are sulfur-containing thioglucosides that don't appear poisonous, pending they are converted to isothiocyanate, nitrile, or thiocyanate by myrosinase [61]. Various nitriles have been established as poisonous to rats, with the product goitrin from isothiocyanate that suppresses the thyroid.

## *4.3.2 Glycosides*

### 4.3.2.1 Cyanogenic Glycosides

A variety of food products widely eaten in Hong Kong contain cyanogenic glycosides. There are at least approximately 2000 plant species in which cyanogenic glycosides are found, some of which are utilized for food such as apricots, bamboo shoots, berries, and cassava [62]. Around 25 cyanogenic glycosides exist; they are amino acid–derived components generated as secondary metabolites in plants. These cyanogenic glycosides are present in the compound form of mono- or disaccharide conjugate cyanohydrins or precursors of hydrocyanic acid, which are known to be natural herbicides, in a variety of cyanogenic food plants (such as linamarin in cassava and taxiphyllin in bamboo shoots). When plant tissues are macerated, hydrocyanic acid (HCN) is produced (disrupted). Cassava and sorghum that are staple foods in many African countries are known to contain cyanogenic glycosides [63]. When preparing the former, great care is taken to ensure proper soaking, grating, and fermenting so as to ensure and allow the release of HCN prior to consumption. Many other edible plants such as bamboo shoot, flaxseeds, kernels of stone fruits (i.e., apricot and peach), and peas, as well as lima beans and shell of soya beans, all contain cyanogenic glycosides. When preparing lima beans, it is important to ensure that the temperature is high enough to destroy the toxic glycosides. Furthermore, cyanogenic glycosides may be found in food items with flavoring features such as ground almonds (powder/paste), marzipan, preserved kernel fruit (apricot, cherry, peach, and plum), and stone fruit drinks (juices and alcoholic), making the sources of hydrogen cyanide. Amygdalin, commonly known as vitamin B17 or laetrile, is a cyanogenic glycoside that may be found in bitter almonds, cherry kernels, and apricot and peach seeds. It was formerly related to the notorious cancer therapy of the 1970s. As little as 12 bitter almonds have been reported to have the ability to kill a toddler [47, 63].

#### 4.3.2.1.1 Toxicity

Although cyanogenic glycosides are chemical compounds contained in foods, on their own, cyanogenic glycoside is not toxic. However, the cyanide that is released upon their ingestion is the reason for the toxicity of cyanogenic glycoside–containing plants [64]. Cyanogenic glycoside–producing plant species normally have a corresponding hydrolytic enzyme (β-glucosidase). When water is available, the nonlethal cyanogenic glycosides are hydrolyzed by the enzyme-producing cyanohydrins, which swiftly disintegrate the deadly hydrogen cyanide. This occurs when the plant is masticated, discharging the deadly cyanide into the mouth of the consumer. Similarly, during food preparation, poisonous cyanide is discharged when plants are chopped or grinded into minute portions, with the produced hydrogen cyanide easily removed through boiling as it is volatile. Collectively cyanohydrins, cyanogenic glycosides, and hydrogen cyanide are referred to as cyanogens.

Cyanide bonds to ferric ions in mitochondrial cytochrome oxidase, which stops cellular respiration. Mental disorientation, muscle paralysis, and respiratory breakdown are all signs of acute cyanide poisoning. The cyanide ion can be processed metabolically to produce thiocyanate, with the rhodanese enzyme catalyzing the synthesis. Although rhodanese is found in a variety of tissues, it is unknown if its key task is to detoxicate cyanide. Thiosulfate is crucial in the detoxification of cyanide [64]. Thiosulfate is a product of the sulfate metabolism.

#### 4.3.2.1.2 Acute Toxicity

In human beings, decline in blood pressure, fast respiration, faintness, abdominal discomfort, rapid pulse, headache, diarrhea, vomiting, stupor, mental disorientation, cyanosis with spasms and convulsions, and terminal coma are all signs of acute cyanide poisoning.

Mortality as a result of cyanide poisoning is possible when the cyanide level exceeds an individual's ability to detoxify. The probability of cyanide poisoning from cyanide-containing food is dependent on what the consumer weighs. For instance, people who are light in body weight (including children) and consume inadequately prepared bamboo shoots and cassava are known to struggle in adequately detoxifying cyanide from their system. For human, hydrogen cyanide's instant fatal dose is acknowledged to be between 0.5 and 3.5 mg/kg body weight. Moreover, about 50–60 mg of free cyanide is considered to be a deadly dose for a male adult.

#### 4.3.2.1.3 Chronic Toxicity

Chronic diseases from dietary cyanide intake are a rare occurrence. However, adverse effects are common in individuals with underlying dietary deficiencies like inadequate iodine and/or protein intake. For instance, people with iodine deficiency and already suffering from goiter and cretinism often suffer aggravation following continued dietary exposure to cyanide.

A number of neurological diseases (including konzo and tropical ataxic neuropathy (TAN)) have been noted in people who lack proteins and iodine in their diet and eat cassava as their main daily food. The upper motor neuron disease known as konzo is defined by an irreversible nonprogressive symmetric spastic paraparesis that set up abruptly, whereas TAN is a severe neurological syndrome characterized by angular stomatitis, optical atrophy, neurosensory deafness, and sensory gait ataxia.

#### 4.3.2.2 Saponins

Forming a soapy foam even at the lowest concentrations, saponins are water-soluble plant constituents that are glycosides composed of a non-sugar aglycone known as sapogenin [65]. Distinguished by their bitter taste, saponins have the ability to hemolyze red blood cells. Saponins are categorized into two major groups based on their chemical composition of sapogenin, namely, steroidal and triterpenoid saponins.

In the plant kingdom, saponins are widely distributed and occur in all parts of plants, with the concentration varying at different stages of growth. They are found in different food plants, including, apples, asparagus, broccoli, eggplants, peanuts, soybeans, sugar beets, alfalfa, ginseng roots, and green tomatoes (major saponin, α-tomatine). Moreover, saponins can also be found in starfish and sea cucumbers.

A lot of saponins have the potential to disrupt the normal bilayer packing of phospholipids in cell membranes, potentially resulting in affected cells becoming abnormal and leaky ions, ultimately leading to death of cells (lysis). In foxglove, the major saponin is digitonin, which is an extremely active detergent.

Ginseng (panax) is a widely used traditional herbal medicine for ailments including cancer, heart disease, sexual impotency, and many others [66]. The root is rich in saponins called glycyrrhizins. Ginseng is usually used orally as tea or in tablet form. However, extended excessive consumption of ginseng has been noted to result in harmful side effects, which include hypertension, diarrhea, and insomnia skin eruptions.

Luckily, majority of the saponins are not easily absorbed by our gastrointestinal tracts as glycosides. Alternatively, intestinal glycosidase enzymes break the sugar groups connected to the 3-B-OH group on the sterol skeleton, which technically gets rid of their toxic properties. The nonpolar aglycones are promptly absorbed and plausibly are pharmacologically active components. The saponins are a sizable, chemically diverse group. Irrespective of immense efforts by chemists to decrypt their complex structures, their pharmacological mechanisms of action remain largely unknown. They probably apply an assortment of actions through multiple cell receptors. Despite the popularity of saponins in traditional herbal medicines, in terms of treatment, their clinical efficacy for most of these disorders is yet to be fully demonstrated.

##### 4.3.2.2.1 Toxicity

Although saponins are known as being typically innocuous to mammals (and other warm-blooded species) unless in high quantities, they, however, have the potential to disrupt red blood cells and result in vomiting and diarrhea. Their toxicities are associated to the decrease of surface tension. The body has the ability to detox when saponin is used in small doses as the intestinal microflora and the blood plasma can destroy and can inhibit the saponin actions, respectively. However, in huge amounts, saponin can irritate the gastrointestinal tract leading to puking and diarrhea.

#### 4.3.2.3 Cardiac Glycosides

Cardiac glycosides are both animal and plant products. These classes of organic compounds are used for medicinal purposes in the West and are traditionally sourced from the foxglove flowering plant that is now cultivated extensively in many countries [67]. The major glycosides of the foxglove are referred to as digitoxin and digoxin. In the Orient, toad venom glands were utilized as a leading source of very similar medicinal compounds known as bufotoxins. Digitalis glycosides *are* primarily used as therapeutic treatment for congestive heart failure, which is a condition defined by loss of myocardial contractility. For a number of different reasons (such as atherosclerosis, kidney failure, long-term hypertension), the heart has to function optimally and pump the blood sufficiently to avoid flooding of the lungs and extremities.

Over three centuries ago, cardiac glycosides, derived from *Digitalis purpurea*, were discovered by Withering who realized that the foxglove leaf was very effective in treating a condition called dropsy [68]. Unfortunately, *Digitalis glycosides* often lead to cardiac arrhythmias at concentrations that are necessary to significantly enhance the cardiac output and, subsequently, are also among the most toxic drugs. Their action site is the sodium-potassium pump (Na,K-activated Mg-ATPase) found in the cell membrane. This transport system is actively liable for keeping the high-potassium, low-sodium intracellular environment of all cells. However, in the heart, it seems that obstruction of a fraction of the pumping sites with digitalis permits the intracellular sodium concentration to transiently rise above normal range during each myocardial action potential, and this raised sodium then is exchanged with calcium from outside the cell through a membrane carrier known as the sodium-calcium exchanger. This leads to elevation in the intracellular calcium when the heart is beating, thereby stimulating the actomyosin system to contract with more force. It is quite extraordinary that these glycosides can be utilized as inotropic drugs at all, noting that all cells have sodium-potassium pumps, which are inhibited by digitalis. Other plants that produce dangerous amounts of digitalis compounds are the oleander bush (*Nerium*), an extremely common ornamental shrub in the Southeastern United States. The lily of the valley ornamental flower (*Convallaria*) is a wildflower and the butterfly weed (*Asclepias*). A single oleander leaf reportedly contains adequate cardiac glycoside to be deadly to an adult human. The danger with foxglove is that during the nonflowering season, its leaves are confused with those of the common comfrey plant, whose leaves are popularly used in the preparation of herbal teas, thus leading to a number of deaths due to inadvertent use of foxglove leaves.

### 4.3.3 Phenolic Substances

More than 800 phenolic substances are known in plants. Such compounds contribute to the bitter taste, flavor, and color of foods.

Most of phenolic compounds are free of acute toxicity now that ways to detoxify them are available. Tannins have developed to make plants less appealing to herbivores, and they defend the plant from microbial and fungal assault. Tannins are divided into two categories: condensed and hydrolyzable substances. Gallic, digallic, and ellagic acids; esters of glucose; or quinic acid are examples of hydrolyzable. Tannic acid is an illustration of a tannin that may be hydrolyzed. Condensed tannins, on the other hand, are flavonoids that polymerize into polymers containing carbon bonds between the monomers. Tannins, such as gallic acid, have the potential to bind metals together. Tannins, on the other hand, are thought to harm the liver (necrosis and fatty liver). Tannins attach to proteins, resulting in protein precipitation and blocking digestive enzymes. Additionally, they impair iron bioavailability. Betel nuts, which are frequently chewed post dinner in the Far East, contain around 26% tannins and are thought to be the cause of alarming prevalence of cheek and esophageal cancers. The consumption of sorghum and a high intake of tannin-rich teas are thought to be the cause of esophageal cancer in South Americans.

There are six subgroups of flavonoids [47]. The majority of the chemicals in this category are glucosides. Flavones are abundant in plants, and most of them contribute to the yellow pigments seen in oil vesicles and citrus fruit peels. The flavone quercetin, which is abundant in onions, has been shown to induce cancer in animals (two strains of rats) when administered orally. Flavonoids, on the other hand, have been lauded for their role in reducing heart disease, notably among French and Asian people who consume a lot of wine and tea, respectively. The French paradox is one situation that has received some attention. As a population, the French from Southern France appears to have a lower cardiovascular disease mortality rate despite consuming a diet high in trans fat and their continued smoking of tobacco. This paradox has been suggested to be due to their intake of wine (providing flavonoids) along with meals. Recently, a Danish study revealed that wine intake was closely linked with a lower risk of heart diseases in comparison to drinking alcohol in other forms.

Consumption of tea has been said to have a protective effect against cardiovascular diseases. Green tea, particularly, has high flavanol content such catechin, epicatechin, epicatechin gallate, gallocatechin, epigallocatechin, and epigallocatechin gallate, all of which have health benefits.

Coumarin, safrole, and myristicin are found as diverse flavor components [45, 47]. Coumarin is a compound present in citrus oils that inhibits blood coagulation and harms the liver. Safrole, a compound present in black pepper and sassafras oil, induces malignancies in the liver of rats. Safrole is a phenolic compound that accounts for around 80% of the essential oils derived from the sassafras tree (roots and bark). Tonics, teas, and different cure-alls include it. It's a common addition to

New Orleans-style gumbo. Myristicin, on the other hand, is present in parsley, celery, black pepper, carrot, and dill, among other herbs and spices.

Gossypol is a poisonous phenolic compound found in cottonseed. Cottonseed is protein-rich, and many plants have been produced with reduced gossypol using selective breeding. Nevertheless, the consequences of selective breeding have made plants to be more susceptible to aflatoxin problems and mold growth. Gossypol causes a loss of appetite and weight, as well as diarrhea, anemia, decreased fertility, pulmonary edema, circulatory failure, and gastrointestinal hemorrhages. Phenolic chemicals reduce the availability of iron by inhibiting the transformation of pepsinogen to pepsin.

### *4.3.4 Cholinesterase Inhibitors*

The cholinesterase enzyme helps in the breakdown of acetylcholine, a chemical released from nerve synapses' vesicles and responsible for nerve impulse transmission throughout the synapse. Acetylcholine needs to be hydrolyzed after the nerve impulse has been sent to guarantee that the neuron is ready for the next impulse. Physostigmine, a cholinesterase inhibitor derived from the West African Calabar bean, was used in African witchcraft as an ordeal poison [69]. An extract of the Calabar bean was later used for medicinal purposes, and in 1864, the active compound was isolated and called physostigmine [70, 71].

Another anticholinesterase toxicant from foods is solanine. Glycoalkaloids (or steroidal alkaloids) such as solanine and chaconine inhibit cholinesterase and are also irritants of the intestinal mucosa. These compounds are encountered in potatoes, eggplants, and tomatoes, which are members of the *Solanum* genus. Huge quantities of these toxicants are found in potatoes, more especially if they've had exposure to light, fungal diseases, or trauma. Solanine is not soluble in water and is heat-stable; thus, the toxicants are neither removed nor destroyed by cooking. The market-sold potato has around 2 mg of solanine per 100 g; however, the greening (which is caused by high chlorophyll content but is not harmful) increases the quantity to about 50–100 mg per 100 g. Exposure of russet potatoes to white fluorescent light has the tendency to increase its overall glycoalkaloid peel content to 70 mg/100 g. Lenape, a potato cultivar designed particularly for potato chips, was phased out due to a total glycoalkaloid concentration of approximately 30 mg/100 g for fresh tuber. As per the FDA, solanine content should not be more than 20 mg/100 g.

#### 4.3.4.1 Clinical Symptoms

Solanine poisoning has caused sickness or death in a few cases. The usual signs are increased stomach discomfort followed by nausea and puking, as well as weakness, respiratory problems, and heat exhaustion. Where human volunteers are used as experimental subjects, 0.3 mg/100 g of solanine leads to drowsiness, itchiness, and hyperesthesia, as well as difficulty breathing. Symptoms of organophosphate poisoning can be seen with higher doses.

### *4.3.5 Biogenic Amines*

Biogenic amines can be found in a variety of plant- and animal-based diets. Dopamine and tyramine are natural elements of foods like avocado, banana, and cheese, while putrescine and cadaverine are produced by bacteria acting on amino acids present in fish and meat. In couple of food poisoning outbreaks, histamine and beta-phenylethylamine were implicated as etiological agents. These amines are shown to have an impact on the vascular system, causing vessel constriction as well as increased blood pressure as a result (pressor amine effects).

Catecholamines such as norepinephrine (noradrenaline), dopamine (adrenaline), and epinephrine (adrenaline) are biologically active molecules (hormones) that are produced by the adrenal glands and function as neurotransmitters in adrenergic nerve cells. They can trigger diet-induced migraine headaches, as well as hypertensive crises in some circumstances.

Normally, the increased dietary load of bioactive amines has little to no consequence; as such, amines are cautiously regulated by the widely distributed enzyme monoamine oxidase (MAO). Consumption of foods loaded in biogenic amines is indicated as contraindications for patients using MAO inhibitors like antidepressants including iproniazid. Examples that are more specific include the fava bean containing dihydroxyphenylalanine (DOPA) which can be decarboxylated to dopamine.

Oriental preserved foods like soybean paste, soy sauce, and various condiments have high tyramine content. Outbreaks of histamine poisoning have been linked with *Lactobacillus buchneri* found in Swiss cheese. A number of wines have been associated with an increase in histamine and tyramine content within the first 5 days of manufacturing.

#### 4.3.5.1 Clinical Symptoms

When monoamine oxidase is inhibited, marked pressor effects are visible. Severe headaches, palpitations, and hypertension are the most common clinical signs. Uncommon signs include intracranial bleeding, which can be fatal.

## 4.3.6 *Alkaloids*

Thousands of compounds of alkaloids have been isolated and investigated, in many cases quite superficially. Sources of alkaloids include seeds, roots, leaves, and/or bark of at least 40% of all plant families, in particular the Amaryllidaceae, Compositae, Leguminosae, Liliaceae, Papaveraceae, and Solanaceae, which are rich in alkaloids [72]. Majority of these substances are also referred to as heterocyclic compounds, since they generally have a halo structure comprising at least one non-carbon atom, usually O or N. Considered to be "the bitter components" of widely found natural plants, alkaloids that mostly act as secondary plant metabolites often possess necessary pharmacological properties. Alkaloids commonly contain basic nitrogen molecules, which can produce salts when reacting with acid. Plants bearing flowers, in particular, are a rich source of alkaloids. Recently, a huge number of alkaloids have been isolated from marine organisms, which are the less likely sources, and in some cases, they tend to be toxins.

#### 4.3.6.1 Pyrrolizidine Alkaloids

Pyrrolizidine alkaloids are a type of alkaloid that is discovered widely in the plant kingdom, with as many as 6000 flowering plant species (3%) distributed in all kids of climatic regions known to have the chemicals globally [73]. Consumption of plants rich in pyrrolizidine alkaloids is normally through contaminated crops as these alkaloid-containing plants often germinate as weeds among crops like corn or wheat and are gathered as part of the grain. This group of toxins may be ingested through the oral ingestion of toxin-filled herbal foods and concoctions. To prevent envenoming, intake of alkaloid-rich foods and herbal preparations can be reduced, through implementing efficient agricultural strategies to prevent and decrease contaminating food crops with pyrrolizidine-containing plants.

Many of these alkaloids are produced by plants from the *Crotalaria*, *Heliotropium*, and *Senecio* species, most of which occur as weeds globally. On a number of occasions, contaminated cereal crops and human consumption of flour-made products have led to poisoning.

This occurred in different countries globally, especially in places with poor agricultural conditions and the natives possibly forced to consume crops that are contaminated. For example, in the 1930s in South Africa, poor whites were exposed to the toxic effects of alkaloids as a result of consuming contaminated staple food (wheat), while their neighbors who are Bantus consumed the maize, which was not contaminated, and thus were not affected. In recent times, Central India (Tashkent) and northern Afghanistan have experienced poisonings. Approximately 1600 poisoning cases reportedly occurred in one incident, where threshed wheat was reportedly contaminated with *Heliotropium popovii* seeds with an alkaloid concentration around a minimum of 0.5%.

In the West Indies specifically, a variety of these plants are utilized in traditional medicines to produce herbal teas. After acute exposure and ingestion of these alkaloids through consumption of herbal teas, veno-occlusive disease that is a form of liver disease usually occurs. Chronic exposure to alkaloids in low doses causes liver cirrhosis. This condition affects about one-third of the West Indian population and is prevalent within the Jamaican population.

Alkaloids, including monocrotaline, a substance extensively studied, go through metabolic activation to a reactive metabolite, which causes damage to cells lining the liver sinusoids and the hepatocytes, consequently resulting with hemorrhagic necrosis and lastly causing the veno-occlusive disease [73]. In the liver, the blockage of the blood vessels ultimately leads to the alteration of the vasculature in a manner that results in the blood supply to the liver being diverted and the growth of new blood vessels.

Animals can also be exposed to alkaloids and thus suffer toxic exposure. Where there is a variety and abundance of grazing fauna, animals often disregard plants that contains the alkaloids such as ragwort (*Senecio jacobaea*). However, in countries such as Australia, widespread losses of livestock animals such as cattle, sheep, and horses have occurred due to *Heliotropium* poisoning. Subsequently, this livestock exposure to alkaloids is considered to be another potential route that humans who consume milk and other dairy products can be introduced to alkaloids.

The glycoalkaloids (like solanine and tomatine) are other types of plant alkaloids commonly detected in food plants, which act as a natural pesticide against pests. Potatoes often contain minute amounts of solanine, whereas tomatoes contain tomatine.

#### 4.3.6.2 Glycoalkaloids

Greened potatoes, sprouting potatoes, and raw potatoes all contain tiny amounts of glycoalkaloids, natural toxins that are heat-stable, meaning when they are cooked, baked, steamed, fried, or microwaved, they do not get decomposed. Moreover, they taste bitter and are not easily soluble in water. Glycoalkaloids contain steroidal alkaloids with one or many monosaccharides attached together. Toxic glycoalkaloids are produced by the entire Solanaceae family. The glycoalkaloids solanine and chaconine are the most common glycoalkaloids found in potatoes. They're present in the parenchyma cells of tuber cortex and periderm parenchyma cells. These chemicals serve to protect plants from predators and thus are toxic to insects and animals. They're found in tiny amounts in all potato tubers but have a higher concentration in the potato skin and places with high metabolic activity, including the eye. Its concentration rises in greened or damaged potatoes, peaking very elevated levels in sprouts. The rapid formation of these toxins can be formed due to factors like improper storage conditions and light exposure, mechanical damages, and decomposition from microorganism such as fungi and/or bacteria [74]. These toxins are not reduced by either soaking, rinsing, or cooking. Cooked potatoes containing high amounts of solanine have a bitter flavor when eaten and cause a burning feeling in the throat.

##### 4.3.6.2.1 Toxicity

Fresh, intact glycoalkaloids commonly present in eatable plants do not generally provoke toxicity. Glycoalkaloids, on the other hand, can be hazardous in excessive concentrations.

##### 4.3.6.2.2 Acute Toxicity

For glycoalkaloid toxicity, the greatest concern is its acute toxicity. A lot of human poisoning incidents (some deadly) have been documented as a consequence of ingesting greened, spoilt, or sprouting potatoes with high amounts of glycoalkaloids. When glycoalkaloid levels exceed 2.8 mg/kg body weight, acute toxicity symptoms in humans have been reported. The incubation period after ingestion of toxic potatoes ranges from 2 min to 2 days, with prolonged incubation time being linked to more grave conditions.

Clinical symptoms of low-grade poisoning are immediate gastrointestinal distress, diarrhea, severe abdominal cramps, and vomiting. On the central nervous system, the *Solanum* alkaloids exhibit a significant anticholinesterase action. Most alkaloids induce acute poisoning by imitating or inhibiting the function of nerve transmitters. In more severe cases, neurological symptoms such as confusion, weakness, drowsiness, apathy, and vision disturbances followed by unconsciousness may be observed, with death occurring in some cases. Glycoalkaloids have saponin-like properties and possibly cause hemorrhagic injury by disrupting the functionality of the membrane in the gastrointestinal tract. The degree of necrosis considerably outweighs the inhibitory effects on acetylcholinesterase function, making this damage potentially lethal.

##### 4.3.6.2.3 Chronic Toxicity

A proper no-observed-adverse-effect level (NOAEL) for potato glycoalkaloids is yet to be made owing to the absence of chronic toxicity data, and a tolerable daily intake (TDI) for humans is also yet to be determined.

##### 4.3.6.2.4 JECFA Evaluation

Joint FAO/WHO Expert Committee on Food Additives (JECFA) considered that, despite the fact that eating of plants containing glycoalkaloids dates far back, the existing evidence (epidemiological and experimental) reported by diverse human and laboratory animal research is yet to establish a safe consumption level [75]. JECFA deems a glycoalkaloid consumption of 3–6 mg per kg body weight to be a potentially fatal dosage for humans based on current evidence. Furthermore, a cumulative glycoalkaloid dosage exceeding 1–3 mg per kg of body weight is regarded hazardous to humans. Children, on the other hand, appear to be more vulnerable than adults. Several variables may be found in suspected potatoes that alter the steroid glycoalkaloids' toxicity.

The JECFA Committee noted that gathering empirical facts to establish a safe intake level takes ample time and work. Nonetheless, the substantial amount of research on the regular and daily eating of potatoes suggested that typical glycoalkaloid levels (20–100 mg/kg) seen in well-produced and well-managed tubers do not pose a serious problem.

## *4.3.7 Pennyroyal Oil*

In the United States, where oil can be bought over the counter, the pennyroyal plant and its oil are usually utilized to induce abortions [76]. The plant can be used as tea or used directly as oil. However, both routes of use can result in toxic effects, particularly to the liver, as well as inducing abortion. The pennyroyal oil contains a number of terpenoid compounds and metabolic activation that is believed to be necessary for the toxicity.

## *4.3.8 Plants Containing Peptide and Protein Toxins*

Various plants contain protein toxins that are deadly when orally ingested or parenterally administered. Rosary bean seeds are captivating red seeds with a black spot and are frequently used in the manufacturing of necklaces. These attractive seeds contain a 70 kDa protein known as abrin, a ribosomal protein synthesis inhibitor.

### 4.3.8.1 Ricin

Castor beans, which are now naturalized in southern California, are also used for manufacturing decorative and aesthetic necklaces. They contain ricin, which is a homologous protein with the same properties and potential lethality. Ricin is considered to be a highly toxic plant product found in the castor oil plant seeds. It became notorious in 1978 when it was reported that it had been used by the Bulgarian secret police to murder the Bulgarian journalist Georgi Markov in London. Even though no poison traces were found in the deceased, the symptoms were consistent with those associated with ricin poisoning. A tiny metal pellet was recovered from a wound on Georgi Markov's leg, apparently inflicted accidentally by an umbrella. The pellet was almost certainly a reservoir for a poisonous substance, but it could only contain a few nanograms of the poisonous substance.

Ricin is a small protein consisting of two polypeptides, a longer B chain and a short chain. The A chain is the active inhibitor of protein synthesis, and the B chain is used to bind to the cell membrane and stimulate internalization of the toxin. Both chains (A and B) are linked via a disulfide bridge. The B chain attaches the ricin molecules to the outside of the mammalian cell by binding to the galactose part of a glycoprotein. The cell membrane invaginates, and the ricin is taken into the cell inside a vacuole. The ricin molecule is discharged from the glycoprotein, and the A and B chains then break at the disulfide bridge. The B chain makes a channel through the vacuole membrane, allowing the A chain to get into the cytoplasm to reach the ribosomes where it stops protein synthesis and kills the cell. One ricin molecule is adequate to kill one cell.

##### 4.3.8.1.1 Toxicity

The poisoning clinical signs of ricin include pain in the mouth, delayed onset of abdominal pain, vomiting, severe diarrhea, hemolysis, and renal failure. After ingestion, the symptoms of poisoning develop rather slowly during the first 24 h; however, if the victim has ingested several seeds, he or she may suffer a lot during the ensuing couple days and later succumb to an awful death. The toxins are implanted inside the fibrous seed pit; in case it isn't broken up by chewing, the individual may not get much poison. Initiated spewing by ipecac syrup followed by gastric lavage is suggested as soon as possible amid the first few hours after ingestion; otherwise, symptomatic treatment is all that can be done, since the toxins are internalized inside the cell.

#### 4.3.8.2 Lectins

Lectins are a nonimmune protein or glycoprotein with a large number of very specific carbohydrate binding sites. Initially, they were only found in the castor bean but currently exist across the plant kingdom, with the inclusion of grain products. Lectins are found in high concentration in legume seeds and have been linked to nausea, gastroenteritis, and diarrhea in males. Lectins may be found in a variety of beans, including green, red, and white kidney beans.

##### 4.3.8.2.1 Toxicity

Intense stomach pain, vomiting, and diarrhea are all symptoms of acute poisoning. Lectins can destroy the gastrointestinal epithelia, inhibit cell mitosis, and cause localized hemorrhages and impairment of the kidneys, liver, and heart, as well as agglutination of the red blood cells.

#### 4.3.8.3 Viscumin

As herbal medicines, mistletoe leaves and berries have been utilized to prepare orally administered extricates and teas for the treatment of conditions that include high blood pressure, tachycardia, sleeping disorders like insomnia, depression, sterility, cancer, and ulcers. While some of these conditions, such as hypertension and tachycardia, might apparently be ameliorated, based on the current knowledge of the contents of mistletoe, presently, there are no medical reports that support the therapeutic usage of mistletoe extracts. Consumption of mistletoe extracts is likely to cause injury to one's health, as a result of the presence of a toxin referred to as viscumin whose action is similar to ricin and abrin, smaller peptide toxin called viscumin whose action is comparative to abrin and ricin, and smaller peptide toxins known as viscotoxins, which depolarize muscle cell membranes and can lead to bradycardia, hypotension, and other problems.

##### 4.3.8.3.1 Toxicity

The poisoning symptoms of viscotoxin are pain in the mouth, delayed onset of vomiting, diarrhea, hypotension, and bradycardia.

### *4.3.9 The Seed of* Ginkgo biloba

*Ginkgo biloba* (ginkgo nuts) seeds are frequently consumed as a household nut in Asian nations such as China, Korea, and Japan. Plant poisons found in *Ginkgo biloba* include cyanogenic glycosides and 4′-methoxypyridoxine (4′MPN). 4′MPN is considered to be the main component in regard to toxicity. Attributed to toxins present in *Ginkgo* seeds, serious poisonous effects (mostly neurotoxic) are associated with this seed.

#### 4.3.9.1 Toxicity

For *Ginkgo biloba* poisoning, acute toxicity is the principal concern. Symptoms of toxicity from *Ginkgo* seeds generally starts 1–12 h after consumption of the seed and include irritability, vomiting, and tonic and clonic convulsions. This form of food poisoning is especially dangerous for children. Unconsciousness and fatalities may ensue in severe situations where large doses have been consumed or in susceptible persons. Ingestion of approximately 10–50 of the cooked seeds at once has been reported to result in intense poisoning in humans. Intake of *Ginkgo biloba* seeds ranging from 15 to 574 seeds has resulted in death in certain people. The pathophysiological processes behind the convulsion-inducing action of *Ginkgo* nuts remain unknown. Because of its pharmacological competition with vitamin B6, a cofactor of glutamate decarboxylase, 4′MPN is regarded as a potential agent. 4′MPN was shown to indirectly block the enzymatic activities of glutamate decarboxylase, resulting in a reduction in the brain's aminobutyric acid (GABA) level, causing convulsions. Seeds that are not ripe or uncooked are said to be more dangerous. Despite the fact that 4′MPN is heat-stable and cannot be neutralized by boiling, heating *Ginkgo* seeds can lessen its potency by deactivating other heat-labile toxins (like cyanogenic glycosides) in the toxin, in the presence of water.

## 4.4 Animal Toxicants

Toxicants are widely distributed and come from a variety of animal and plant kingdoms, from the unicellular to the multicellular organisms. The chemical composition of hazardous substances varies dramatically in the natural environment, making categorization based on structure difficult. Poisonous compounds in food may have developed as a result of animals or plants evolving to produce chemicals to protect themselves from predators, insects, nematodes, microbes, or even humans. There are roughly 1200 species of venomous and poisonous creatures in the animal world. Venomous creatures have the ability to produce poison via a relatively sophisticated secretory gland or collection of cells and then transmit the poison to its prey via a sting or bite. The venom can serve a variety of purposes, including attack and defense. Normally, humans do not ingest venomous animals for food purposes, but when they do, special caution must be exercised to prevent dangerous glands harboring toxicants.

Some poisonous animals possess poisonous chemicals in their tissues, but they lack a system for delivering the poison; therefore, poisoning only happens by ingestion of the animal material, particularly the toxin-containing tissues. Poison has a little part in the aggressive and defensive behaviors of toxic animals. The toxin might be found as a by-product, a metabolic product, or a substance that passes through the food chain. Barracuda, snapper, and other grouper fish are examples of the latter that might pose a health risk to humans by devouring smaller poisonous fish and other marine creatures such as dinoflagellates. In the food chain, little fishes with ingested poison are eaten by bigger ones, which are in turn eaten by humans, consequently resulting to illness in humans. While poisoning after consumption of terrestrial animals is to a certain degree not common, poisoning as a result of marine toxins happens globally. Marine toxins from toxic microalgae are amassed in shellfish, crustacean, and finfish after their consumption. Tetrodotoxin, which is a potent marine neurotoxin, is believed to be from certain bacteria and found in more than 90 species of puffer fish, with the ability to cause casualties after consumption, even in small quantities. Poisoning from seafood normally reported in coral reef fish is as a result of the presence of ciguatoxin, which is present in over 300 species of fish. Histamine from bacterial spoilage of scombroid fish leads to another kind of seafood poisoning.

### 4.4.1 *Marine Animals*

Generally, for animals, seafood poisoning must be differentiated from marine venoms. Many seafood toxins are not exclusive to one species and are susceptible to environmental influences. Ichthyotoxins, for example, are exclusive to a single species or genus of fish. The organs in which the toxins are located in the animal can be used to classify seafood poisons [77]. Toxins discovered in the skin, muscles, liver, or gut of animals are known as ichthyosarcotoxins, whereas toxic compounds found in the testis or ovaries of animals are known as toxic chemicals. Ichthyohemotoxins are toxic substances present in an animal's circulatory system, while ichthyohemotoxins are present in the hepatic system.

### 4.4.2 *Scombroid Poisoning*

Following the intake of bonito, tuna, skipjack, and mackerel-like fish, the kind of envenoming that occurs is known as scombroid poisoning [77]. The clinical signs of this type of poisoning differ from that of the ciguatera toxin. The toxins are formed in the muscular structure of the fish, if the scombroid is immature. Some items like cheese (prominently Swiss cheese) have also experienced the impacts of the toxin.

When bacteria are present and conditions are favorable (i.e., time and temperature), toxins are formed in food. Toxin dispersion within a single fish fillet or across containers in a case parcel can be irregular, resulting in certain parts of the food causing sickness and some don't. These harmful effects are not reduced by cooking, canning, or freezing. Individual sensory tests by itself are insufficient to establish if toxins are present or absent. Chemical test is considered to be the only reliable test for evaluating products. In the United States, scombroid poisoning is among the most prevalent types of fish poisoning; however, poisoning incidents often go unreported due to a shortage of mandatory reporting and information from some medical personnel and lack of clear symptoms, which are often confused with other illnesses [77]. This leads to underreporting of the exact count of incidences, which is a global challenge.

#### 4.4.2.1 Mode of Action

The poisonous effects of scombroid poisoning are partly due to the histamine produced by the enzymatic activity and microorganisms on the fish after it dies.

Scombroid poisoning has clinical symptoms that are comparable to allergic responses caused by histamine production and is caused by bacterial decarboxylation of histidine as the fish degrades. Other toxicants identified in fish that partake in the toxic process include saurine, cadaverine, and putrescine. Maintaining fish on cold storage can slow spoilage. Fresh salmon stored at 32 °F (0 °C) stays unspoilt for 12 days, whereas fresh salmon kept at 60 °F (15 °C) stays just 24 h. Any meal having the proper amino acids that have been contaminated with bacteria, the bacterial growth can cause scombroid toxicity if consumed.

#### 4.4.2.2 Clinical Symptoms

Diagnosis of scombroid poisoning is normally based on the symptoms of the patient, time of onset, and the treatment effect of antihistamine medication. To confirm diagnosis, the suspected foodstuffs must be analyzed within a few hours for elevated levels of histamine. Most people normally recall the infected fish having a fiery or poignant taste. Following consumption (roughly 2 h), these gastrointestinal discomforts are observed: abdominal cramps, obnoxiousness, vomiting, diarrhea, headache, epigastric distress, and fiery sensation in the throat region. In most cases, hypoesthesia, tingly sensations, rash, skin flushing, and urticaria (generally widespread erythema, usually lacking wheals) also follow the first set of symptoms. The onset of intoxication symptoms is rapid, ranging from immediate to 30 min. The duration of the illness is normally 3 h, but it may last for several days. Clinical signs settle down in 16 h, and generally, no lasting ill effects are experienced.

### *4.4.3 Saxitoxin*

Dinoflagellate species, especially *Gonyaulax* species, release a toxin that accumulates in butter clams, cockles, mussels, scallops, soft-shell clams, and some shellfish that consumes algae. The occurrence of red tides or blooms can ensue, causing the dinoflagellates to proliferate rapidly. The North American West Coast Indians were the first to notice the dangers of red tides, and they shunned eating shellfish when the water became red. The toxin has no adverse effects on shellfish, but it can be hazardous to people. Consumption of such fish can pose a serious health risk, which is known as paralytic shellfish poisoning (PSP). Paralytic shellfish poisoning happens globally, especially in regions 30° or those higher in latitude. The toxin is an alkaloid and cannot be removed or destroyed through washing or heating. The body stores toxic substances in different parts of a shellfish. These body parts (the digestive system, gills, liver, and siphons) contain the greatest quantities of poison in the warmer season. Shellfish (mollusks that filter water) are all possibly poisonous. PSP, on the other hand, is often commonly associated with clams, cockles, mussels, and scallops.

#### 4.4.3.1 Mode of Action

Saxitoxin is the primary toxin, with 10 g/kg being the $LD_{50}$ of saxitoxin injection (IP) in mice. Saxitoxin is made up of 20 toxin derivatives, and it is fundamental in paralytic shellfish poisoning (PSP). It has a curare-similar action preventing muscles from responding to acetylcholine. The toxin has the ability to prevent neural transmission at the neuromuscular junction through binding to the surface of the sodium channels and interrupting the flow of $Na^+$ ions; atrioventricular nodal conduction may be suppressed, and there may be direct suppression of the respiratory center and progressive reduction of peripheral nerve excitability. Saxitoxin affects the heart directly by inhibiting nerve impulses in muscles and nerves.

#### 4.4.3.2 Clinical Symptoms

Approximately 120–180 μg PSP toxins per person produce moderate symptoms, which are reversible within hours or even days, while 80 μg of purified PSP toxin per 100 g of tissue (i.e., 0.5–2 mg per person) can be deadly because of asphyxiation, which usually occurs inside 12 h of ingestion. Clinical signs of the disease occur fairly rapidly, between 30 min and 2 h after consumption of the shellfish, depending on the quantity of the toxin ingested. Respiratory paralysis is common in severe cases, and if no respiratory support is provided, death may occur. When respiratory support is provided within 12 h of exposure to the toxin, recovery is usually complete, without any permanent side effects. In uncommon cases, death can result from cardiovascular collapse, despite respiratory support as a result of weak hypotensive action of the toxin.

### 4.4.4 *Pyropheophorbide-A*

The toxin pyropheophorbide-A is discovered in the liver or digestive gland of some abalones (like *Haliotis*). This toxin is a metabolic by-product that is a chlorophyll precursor with a blue-green color. The poison is hypothesized to be generated from the chlorophyll in the seaweed ingested by the abalone. Pyropheophorbide-A is stable to freezing, boiling, and salting.

#### 4.4.4.1 Mode of Action

When people consume organs containing pyropheophorbide-A and are exposed to sunlight, toxic reaction occurs. Pyropheophorbide-A is photoactive, and the photosensitization in the body promotes the production of amine compounds from amino acids, histidine, tryptophan, and tyrosine. Such compounds result in inflammation and other toxic reactions. Murexine and enteramine (5-hydroxytryptamine) are two other compounds that when isolated appears to have muscarine- and nicotine-like activities.

#### 4.4.4.2 Clinical Symptoms

The development of face and extremity redness, as well as dermatitis and edema, is all clinical indications of photosensitized pyropheophorbide-A poisoning. The muscarine- and nicotine-like actions cause low blood pressure and increased breathing, resulting in cardiovascular alterations.

### *4.4.5 Tetrodotoxin*

The balloonfish, blowfish, burrfish, globefish, porcupine fish, molas, fugu (served in Japanese restaurants), and toadfish are among 90 species of puffer fish. The California newt (*Taricha torosa*), some members of the Salamandridae family, frogs, octopus, starfish, flatworms, various crabs, and gastropods also produce tetrodotoxin. Except in an alkaline solution, tetrodotoxin is water-soluble and is stable to boiling. In certain societies, the fish is viewed as a delicacy, and its utilization brings about a shivering sensation.

Individuals who have undergone special training normally prepare the fish; however, slipups can occur. Poisoning can result from improperly preparing and removing of the ovaries, liver, roe, intestines, and skin, which have elevated amounts of tetrodotoxin. As a matter of fact, except when turned alkaline, the toxin is present in virtually all tissues and is resilient to boiling temperatures. During the spawning period of the fish, the highest concentrations of the poison are found.

#### 4.4.5.1 Mode of Action

Tetrodotoxin, a neurotoxic, paralyzes the central nervous system and peripheral neurons by impeding the normal flow of all monovalent cations, consequently intensifying the nerve's early transient ionic permeability. This causes a tingling (prickly) sensation in both the toes and fingers. Tetrodotoxin blocks the skeletal muscle membrane to a minor extent. It causes hypotension and inhibits breathing. In mice, the oral $LD_{50}$ is approximately 322 g/kg, and the dose for fatality per person starts from as little as 1–4 mg. Certain bacteria have been shown to produce the toxin, suggesting that they are the poison's source. Both red algae and puffer fish have been used to isolate *Alteromonas* and *Shewanella*.

#### 4.4.5.2 Clinical Symptoms

Clinical signs usually appear in 10–45 min; however, a lag period of about 3 h or more has been noted. Initially, tingling feelings can be experienced, after which malaise, dizziness, paleness of the skin, and hypoesthesia of the limbs, tongue, and lips occur, and subsequently, obnoxiousness, puking, ataxia, and diarrhea are observed. Other clinical signs include respiratory distress, subcutaneous bleeding, scaling, tremors, muscular twitching, muscle paralysis, loss of coordination, cyanosis, and convulsions, and approximately 40–60% death may occur.

### *4.4.6 Ciguatoxin*

In the beginning, ciguatera was administered to treat poisoning in the Caribbean due to consumption of the marine snail *Livona pica* (cigua). Ciguatoxin and associated toxins (such as maitotoxin and scaritoxin) are ichthyosarcotoxin neurotoxins (anticholinesterase) and found in 11 orders, 57 families, and more than 400 species of fish as well as in oysters and clams. Because of their eating patterns, many fish captured for food have ciguatoxin. Plant components containing the toxin are consumed by fish. Ciguatera poisoning is most likely caused by photosynthetic benthic dinoflagellates.

#### 4.4.6.1 Mode of Action

Ciguatoxin hinders cholinesterase, causing acetylcholine to accumulate in synapses and neuronal function to be disrupted. The respiratory system is usually paralyzed, which results in death. Ciguatoxin is a heat-stable hydroxylated lipid molecule with an unknown structure that is soluble in water. Physostigmine has been discovered to inhibit the toxin.

#### 4.4.6.2 Clinical Symptoms

The asymptomatic period is 3–5 h post consumption but then may last up to 24 h. Ciguatoxin poisoning emerges suddenly and starts with tingling in the lips, tongue, and throat, quickly followed by numbing in the same areas. Soon after, abdominal discomfort (along with intestinal spasms), nausea, vomiting, and diarrhea occur. Patients have complained of headaches, which can be followed by vision disturbances, muscular discomfort, convulsions, and even rashes. Poisoning is usually treated symptomatically. Recovery usually occurs within 24 h, but tingling may continue for a week or more. About 7% of the cases result in fatalities.

### 4.4.7 Amnesic Shellfish Poisoning (Domoic Acid)

Domoic acid (DA) was first reported and isolated from the red algae *Chondria armata* in 1958 [78]. In 1987, eating mussels harvested from Prince Edward Island leads to gastroenteritis, and numerous elderly people were affected, and/or those with underlying chronic diseases developed neurological symptoms such as loss of memory [79, 80]. In spite of the treatment, three patients (between ages 71 and 84 years) passed inside 11–24 days of exposure. The poisoning was connected to the domoic acid from the diatom *Nitzschia pungens* f. multi-series, which was consumed by the mussels during their feeding. Presence of DA has been rumored in the California shellfish in 1991, produced by *Nitzschia pseudodelicatissima*, and in anchovies (leading to pelican deaths), from *Nitzschia pseudoseriata* (now referred to as *Pseudo-nitzschia australis*). Domoic acid has been identified in shellfish in a number of provinces in Canada, Alaska, Washington, and Oregon; it may be as regular as PSP toxins [81]. Moreover, DA has also been found in seaweed.

In the Canadian outbreak, mice were injected with extracts (as in the PSP assay) and died within 3.5 h. The mice presented a scratching syndrome uniquely typical to DA, followed by progressively uncoordinated movements as well as seizures leading to the collapse of the mice on their sides, rolling over, and dying. Domoic acid at amounts >40 μg/g wet weight of mussel meat led to the mouse symptoms, while the Canadian authorities requires stopping of harvesting when DA levels approach 20 μg/g. Under normal circumstances, mice and rats can tolerate DA levels of approximately 30–50 mg/kg (mouse NOEL via intraperitoneal injection is 24 mg/kg). In human, DA is dose-responsive, without any effects between 0.2 and 0.3 mg/kg, with mild gastrointestinal symptoms noted at 0.9–1.9 mg/kg and the most serious symptoms observed at 1.9–4.2 mg/kg. In spite of the fact that rodents seem to be more tolerant to DA, the fatalities in human beings were related to other underlying illness or comorbidities. Domoic acid is an analog of kainic acid and glutamine, a neurotransmitter; the toxicity of all three is pretty much similar as they are excitatory and, thus, act on three types of receptors in the central nervous system, the most sensitive of which is the hippocampus. Domoic acid is potentially a more potent activator of tannic acid receptors than tannic acid itself. The stimulatory action may result in extensive damage of the hippocampus, however with less severe injury to the thalamic and forebrain regions.

### 4.4.8 Moray Eel Poisoning

Even though the moray eel (*Gymnothorax javanicus*) and other carnivorous fish can accumulate ciguatoxin from consuming other contaminated fish such as the Indo-Pacific moray eel (*Lycodontis nudivomer*), they have been reported to pose a mucous skin secretion with hemolytic, toxic, and hemagglutinating properties. These hemolytic properties can be detached from the hemagglutinating properties. Upon treatment with trypsin, the hemolytic characteristic is lost and is unstable when exposed to heat and acidic or alkaline media. Skin mucus of other species of eels, the common European eel (*Anguilla anguilla*) and pike eel (*Muraenesox cinereus*), was found to have proteinaceous toxins immunologically alike to that of the skin mucous toxin from the Japanese eel (*Anguilla japonica*).

### 4.4.9 Sea Urchin Poisoning

During the reproductive season, the etiological agent forms and is limited to the gonads. The sea urchins involved includes *Paracentrotus lividus*, *Tripneustes ventricosus*, and *Centrechinus antillarum*. The symptoms experienced comprise abdominal pain, diarrhea, nausea, vomiting, and migraine-like attacks. The toxin has been reported to result in interference with calcium uptake in nerve preparations.

### 4.4.10 Sea Turtle Poisoning (Chelonitoxin)

Chelonitoxin is the causative agent with its greatest concentration found in the liver but also found in the flesh, blood, viscera, and fat layers. The toxicity is sporadic (or seasonal), suggesting the poison may be resulting from toxic marine algae. Majority of incidents are prevalent in the Indo-Pacific region. There are several turtles involved in the outbreaks, including the green sea turtle, the hawksbill, and the leatherback turtles. In Sri Lanka, the natives offer the liver first to crows to determine its safety and consume it based on whether the birds consumed it or not. As a result of symptoms appearing after several hours to days, thus, the bioassay requires patience. In human, symptoms of poisoning comprise sore throat, tongue, and lips; trouble in swallowing; a white coating on the tongue (which may become covered with pin-sized, pustular papules); foul breath; tightness of the chest; coma; vomiting; diarrhea; and fatality. The toxin has reportedly been transferred from intoxicated to nursing infants. Congestion of internal organs, interstitial pulmonary edema, and necrosis of myocardial fibers have been revealed in postmortem examinations. Fatality rates that range from 7% to 25% have been reported.

## References

1. Ostroff S. Yersinia as an emerging infection: epidemiologic aspects of Yersiniosis. Contrib Microbiol Immunol. 1995;13:5–10.
2. Brunelle T, Dumas P, Souty F. The impact of globalization on food and agriculture: the case of the diet convergence. J Environ Dev. 2014;23(1):41–65. https://doi.org/10.1177/1070496513516467.

3. SADAOC. Food hygiene and the problem of street food in West Africa. Six monthly bulletins on food security policies and strategies in West Africa. 2002;6(1). http://www.sadaoc.bf/anglais/sadaocinfo6.htm. Accessed 15 Apr 2020.
4. FAO. The future of food and agriculture: trends and challenges. Rome: FAO; 2017. http://www.fao.org/3/a-i6583e.pdf. Accessed 13 Apr 2020.
5. WHO. Food borne disease: a focus for health education. Geneva: WHO; 2000. https://apps.who.int/iris/bitstream/handle/10665/42428/9241561963.pdf. Accessed 10 Apr 2020.
6. Bintsis T. Foodborne pathogens. AIMS Microbiol. 2017;3(3):529–63. https://doi.org/10.3934/microbiol.2017.3.529.
7. Bacon RT, Sofos JN. Characteristics of biological hazards in foods. In: Schmidt RH, Rodrick GE, editors. Food safety handbook. New Jersey: Wiley; 2003. p. 157–95.
8. Celandroni F, Vecchione A, Cara A, et al. Identification of *Bacillus* species: implication on the quality of probiotic formulations. PLoS One. 2019;14(5):e0217021. https://doi.org/10.1371/journal.pone.0217021.
9. Hauge S. Food poisoning caused by aerobic spore-forming bacilli. J Appl Bacteriol. 1955;I8:591.
10. Ormay L, Novotny T. The significance of *B. cereus* food poisoning in Hungary. In: Kampelmacher EH, Ingram M, Mossel DAA, editors. The microbiology of dried foods. Proceedings of the sixth international symposium on food microbiology, Bilthoven, The Netherlands, Jone 1988. Haarlem: Grafische Industrie; 1969. p. 279.
11. Artenstein AW. Biological attack. Ciottone's Disaster Med. 2016:480–8. https://doi.org/10.1016/B978-0-323-28665-7.00079-0 Accessed 11 Apr 2020.
12. McDowell RH, Sands EM, Friedman H. *Bacillus cereus.* [Updated 2019 Jul 5]. In: StatPearls [Internet]. Treasure Island: StatPearls Publishing; 2020. https://www.ncbi.nlm.nih.gov/books/NBK459121/. Accessed 8 Apr 2020.
13. Bell JH, Fee E, Brown TM. Anthrax and the wool trade. 1902. Am J Public Health. 2002;92(5):754–7. https://doi.org/10.2105/ajph.92.5.754.
14. Espelund M, Klaveness D. Botulism outbreaks in natural environments—an update. Front Microbiol. 2014;5:287. https://doi.org/10.3389/fmicb.2014.00287.
15. Sobel J, Tucker N, Sulka A, et al. Foodborne botulism in the United States, 1990-2000. Emerg Infect Dis. 2004;10(9):1606–11. https://doi.org/10.3201/eid1009.030745.
16. Barber MA. Milk poisoning due to a type of *Staphylococcus albus* occurring in the udder of a healthy cow. Philipp J Sci. 1914;9:515–9.
17. Dack GM, Cary WE, Woolpert O, et al. An outbreak of food poisoning proved to be due to a yellow hemolytic *Staphylococcus*. J Prev Med. 1930;4:167–75.
18. Paparella A, Serio A, Rossi C, et al. Food-borne transmission of staphylococci. In: Pet-to-man travelling staphylococci: a world in progress. 1st ed. London: Elsevier; 2018. p. 71–94.
19. Scallan E, Hoekstra RM, Angulo FJ, et al. Foodborne illness acquired in the United States—major pathogens. Emerg Infect Dis. 2011;17(1):7–15.
20. EFSA. The European Union summary report on trends and sources of zoonoses, zoonotic agents and food-borne outbreaks in 2016. EFSA J. 2017;15:5077.
21. Wang W, Baloch Z, Jiang T, et al. Enterotoxigenicity and antimicrobial resistance of *Staphylococcus aureus* isolated from retail food in China. Front Microbiol. 2017;8:2256.
22. GBD 2015 Mortality and causes of death, collaborators. Global, regional and national life expectancy, all-cause mortality, and cause-specific mortality for 249 causes of death, 1980-2015: a systematic analysis for the global burden of disease study 2015. Lancet. 2016;388(10053):1459–544. https://doi.org/10.1016/s0140-6736(16)31012-1.
23. Williams LP Jr, Helsdon HL. Pet turtles as a cause of human salmonellosis. JAMA. 1965;192:347–51.
24. Crump JA. Progress in typhoid fever epidemiology. Clin Infect Dis. 2019;68(Suppl 1):S4–9. https://doi.org/10.1093/cid/ciy846.

25. Soofi SB, Habib MA, Seidlein LV, et al. A comparison of disease caused by *Shigella* and *Campylobacter* species: 24 months community-based surveillance in 4 slums of Karachi, Pakistan. J Infect Public Health. 2011;4:12–21.
26. Vogt RL, Sours HE, Barrett T, et al. *Campylobacter enteritis* associated with contaminated water. Ann Intern Med. 1982;96(3):292–6. https://doi.org/10.7326/0003-4819-96-3-292.
27. Lund BM, Peck MW. A possible route for foodborne transmission of *Clostridium difficile*? Foodborne Pathog Dis. 2015;12(3):177–82. https://doi.org/10.1089/fpd.2014.1842.
28. Centers for Disease Control and Prevention (CDC). Clostridium perfringens. Centers for Disease Control and Prevention. 2018. https://www.cdc.gov/foodsafety/diseases/clostridium-perfringens.html. Accessed 21 Apr 2020.
29. Leung JM, Gallant CV. Infections due to *Escherichia* and *Shigella*. In: Reference module in biomedical sciences. Amsterdam: Elsevier; 2014. https://doi.org/10.1016/B978-0-12-801238-3.05090-X.
30. Taneja N, Mewara A. Shigellosis: epidemiology in India. Indian J Med Res. 2016;143(5): 565–76. https://doi.org/10.4103/0971-5916.187104.
31. Mitscherlich E, Marth EH. Microbial survival in the environment: bacteria and rickettsiae important in human and animal health. Berlin: Springer-Verlag; 1984.
32. Ameer MA, Wasey A, Salen P. *Escherichia coli* (*E coli* 0157 H7). [Updated 2020 Mar 17]. In: StatPearls [Internet]. Treasure Island: StatPearls Publishing; 2020. https://www.ncbi.nlm.nih.gov/books/NBK507845/. Accessed 21 Apr 2020.
33. Ochoa TJ, Contreras CA. Enteropathogenic *Escherichia coli* infection in children. Curr Opin Infect Dis. 2011;24(5):478–83. https://doi.org/10.1097/QCO.0b013e32834a8b8b.
34. Ekici G, Dümen E. *Escherichia coli* and food safety. In: The universe of *Escherichia coli*. London: IntechOpen; 2019. https://doi.org/10.5772/intechopen.82375.
35. Konowalchuk J, Speirs JI, Stavric S. Vero response to a cytotoxin of *Escherichia coli*. Infect Immun. 1977;18:775–9.
36. Harkins VJ, McAllister DA, Reynolds BC. Shiga-toxin E. coli hemolytic uremic syndrome: review of management and long-term outcome. Curr Pediatr Rep. 2020;8:16–25. https://doi.org/10.1007/s40124-020-00208-7.
37. Islam MA, Mondol AS, Azmi IJ, et al. Occurrence and characterization of Shiga toxin–producing *Escherichia coli* in raw meat, raw milk, and street vended juices in Bangladesh. Foodborne Pathog Dis. 2010;7:1381–5.
38. Vivant AL, Garmyn D, Piveteau P. *Listeria monocytogenes*, a down-to-earth pathogen. Front Cell Infect Microbiol. 2013;3:87. https://doi.org/10.3389/fcimb.2013.00087.
39. Ivanek R, Gröhn YT, Wiedmann M. *Listeria monocytogenes* in multiple habitats and host populations: review of available data for mathematical modeling. Foodborne Pathog Dis. 2006;3(4):319–0336. https://doi.org/10.1089/fpd.2006.3.319.
40. Kokashvili T, Whitehouse CA, Tskhvediani A, et al. Occurrence and diversity of clinically important *Vibrio* species in the aquatic environment of Georgia. Front Public Health. 2015;3: 232. https://doi.org/10.3389/fpubh.2015.00232.
41. Ojeda Rodriguez JA, Kahwaji CI. Vibrio Cholerae. [Updated 2020 Feb 5]. In: StatPearls [Internet]. Treasure Island: StatPearls Publishing; 2020. https://www.ncbi.nlm.nih.gov/books/NBK526099/. Accessed 23 Apr 2020.
42. Sabina Y, Rahman A, Ray RC. *Yersinia enterocolitica*: mode of transmission, molecular insights of virulence, and pathogenesis of infection. J Pathog. 2011;429069:1–10. https://doi.org/10.4061/2011/429069.
43. Zain ME. Impact of mycotoxins on humans and animals. J Saudi Chem Soc. 2011;15(2): 129–44. https://doi.org/10.1016/j.jscs.2010.06.006.
44. Marcellino OSBSN, Benson DR. The good, the bad, and the ugly: tales of mold-ripened cheese. Microbiol Spectr. 2013;1(1). https://doi.org/10.1128/microbiolspec.CM-0005-12.
45. Eskola M, Kos G, Elliott CT, et al. Worldwide contamination of food-crops with mycotoxins: validity of the widely cited 'FAO estimate' of 25%. Crit Rev Food Sci Nutr. 2019;60(16): 2773–89. https://doi.org/10.1080/10408398.2019.1658570.

46. Squire RA. Ranking animal carcinogens: a proposed regulatory approach. Science. 1981;214: 877–80.
47. European Food Safety Authority (EFSA). Opinion of the scientific panel on contaminants in the food chain [CONTAM] related to the potential increase of consumer health risk by a possible increase of the existing maximum levels for aflatoxins in almonds, hazelnuts and pistachios and derived products. EFSA J. 2007;5(3):446–53. https://efsa.onlinelibrary.wiley.com/doi/pdf/10.2903/j.efsa.2007.446. Accessed 15 Apr 2020.
48. Pfohl-Leszkowicz A, Petkova-Bocharova T, Chernozemsky IN, et al. Balkan endemic nephropathy and associated urinary tract tumours: a review on aetiological causes and the potential role of mycotoxins. Food Addit Contam. 2002;19(3):282–302. https://doi.org/10.1080/02652030110079815.
49. Anfossi L, Giovannoli C, Baggiani C. Mycotoxin detection. Curr Opin Biotechnol. 2016;37: 120–6.
50. Hocking AD. Common mycotoxigenic species of *Fusarium*. In: Semple RL, Frio AS, Hicks PA, Lozare JV, editors. Mycotoxin prevention and control in food grains. Regnet/AGPP; 1991.
51. European Food Safety Authority (EFSA). Scientific opinion on the risks for public health related to the presence of zearalenone in food. EFSA J. 2011;9:2197. https://efsa.onlinelibrary.wiley.com/doi/pdf/10.2903/j.efsa.2011.2197.
52. Haschek WM, Voss KA. Chapter 39—Mycotoxins. In: Haschek WM, Rousseaux CG, Wallig MA, editors. Haschek and Rousseaux's handbook of toxicologic pathology. 3rd ed. Boston: Academic Press; 2013. p. 1187–258.
53. Sydenham EW, Shephard GS, Thiel PG, et al. Fumonisin contamination of commercial corn-based human foodstuffs. 1. Agric Food Chem. 1991;39:2014–8.
54. Santini A, Meca G, Uhlig S, et al. Fusaproliferin, beauvericin and enniatin: occurrence in food—a review. World Mycotoxin J. 2012;5:71–81.
55. Babaali E, Abbasi A, Sarlak Z. Risks of Patulin and its removal procedures: a review. Int J Nutr Sci. 2017;2(1):10–5.
56. Burnside JE, Sipple WL, Forgacs J, et al. A disease of swine and cattle caused by eating moldy corn. II. Experimental production with true cultures of molds. Am J Vet Res. 1957;69:817–24.
57. Haarmann T, Rolke Y, Giesbert S, et al. Ergot: from witchcraft to biotechnology. Mol Plant Pathol. 2009;10(4):563–77. https://doi.org/10.1111/j.1364-3703.2009.00548.x.
58. Urga K, Debella A, W'Medihn Y, et al. Laboratory studies on the outbreak of gangrenous ergotism associated with consumption of contaminated barley in Arsi, Ethiopia. J Health Dev. 2002;16:317–23.
59. Wink M. Plant secondary metabolites modulate insect behavior-steps toward addiction. Front Physiol. 2018;9:364. https://doi.org/10.3389/fphys.2018.00364.
60. Sinclair ARE, Leakey MD, Norton-Grieffiths M. Migration and hominid bipedalism. Nature. 1986;324:307–8.
61. Ishida M, Hara M, Fukino N, et al. Glucosinolate metabolism, functionality and breeding for the improvement of Brassicaceae vegetables. Breed Sci. 2014;64(1):48–59. https://doi.org/10.1270/jsbbs.64.48.
62. World Health Organization. Cyanogenic glycosides. Toxicological evaluation of certain food additives and naturally occurring toxicants WHO food additive series 30. Geneva. 1993. http://wwwinchemorg/documents/jecfa/jecmono/v30je18htm. Accessed 20 Apr 2020.
63. FAO. Why cassava? http://www.fao.org/ag/AGP/agpc/gcds/index_en.html. Accessed 19 Apr 2020.
64. Kwok J. Cyanide poisoning and cassava. Food safety focus (19th issue February, 2008). Incident focus 2008. http://wwwcfsgovhk/english/multimedia. Accessed 13 Apr 2020.
65. Güçlü-Üstündağ Ö, Mazza G. Saponins: properties, applications and processing. Crit Rev Food Sci Nutr. 2007;47:231–58. https://doi.org/10.1080/10408390600698197.
66. Leung KW, Wong AS. Ginseng and male reproductive function. Spermatogenesis. 2013;3(3): e26391. https://doi.org/10.4161/spmg.26391.

67. Norn S, Kruse PR. Hjerteglykosider: Fra oldtiden over Witherings digitalis til endogen glykosider [Cardiac glycosides: from ancient history through Withering's foxglove to endogeneous cardiac glycosides]. Dan Medicinhist Arbog. 2004:119–32.
68. Silverman ME. William withering and an account of the foxglove. Clin Cardiol. 1989;12(7): 415–8.
69. Koelle GB. Anticholinesterase agents. In: Goodman LS, Gilman A, editors. The pharmacological basis of therapeutics. 5th ed. New York: Macmillan; 1975. p. 445.
70. Davis W. The serpent and the rainbow. New York: Warner Books; 1985. p. 36–7.
71. Fraser TR. On the characters, actions, and therapeutic use of the ordeal bean of Calabar. Edinb Med J. 1863;9:124–32.
72. Sinha K, Khare V. Review on: ant nutritional factors in vegetable crops. Pharma Innov J. 2017;12:353–8.
73. Moreira R, Pereira DM, Valentão P, et al. Pyrrolizidine alkaloids: chemistry, pharmacology, toxicology and food safety. Int J Mol Sci. 2018;19(6):1668. https://doi.org/10.3390/ijms19061668.
74. Sharma RP, Salunkhe DK. Solanum glycoalkaloids. In: Cheeke PR, editor. Toxicants of plant origin, vol. 1. Boca Raton: CRC Press; 1989. p. 179–236.
75. Joint FAO/WHO Expert Committee on Food Additives. Evaluation of certain food additives and naturally occurring toxicants. WHO technical report series. Geneva. 1992. https://apps.who.int/iris/bitstream/handle/10665/40033/WHO_TRS_828pdf?sequence=1. Accessed 14 Apr 2020.
76. Riddle J. Contraception and abortion from the ancient world to the renaissance. Cambridge: Harvard University Press; 1994. ISBN 9780674168763.
77. Traylor J, Mathew D. Histamine (scombroid toxicity, Mahi-Mahi flush) toxicity. [Updated 2020 Feb 18]. In: StatPearls [Internet]. Treasure Island: StatPearls Publishing; 2020. https://www.ncbi.nlm.nih.gov/books/NBK499871/. Accessed 5 Apr 2020.
78. Takemoto T, Daigo K. Constituents of Chondria armata. Chem Pharm Bull. 1958;6(578):580.
79. Wright JLC, Boyd RK, de Freitas ASW, et al. Identification of domoic acid, a neuroexcitatory amino acid, in toxic mussels from eastern Prince Edward Island. Can J Chem. 1989;67:481–90.
80. Perl TM, Bédard L, Kosatsky T, et al. An outbreak of toxic encephalopathy caused by eating mussels contaminated with domoic acid. N Engl J Med. 1990;322:1775–80.
81. Bates SS, Bird CJ, de Freitas ASW, et al. Pennate diatom as the primary source of domoic acid, a toxin in shellfish from eastern Prince Edward Island, Canada. Can J Fish Aquat Sci. 1989;46: 1203–15.

# Chapter 5
# Food Contaminants

**Yi Shuai, Haixia Sui, Gonghua Tao, Qian Huo, Chen Li, and Naimin Shao**

**Abstract** Nowadays, with food safety incidents occurring repeatedly in many countries, consumers and organizations are increasingly concerned about the quality and safety of foodstuffs. The presence of food contaminants in humans has become subject of intense research for human exposure and health risk assessment. The release of an alarming number of toxic polluting agents such as volatile organic compounds, dyes, heavy metals, pharmaceuticals, pesticides, industrial wastes, radiation contamination, and personal care products due to natural or anthropogenic activities poses direct adverse effects on human health and living entities. Institutions like the World Health Organization and various nations have categorized food contamination into several groups: chemical, biological, physical, and cross-contamination. All these make them a threat to human health and the environment across the globe. Here, we describe various concepts related to food contaminants, such as their sources, risk characterization, formation mechanism, and health effect. And a thorough understanding of these concepts can help researchers gain better insight into food contaminants.

**Keywords** Pesticide · Veterinary drug · Heavy metals · Persistent organic pollutants · Process-induced toxicants · Food irradiation

Y. Shuai (✉)
Syngenta (China) Investment Co., Ltd, Shanghai, People's Republic of China
e-mail: yi.shuai@syngenta.com

H. Sui
Division III of Risk Assessment, China National Center for Food Safety Risk Assessment, Beijing, People's Republic of China
e-mail: suihaixia@cfsa.net.cn

G. Tao · Q. Huo · C. Li · N. Shao
Shanghai Municipal Center for Disease Control and Prevention, Shanghai, People's Republic of China



L. Zhang (ed.), *Nutritional Toxicology*,
https://doi.org/10.1007/978-981-19-0872-9_5

## 5.1 Pesticides and Veterinary Drug Residues in Food

Natural or synthetic pesticides and veterinary drugs are a diverse group of chemicals used to improve crop productivity and prevent major biological disasters involving carriers of human or animal disease and harmful plants or animals to protect human and animal health. Pesticide residue refers to any specific substances in or on the surface of food, agricultural commodities, or animal feed resulting from the use of pesticides. Pesticide and veterinary residues include residues from unknown or inevitable sources (e.g., environmental) as well as known uses of the chemical, their derivatives, metabolites, by-products, and impurities of the pesticide and veterinary considered to be toxic.

Pesticide residues in soil may affect the nature and level of residues in crops, especially for soil or seed treatments. Rotational crop studies aim to determine the characteristics and levels of pesticide residues in new crop after the original crop applied with the targeted pesticides was harvested.

There are different definitions of pesticide residues for enforcement or dietary exposure assessment purposes. Residues and their transformations of parent pesticides are usually expressed as equivalents of the parent pesticide.

For purposes of dietary exposure, definitions of pesticide residues should also include pesticide metabolites, photolysis, or other degradation products with toxic properties similar to those of their parent pesticides. For purposes of enforcement (analyzed food for compliance with maximum residue limits (MRLs)), transformations may not be included in the definition of residue, if they are only the minorities of the residue or if they are difficult or expensive to conduct an analysis on. It is usually possible to avoid transformations in the residue definition, which are identical to other pesticides if these pesticides have separate sets of MRLs. Otherwise, the law enforcement work will be abnormal.

Pesticide residues are a concern for consumers because pesticides are known to have potential harmful effects to other nontargeted organisms than pests and diseases. People are mainly exposed to pesticides through diet. Human exposure to pesticide residues through food is usually much higher than those from drinking water and air inhalation (except for occupational exposure or at-home application, e.g., home gardens or animal handling). Nowadays, risk assessment of pesticide residues is the most important to ensure food quality and protect human health [1].

The definition of residues used to estimate dietary exposure depends on the results of metabolism, toxicological studies, and their general applicability in comparing the estimation of dietary intake with the ADI and acute reference dose (ARfD) [2]. Depending on the identified relevant end point of toxicity, two dietary exposure approaches could be relevant, namely, acute and chronic exposure assessment.

The general methodology used to estimate acute dietary exposure to food chemicals at the global level is the International Estimated Short-Term Intake (IESTI) method, which is currently based on consumption of a single food product at the highest available 97.5th percentile for a short period of time (i.e., within

1 day), combined with the highest residues (HR) of the chemicals assessed assuming its presence in a single food eaten by consumers of the population group studied. There are four different cases to calculate acute dietary exposure.

For Joint FAO-WHO Meeting on Pesticide Residues (JMPR), the HR is obtained from the supervised field trials conducted in accordance with good agricultural practices (GAP). For the Joint FAO/WHO Expert Committee on Food Additives (JECFA), depending on the chemical, the HR is obtained from depletion studies in respective tissue according to good veterinary practices (GVP).

JMPR uses the international estimated daily intake (IEDI) method to estimate chronic dietary exposure. The underlying assumption is average daily per capita consumption of raw and semi-processed agricultural commodities based on the FAO food balance sheet (food supply data, GEMS/Food consumption cluster diets) combined together with the median of the residues from supervised field trials (STMR) conducted according to good agricultural practices (GAP).

For comparison, the European Food Safety Authority (EFSA) uses the average consumption data from the EFSA Comprehensive European Food Consumption Database in combination with the STMR from dossier provided by applicants (pesticide residue intake model (PRIMo) exposure model, available from the EFSA website).

However, the use of the GEMS/Food consumption cluster diets tends to overestimate the average consumption, underestimate specific food consumption for the average consumers, and underestimate food consumption of high consumer's patterns. Moreover, cluster diets do not cover sensible populations such as toddlers, children, and adolescents.

For veterinary drugs, a theoretical modelled diet intended to represent the high daily consumption of animal products (meat tissues, fish, milk, eggs, and honey) is used. The modelled diet is combined with the median of the residue distribution from depletion studies in respective tissues obtained from studies reflecting GVP.

For comparison, the Committee for Medicinal Products for Veterinary Use (CVMP) from the European Medicines Agency (EMA) uses the same exposure modelling approach as JECFA (theoretical food basket and median residues from depletion studies).

However, the Joint FAO/WHO Expert meeting on dietary exposure assessment methodologies for residues in veterinary drugs in 2011 showed that the use of the theoretical JECFA modelled diets in comparison to the highest consumption level (P95th) reported from individual food consumption survey tends to be conservative enough for some broad food commodities like milk, fish, seafood, and poultry but not conservative enough for some other food commodities like beef, pork, sheep, goat, mammalian offal, eggs, fish, crustaceans, molluscs, and honey. Based on the limited examples provided in the report, the expert meeting showed that the current JECFA model could underestimate exposure by 30% for adult population and up to 100% for infant population compared with the approach using individual food consumption data. It concluded that the current JECFA modelled diet may not be as conservative as previously assumed.

More recently, the 78th JECFA meeting has conducted a pilot study with a limited number of case studies to explore new approaches for estimating exposure to veterinary drug residues aiming for routine use in its safety assessment. The calculations of dietary exposure assessment were calculated for four veterinary drug residues using the current JECFA modelled diet approach as well as the new proposed exposure model (called GECDE for Global Estimate of Chronic Dietary Exposure) with the calculation assumptions described above (highest high level of exposure from one food category + the mean exposure values for the remaining categories). Overall, the JECFA at its 78th meeting agreed that the new exposure approach is preferable to the current model because it is based on more accurate consumption data, which provides improved estimates of dietary exposure to veterinary drugs. It also recommended to use the new approach in parallel with the current model diet approach in future committee meetings.

People may be exposed to pesticides through occupational or environmental pathway. Biomonitoring was recognized as the best approach to assess total exposure. It provides information on internal doses and reflects actual exposures rather than estimated exposures [3]. In addition, biomonitoring exposure provides early warning signs before irreversible changes occur and allows inter- and intraindividual variability to be recorded. Urine biomarkers are commonly used as exposure indicators for many pesticides that have short biological half-lives and are largely eliminated in urine [4]. Meanwhile, in order to understand the results of biomonitoring properly, it is necessary to understand the kinetics of target pesticide and its metabolites and sample collection strategy [5]. Once the pesticide kinetics has been documented in human, biological guidance values designed to prevent harmful effects may then be obtained by correlating the safe exposure levels with corresponding urinary biomarker values [6]. For the general population, governmental organizations have not yet proposed biological limit values for pesticides [7]. Biomonitoring values are preferred to be compared with those obtained in large-scale longitudinal epidemiological studies and considered as a reference for baseline or control value [8].

### *5.1.1 Pyrethroids*

The use of synthetic pesticides had a significant impact on agriculture and human health. In the past few decades, attention has turned to pesticide residues in foods. Pesticide residues in foods have become a global concern due to the adverse effects of pesticide residues on human health. Pesticide contamination of surface water and groundwater is harmful to agriculture. Most of the pesticides used nowadays degrade rapidly in the environment or are too insoluble to move through the environment. Some of the pesticides used in the past years are either persistent, mobile, or both.

Pyrethroid pesticide is a class of synthetic insecticides based on the design and optimization of the pyrethrin structure. Pyrethroids are found in natural pyrethrum extracted from chrysanthemum flowers [9]. Because pyrethroids are relatively safe for humans and have high insecticidal potency, they are used as insecticides in agriculture to control insects in vegetables, fruits, and field crops and in public health to control disease caused by vectors and as a veterinary drug against ectoparasites [10]. Synthetic PIs have high insecticidal efficacy and relatively low mammalian toxicity, which are similar to natural esters. In addition, compared with the prototype compounds, the synthetic PIs have the advantage of being less susceptible to hydrolysis and photodegradation [11]. In addition to controlling agricultural pests, PIs are also effective ingredients in household insecticides.

Pyrethroid pesticides do not persist in the environment and have weak resistance to insect. Pyrethroids are more effective against a wider range of insect pests. Furthermore, pyrethroids are much less toxic to mammals than organochlorine, organophosphate, and carbamate counterparts. Thus, the use of pyrethroids has continued to increase over the past two decades [12–14]. However, the widespread use of pyrethroids has led to an increased exposure of workers and the ecosystem and increased possibility of pyrethroids entering the food chain through residues in meat, fruit, vegetables, and water. The widespread use of pyrethroids has led to population resistance with different levels worldwide, albeit with substantial geographical variation.

According to the chemical structure and toxicity of acute intravenous or intracerebral administration, pyrethroids are divided into two subclasses (type I and type II) based on exposure to rodents [15]. Type I PIs (e.g., allethrin or permethrin) involve a variety of structurally non-cyano compounds and cause poisoning syndromes including general tremor and convulsion. Meanwhile, type II PIs (e.g., deltamethrin and cypermethrin) display the cyano and trigger a poisoning syndrome, which includes salivation and choreoathetosis.

Cismethrin (T-syndrome) and deltamethrin (CS-syndrome) pyrethroids were shown to activate and inactivate membrane depolarization; each type of PI exhibits secondary target relative to its acute neurotoxicity.

Although PIs are rapidly metabolized, there is also evidence that it has toxic effects even under low doses of exposure and increases the sensitivity of developmental organisms. Due to the limited metabolic capacity of immature rodents in the neonatal period, the differences in toxicokinetics between juvenile and adult rodents are the causes of age-related differentiation [16]. For example, the deltamethrin of type II PI is detoxified by cytochrome P450s and carboxylesterases, which increase the ability of these metabolizes to this PI during the postnatal development of rats [17]. According to these data, the severity of neurotoxicity caused by deltamethrin is negatively corrected with age [18]. The oral toxicity of pyrethroids to mammals is relatively low. Generally, the toxicity ratio of insects (partial) to mammals (oral) is much higher than that of other major types of insecticides [19]. The LD50 values of most compounds after application combined with vegetable oil are more than 50 mg/kg per body weight, so they are classified as moderate toxicity and equivalent to Globally Harmonized System of Classification and Labelling of Chemicals (GHS) category 3 or above. In conclusion, pyrethroids have very low acute dermal toxicity and moderate acute inhalation toxicity [20]. The acute toxicity of pyrethroids with a cyano is significantly greater than the toxicity of pyrethroids combined with vegetable oils.

In most instances, the results observed under different experimental conditions are inconsistent, which complicates the analysis of pyrethroid potency. In many reports, similar animal models, test compounds, and dosing vehicles have been used to study the comparable neurobehavioral areas. However, due to inconsistent experimental conditions or incomplete description of experimental methods, the experimental results are not comparable. The neurotoxicity induced by pyrethroids may be affected by a variety of biological factors and/or experimental conditions [21–23], which include vehicle, exposure pathway [24, 25], dose [26], composition of isomers [20], pesticide formulation [27], body size [28], and age [29]. There is also new evidence showing that different testing tools [30] and lack of food [31] may also determine differences in results after acute exposure to pyrethroids. Finally, circadian rhythms [32, 33], ambient temperature, and housing conditions [34, 35] are additional factors that may influence the neurobehavioral end point to estimate pyrethroid potency. Determining the biological and experimental determinants of pyrethroid neurotoxicity and their actual impact on estimating pyrethroid potency can provide valuable guidance for interpreting neurobehavioral data collected in various laboratory environment. Based on factors that affect the neurotoxicity end points of pyrethroids, it is essential to consider experimental conditions carefully before conclusions are elaborated within or across studies.

With the increasing use of pyrethroids and recent studies showing exposure to pregnant women and children, there is growing concern about potential developmental neurotoxicity. Pyrethroids at higher doses had mildly teratogenic effects. Exposure of zebra fish embryos to pyrethroids increased mortality and the development of pericardial effusion that is dose-dependent. Type II compounds are the most potent. Meanwhile, permethrin and deltamethrin cause craniofacial deformities at doses close to LC50. However, body axis curvature and spasms were observed at lower doses, which was known as a typical syndrome associated with pyrethroid toxicity. Other studies have shown that pyrethroid-induced neurotoxicity is similar in zebra fish and mammals, including treatment with diazepam to improve spasms and bendiness and treatment with the sodium channel antagonist MS-222 for ameliorated spasms and body curvature.

Pyrethroids can possess one of three chiral centers, which results in two to eight isomers. Acute toxicity of pyrethroids and insecticidal action differ from stereospecificity of pyrethroids [36]. The toxicity of pyrethroids depends on the stereochemical configuration of cyclopropane C-1 or the homologous position of compounds lacking cyclopropane carboxylate moiety. Besides, ester of 1*R*-cyclopropane carboxylate isomeric and 2*S* isomers of non-cyclopentanecarboxylate esters of primary alcohols also strongly influence toxicity. Compounds with the 1*R* and *cis* conformation are also toxic. Furthermore, the acute toxicity is enhanced due to the presence of α-cyano substituent in the *S* configuration of 3-phenoxybenzyl alcohol moiety.

There are little data on the indications of pyrethroid-induced toxicity in larger mammal species. Some clinical veterinary reports have described the pyrethroid poisoning in dogs and cats, which are mainly concerned dermal exposures. The characteristics of pyrethroid poisoning were similar in the mammalian species investigated. According to veterinarian observations, acute and sublethal exposure to pyrethroids causes restlessness and hyperexcitability followed by "drunken movements" (i.e., locomotor ataxia), mydriasis, diarrhea, and general depression. In some cases, motor incoordination, paresis, head bobbing, chewing, hypersalivation, and/or whole-body tremors have also been reported [37, 38].

There is little data available from quantitative assessment of acute neurotoxicity in humans and primates following acute pyrethroid exposures, which precludes direct comparisons between the available laboratory rodent studies in peer-reviewed literature and human clinical reports. In short, although mammalian species appear to be less sensitive to pyrethroid-induced neurotoxicity than insects, safety concerns need to be continued as these compounds have similar cross-species neurotoxicity when doses are reached in active target tissues.

Two different toxicity syndromes have been described in the previous studies. According to the results of the acute intravenous toxicity of 36 pyrethroids in rats, the toxicological syndrome trigged by pyrethroids is classified into Verschoyle and Aldridge, tremor (T) and choreoathetosis with salivation (CS). T syndrome involves aggressive sparring, sensitivity to external stimuli, whole-body tremor, and prostration, which is mainly observed in the compounds with α-non-cyano-containing pyrethroids, such as bifenthrin and permethrin. Meanwhile, CS syndrome involves twitch (athetosis) and clonic seizures, which is mainly observed in the compounds containing alpha-cyano pyrethroids, such as deltamethrin. In addition, there are exceptions to moderate symptoms. Lawrence and Casida confirmed the classification by administration of 29 pyrethroids to mice via intracerebral injection [39].

### *5.1.2 General Remarks and Conclusions*

Residue occurred when a drug or pesticide is applied to animals or plants that produce food. Residue compliance is defined as a combination of pesticides and their metabolites, derivatives, and related compounds for which MRL is applicable.

The definitions of pesticide residues and veterinary drug residues are equivalent. For the definition of "veterinary drug residue," the expression "considered to be of toxicological significance" was added to the definition of "pesticide residue." In the same case, the definition of residues does not include other substances that may appear as adjuvants in formulated products, carriers, or delivery devices.

Whether it is a pesticide or a veterinary drug, it is of utmost importance to identify the composition of the active substance, especially when stereoisomers are involved. Under such circumstances, relative proportions of the isomers should be given. Sometimes, only one active isomer is more biologically active than others.

Pyrethroid insecticides are widely used to control insects in agriculture and control disease caused by vectors in public health. The implication of increasing use of pyrethroids applications is a significant challenge for the environmental and biomedical research communities. The principal effect of pyrethroids is its neurotoxicity. Pyrethroids are classified into two subclasses (types I and II) based on chemical structure and the intoxication syndrome via rodents' intravenous or intracerebral exposure. Dietary exposure to pyrethroids is a major source for nonoccupational population. Assessment is an effective strategy for dealing with pesticide contamination in the environment.

The toxic and pharmacological effects of pyrethroids suggest that a simple additivity model based on a combined effect on a single target is not appropriate for assessing the risks of cumulative exposure to multiple pyrethroids.

## 5.2 Heavy Metals

Food is the major source of exposure to both essential and nonessential metals. Some metal elements are necessary for the normal functions of organisms. They play key roles in numerous important biological processes, such as enzymatic reactions, erythropoiesis, electron transport, hormone and vitamin synthesis, and immune system functioning. Their lack or excess in the human body may be a cause of many serious diseases, while some metal elements, also called trace elements, are not essential for the human body and may cause severe poisonings if accumulated equal to or higher than the minimal dose in amounts.

It's difficult to determine the roles of metal elements and human daily requirements due to their low concentrations in the human body, also as effects that connected with the elimination of their constant inflow. Biologically, the human body is a coordinated organism that is in a dynamic balance. The relationship among various elements in the body is in an appropriate proportion, especially trace elements, and any excessive or insufficient element will make the proportion out of the balance. However, the absorption of trace elements from the environment is frequently and widely disturbed by human industrial and economic activities, which are leading to contamination of the environment, including foods, with heavy metals. Heavy metals are part of trace elements, and some of which have recently received particular attention.

Heavy metal toxicity is a dominant environment health problem in modern society, which is potentially dangerous due to bioaccumulation through the food chain. Moreover, it can cause risky effects on livestock and human health. Typically, "heavy metals" are described as metal elements with a density greater than 4.5 $g/cm^3$ under standard conditions or the specific weight over 5, which are distinguished from light metals (such as aluminum and magnesium). But what matters most are the chemical properties of the heavy metals comparing to their density. Owing to the toxicity of some heavy metals and the possibility of environmental contamination, the potential for high risk is associated to lead (Pb), cadmium (Cd), arsenic (As), mercury (Hg), nickel (Ni), iron (Fe), zinc (Zn), chromium (Cr), cobalt (Co), silver (Ag), and the platinum group elements [40].

### 5.2.1 *Exposure to Heavy Metals of Food from the Environment*

Although some sporadic cases of serious metal poisoning originated from contaminated food, the main worry is population risk related with exposures at levels that is lower than those problems causing overt or clinical signs of toxicity.

Exposure is considered throughout the whole lifetime for the population, including vulnerable groups. Food can be contaminated via air, soil, and water, as well as food processing and food packaging, which will be discussed in the following passages. In developing countries, the risk of exposure to metals via food is an increasingly growing problem as a result of, for instance, overused cadmium-containing fertilizers, contaminated water used to irrigate agricultural areas where staple food crops grow, and the location of metal-recycling facilities in these areas where food crops may be metal-contaminated by air and water releases [41]. The heavy metals in the environment can be from different sources (Fig. 5.1). And heavy metal pollution can be rooted in both natural and anthropogenic sources [42].

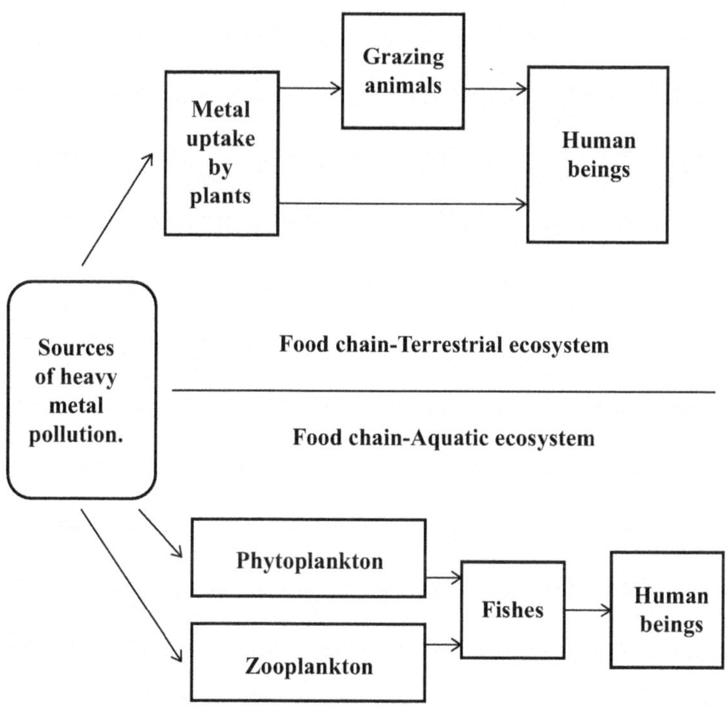


**Fig. 5.1** Human beings are exposed to heavy metal pollution via both food chain-terrestrial ecosystem and food chain-aquatic ecosystem

#### 5.2.1.1 Nature Source of Heavy Metals

Heavy metals root within the Earth's crust; for that, their occurrence in soil is naturally from weathering process. The most important natural source of heavy metals is geologic parent material and rock outcroppings. The different rock type and environmental conditions can cause the differences in composition and concentration of heavy metals. The geologic plant materials usually have relatively high concentrations of Hg, Pb, Cr, Ni, Cu, Zn, Cd, Mn, Co, and Sn. Class-wise, the heavy metal concentrations vary in the rocks, and soil formation occurs mostly from sedimentary rock. However, it is only a small source of heavy metals due to the difficult weathering processes.

#### 5.2.1.2 Agricultural Sources of Heavy Metals

Historically, agriculture was the dominant human effect on the soil. Plants need not only macronutrients (N, P, K, S, Ca, and Mg) but also essential micronutrients to grow and complete the lifecycle. Some heavy metals (such as Co, Cu, Fe, Mn, Mo, Ni, and Zn) are deficient in some soils, which are necessary for healthy plant growth [43]. In the meantime, crops may obtain these heavy metals from the soil or a foliar spray. To provide adequate N, P, and K for crop growth, a large amount of fertilizers is often added to soils in intensive farming systems. The compounds providing these elements contain trace amounts of heavy metals as impurities. Due to continued fertilizer, application may significantly increase their content in the soil. The content of heavy metals may be significantly increased by continuously fertilizing [44, 45].

The main sources of heavy metals in agricultural soil are fertilizers (inorganic and organic) including liming, sewage sludge, irrigation waters, and pesticides [46]. For example, lead arsenate was used in fruit orchards for decades to avoid from some parasitic insects. Cadmium accumulates in leaves of plants at very high levels, which is of particular concern, and these plants may be misused by animals or human. Cadmium enrichment also occurs as a result of the application of sewage sludge, manure, and limes. Although the contamination levels of heavy metals in agricultural soil account for a very small part, some metals could be still dangerously accumulated at a high risk if overusing the phosphate fertilizer due to its long persistence [47].

#### 5.2.1.3 Industrial Sources of Heavy Metals

Mining and refinement, including spoil heaps and tailings, smelting and metal finishing, transport of ores, and recycling of metals, are two main industrial sources of heavy metals. Mining operation releases different kinds of heavy metals up to the type of mining. For instance, As, Cd, and Fe enrich the soil around the coalfield directly or indirectly [48], which proves that coalmines are sources of these heavy metals.

Mining and milling of metal ores along with industries have caused a serious consequence in many countries, which means heavy metals accumulated from industries have contaminated the soil. During mining, tailings (heavier and larger particles settled at the bottom of the flotation cell during mining) are directly released into natural depressions, which causes the elevated concentrations in on-site wetlands. In order to mine and smelt considerable Pb and Zn ore, human and ecological health have been exposed to the risk of contaminations [49]. And, it is certain that this kind of mining has already caused contamination of soil. However, the possible ways to restore the soil productivity used to reduce the contaminations need a long period and a large amount of money. There are many researches that show that the bioavailability is associated with the soil heavy metal environmental risk to humans. Assimilation pathways include the ingestion of plant material grown in (food chain) contaminated soil or the direct ingestion (oral bioavailability) of contaminated soil.

Another way to release metals is in vapor forms, which is mainly from high-temperature processing of metals like smelting and castings. The water in the atmosphere could be combined with vapor form of heavy metals including As, Cd, Cu, Pb, Sn, and Zn to form aerosols. In that way, the heavy metals could be dispersed by wind (dry deposition) or enriched in rainfall (wet deposition), which result in the contamination of soil or water bodies [50].

In addition, other industrial sources like processing of plastics, textiles, microelectronics; wood preservation; and paper processing could also disperse too many heavy metals in the soil or water causing contaminations.

#### 5.2.1.4 Domestic Effluents

It has been 400 years that people attempt to reuse the municipal and industrial wastewater and related effluents to land. So far, it is estimated that wastewater is used to irrigate appropriately 20 million hectares of arable land. And in several Asian and African cities, studies suggest that wastewater irrigation becomes the main way in agriculture, which take up for 50% of the vegetable supply to urban areas [51]. Although the metal concentrations remained in wastewater effluents are usually comparatively low, the heavy metals still can enrich in the soil if using wastewater irrigation for a rather long time. Moreover, the wastewater, which is the largest single source, possibly increases metal values in rivers and lakes.

Domestic effluents, through different sewage disposal process, may contain untreated or solely mechanically treated wastewaters; substances from biological treatment plants and waste substances passed through sewage outfalls and discharged to receiving water bodies often end up into the sea from coastal residential areas [52]. In terms of pollution stemming from urbanized areas, there is an increasing trend that people start to pay more attention to the heavy metal contaminations, which means urban runoff presents a serious problem of heavy metal contamination.

#### 5.2.1.5 Atmospheric Sources

Some researches show that both natural and man-made processes can bring about metal-containing airborne particulates. Hinging on prevailing climatic conditions, these particulates may become windblown over great distances, although they are doomed to return to the lithosphere as precipitations by rain or snowfall eventually.

Since some metals can volatilize under a high temperature, they are released in the form of particulates contained in the gas stream, which mainly comes from automobile exhaust, fuel combustion, industrial exhaust emissions, and other sources. And airborne sources of the inhalable particles can also be classified into two categories: stack or duct emissions of air, gas, or vapor streams and fugitive emissions such as dust from storage areas or waste piles [53]. If the atmosphere is reductive continuously, these metals will turn to oxides and condense as fine particulates. Stack emissions can be dispersed through a wide area by natural air currents after they are removed from gas stream by dry and/or wet precipitation mechanisms. Due to the limitation of emission, fugitive emissions are often distributed over a much smaller area near the ground. Normally, contaminant concentrations are relatively lower in fugitive emissions than in stack emissions.

The site-specific conditions determine the type and concentration of metals from both types of sources. All solid particles in smoke are mainly from fires and emissions from factory chimneys. And, they are eventually deposited on land or sea. It has been a long time that the contaminants of heavy metals from fossil fuels account for a large scale, since the industrial revolution began. For instance, scientists have tested the environmental conditions adjacent to smelting works. And, the test shows there is a rather high concentration of Cd, Pb, and Zn in plants and soils nearby. The combustion of petrol containing tetraethyl lead, which discharges a great deal of Pb into the air, is another main source of soil contamination [54]. Therefore, the concentration of Pb in soils in urban areas and in those adjacent to major roads is especially high. Moreover, the sources being tires and lubricant oils may account for the reason Zn and Cd may also be found in the soils adjacent to roads.

In addition, geothermal sources such as volcanic eruptions have resulted in significant air pollutions by enriching metals in atmosphere.

#### 5.2.1.6 Other Sources

Other sources of heavy metals comprise refuse incineration, landfills, and some anthropic causes. Fly ash produced by burning coal and the corrosion of commercial waste products are two dominant anthropogenic sources that contaminate the soil to some extent by releasing Cr, Cu, Pb, and galvanized metals (primarily Zn) into the environment directly.

## 5.2.2 *Risk Characterization of Heavy Metals*

It is certain that ingesting contaminated food with heavy metals will cause a series of problems that endanger human health, which have the following common characteristics:

1. Strong cumulative toxicity
   Due to the long biological half-life of heavy metals, the heavy metals ingested by the human body will take a long time to discharge slowly.
2. High concentration in food chain
   Heavy metals can reach a high concentration in the organism and in the human body due to the bioaccumulation of the food chain, like the concentration of mercury, cadmium, and other metal poisoning content in aquatic products, which can be as high as hundreds to thousands times of the concentration in their living environment.
3. Health hazards to the human body
   Contaminated food with toxic heavy metals often causes chronic poisoning and long-term effects to the human body, such as carcinogenicity, teratogenicity, and mutagenicity.

## 5.2.3 *Main Toxic Heavy Metals*

The point of this chapter is to expound the four "classical" toxic metals in food, mercury, cadmium, lead, and arsenic, which are mentioned in the ten chemicals of major public health concern published by FAO/WHO.

### 5.2.3.1 Lead (Pb)

Lead is a metal pertaining to group IV and period 6 of the periodic table with atomic number 82, atomic mass 207.2, density 11.4 g/cm$^3$, melting point 327.4 °C, and boiling point 1725 °C. It is a natural element that usually is found as a mineral attached to other elements, such as sulfur (i.e., PbS, $PbSO_4$) or oxygen ($PbCO_3$), and ranges from 10 to 30 mg/kg in the Earth's crust.

Lead in foods may be obtained from the environment in which the food is grown or from food manufacturing process. Usually, due to the growing environment of agricultural crops such as near heavily traveled roads or industrial sources of lead, the concentration of lead is significantly high, which is caused by airborne lead emissions on them or in the soil. Among cases of lead poisoning written in the literature, it is the most often reported that source of high lead concentrations in foods is mainly from ceramic-glazed storage vessels containing acid foods. As many as 5% of children are at risk of lead poisoning because they have taken a rather high enough lead through water and food [54].

#### 5.2.3.1.1 Distribution in Food

Compared to adults, if taking higher food relative to size, higher metabolic levels, and greater motor activity, it will bring about higher dietary lead consumption. It is reported that the ways of getting lead from food are very many. However, it is still a challenging problem to establish a reasonable standard of lead in the diet due to the following reasons: (1) methodological limitations in the accurate analysis for lead in food and (2) the lack of good dietary survey data. Generally, the lead concentration levels in canned foods may be more than twice as much as in uncanned food.

According to the recent scientific studies conducted in 2010, the Joint FAO/WHO Expert Committee on Food Additives (JECFA) estimated that the standard established before, which showed the provisional tolerable weekly intake (PTWI) was 25 μg/kg body weight per week, could no longer be referred as healthy instructions. The weekly intake of lead might be lower than the standard, which showed this standard should be withdrawn. Since the dose-response analyses did not provide any indication of a threshold for the key adverse effects of lead, the Committee concluded that it was unlikely to set forth a new PTWI, which would be considered as a healthy instruction [55–57]. The limitations for lead in food are determined by the national food safety standards of China (GB2762-2012): the maximum concentration is 0.2 mg/kg in grain and grain products, 0.1 mg/kg in fresh vegetables and fruit, 0.2 mg/kg in meat and beans, 0.5 mg/kg in meat products, 0.5 mg/kg in fish and shellfish, 1.0 mg/kg in aquatic products, and 0.2 mg/mL in eggs and egg products.

#### 5.2.3.1.2 Uptake and Metabolism in Humans

The human body may absorb lead by ingestion and inhalation or through the skin. Therefore, the absorption of lead is quite different among the ways of intake. There are many factors that could affect the absorption of lead such as amount of lead presented to portals per unit time and the physical and chemical states in which lead is presented. Moreover, the physiological status and age could also influence it. For most individuals, the gastrointestinal tract is the major path of absorption. There are some researches that indicate that less than 10% of ingested lead are absorbed from the gastrointestinal tract. As for the remaining lead, most of it can be expelled through feces. Lead absorption through the skin is quite different compared with gastrointestinal lead absorption. When lead is absorbed through the skin or respiratory tract, it will be transported first into the plasma. Then, it will rapidly accumulate into the extra cellular fluid pool of sweat and saliva without significant uptake by erythrocytes [58].

Lead in blood is a type of nondiffusible form combined with erythrocytes, while it is a diffusible form in plasma. Plasma inhabits a central position in the distribution equilibrium. So, it would be considered as an index, which can reflect the concentration of lead in all the body tissues. The process is mainly as follows: When lead is absorbed into the body, it enters the blood stream where it rapidly binds to red blood cells or is carried to soft tissues, such as the kidney, liver, and heart. At last, it is transferred to the bone where the reservoir is large with long half-life.

Inorganic lead is not expelled through metabolism but primarily through the urinary tract. Most of absorbed lead (about 76%) is eliminated mainly via the kidney in the urine, while the remaining is via biliary secretion in the gastrointestinal tract [58]. In short, lead is excreted very slowly from the body due to its biological half-life, which is estimated to be nearly 10 years, thus promoting accumulation in the body.

##### 5.2.3.1.3 Toxicity and Clinical Effects

Lead does not do any good to biological function in the human body. It is common knowledge that it is a kind of toxic metal whose effects have been more extensively reviewed than the effects of other trace metals.

Lead can lead to significant injury to the kidneys, red blood cells, nervous system, and brain. Exposure to lead under different conditions, for example, the level and duration of exposure, can bring about a wide range of biological effects. As the teenagers and infants are more sensitive than adults, different range of doses can cause various effects. If children are exposed to lead, especially those who are under the age of 6, they are more easily to suffer from impaired development, shortened attention span, lower IQ, hyperactivity, and mental deterioration. As for adults, if they are exposed to lead, they usually show symptoms like loss of memory, decreased reaction time, nausea, insomnia, anorexia, and weakness of the joints [59].

Lead poisoning, which can cause severe illness, is now very rare. Over the past decades, the whole society has begun to pay more attention to the low-level lead exposure and the maximum dose of lead a "normal" body can take, which could protect our bodies from diseases caused by lead. In the occupational setting, the researchers are re-evaluating the present safe level for lead exposure because they can obtain much more accurate data of the lead's physiological effects through clinical investigations.

Blood sampling is the main way to test the level of human exposure to lead [59]. However, this method does not always accurately show the intoxication level of the individual because lead can move from the vascular system and accumulate in the bones of the human body once it is absorbed. Therefore, if a man was exposed to lead before, the results of tests could not be considered to be absolutely accurate as the historic exposure to lead had already gathered in the bones. Some symptoms of lead exposure could recur in the late life of the individual due to the lead stored in the bones.

#### 5.2.3.2 Cadmium (Cd)

Cadmium is located at the end of the second row of transition elements with atomic number 48, atomic weight 112.4, density 8.65 g/cm$^3$, melting point 320.9 °C, and boiling point 765 °C. Cadmium is very toxic and is not known for any essential biological function of the human body. Both natural and various anthropogenic sources allow cadmium to have the access to the environments. However, the accumulation of cadmium in the soil or plant is mainly through anthropogenic activities such as application of phosphate fertilizers, wastewater, sewage sludge, and manures. Due to the high mobility of cadmium in soils, the accumulation of it in plants has gradually become a threat to animal and human health.

##### 5.2.3.2.1 Distribution in Food

As the result of the high transfer factor properties of plants, cadmium is a usual contaminant, which could be found in most human foodstuffs. The foodstuffs containing cadmium are the main source of cadmium exposure among nonsmoking, nonoccupationally exposed populations. Ordinarily, cadmium selectively accumulates in the kidney and liver of mammals fed with cadmium-rich diets. And high cadmium concentrations are found in certain species of oysters, scallops, mussels, and crustaceans, while lower cadmium levels are found in vegetables, cereals, and starchy roots [60]. Due to the larger consumption of food containing cadmium, most people are absorbing cadmium daily. Some crops, such as rice, can accumulate high concentrations of cadmium if grown on cadmium-polluted soil or irrigated with contaminative water. Moreover, those crops growing in the acidification of cadmium-containing soils may contain higher cadmium concentrations.

The Joint FAO/WHO Expert Committee on Food Additives (JECFA) established a provisional tolerable monthly standard of intake for cadmium of 25 μg/kg body weight (in 2010) [61, 62]. The national food safety standards of China (GB2762-2012) specify the allowable limit for cadmium in food: the maximum concentration is 0.1 mg/kg in grain, 0.05 mg/kg in fresh vegetables and fruit, 0.2 mg/kg in beans, 0.1 mg/kg in meat and meat products (except entrails), 0.5 mg/kg in mushrooms, 0.1 mg/kg in fish, and 0.05 mg/mL in eggs and egg products [62, 63].

##### 5.2.3.2.2 Uptake and Metabolism in Humans

The cadmium uptake by adults is approximately 5% through oral ingestion and is stored primarily in the kidneys. Several studies have shown that the uptake rate of ingested cadmium may be influenced by many factors, such as the daily amount, the composition of the food, and the individual's nutritional status [58]. Cadmium is hard to be excreted out of the body, which causes the obvious accumulation. Since cadmium has a long biological half-life, which is about 33 years, it is accumulated in humans' bodies throughout their whole lifetimes. These factors cause the high level of concern regarding the risks for human health posed by cadmium—cadmium is considered more dangerous than mercury and lead. And the vulnerable groups are mining workers and people with low immunity.

#### 5.2.3.2.3 Toxicity and Clinical Effects

Cadmium exposure in humans is predominantly through food ingestion where fish, meat, and fruit can contain 1–50 μg/kg cadmium. Cadmium is extremely toxic to humans due to its long biological half-life preventing the reduction of the accumulated body burden.

Cadmium poisoning is mainly due to its inhibition to sulfhydryl enzymes. It does harm to the kidney, bone, and digestive system, especially the damage to the renal proximal convoluted tubule epithelial cells, which makes its reabsorption dysfunction and causes proteinuria, aminoaciduria, hypercalciuria, and glycosuria. It may lead to negative calcium balance in the body, and as a result of bone calcium precipitation, osteoporosis and pathologic fracture may come up. "Itai-itai" disease in Japan is a particular syndrome of chronic cadmium poisoning caused by contaminated food with cadmium from the environment, which is characterized by renal tubular proteinuria, osteoporosis, and pseudofracture, and it appears more often among women above 50 years old with the incubation period for 10–30 years [63].

Other effects of cadmium on human health also include hypertension, cardiovascular function, neurological disorders, and carcinogenic effects. Cadmium and its compounds are not only the major cause of lung cancer but also positively associated with kidney and prostate cancers.

### 5.2.3.3 Mercury (Hg)

Mercury is in the same group of the periodic table with zinc and cadmium. The basic characteristic of it is atomic number 80, atomic weight 200.6, density 13.6 g/cm$^3$, melting point −13.6 °C, and boiling point 357 °C and is usually recovered as a by-product of ore processing. Hg contamination is mainly caused by coal combustion, which would release Hg metals. Hg usually exists in the form of mercuric (Hg2+), mercurous (Hg22+), elemental (HgO), or alkylation (methyl/ethyl mercury) after it is released to the environment. Compared to inorganic mercury species, methylmercury (MeHg) is well-known for its efficient bioaccumulation and long-term retention in biological tissues.

#### 5.2.3.3.1 Distribution in Food

A wide variety of foods, including dairy products, poultry, meats, eggs, fruits, and vegetables, should be tested whether they have high concentration of mercury. However, the levels of mercury in these foods are relatively low compared to the levels found in fish. The main pathway for human exposure to methylmercury is the consumptions of fish and shellfish, which could contain a high concentration of mercury [64, 65].

Methylmercury is generated by anaerobic organisms (mainly bacteria), which reside in the anoxic deep seabed. Generally, they are considered as a huge threat to human health. Almost all fish species contain a certain amount of mercury levels. Among them, the levels in swordfish and shark are usually higher than other fishes [66]. Normally, methylmercury is accumulated in the muscle tissues of fish due to the consumption of organisms containing methylmercury. The concentration of mercury levels in food products, excluding fish, varies from a few micrograms to 50 μg/kg. Thus, the daily intake of methylmercury is mainly determined by fish consumption and the concentrations of methylmercury in consumed fish [67].

The Joint FAO/WHO Expert Committee on Food Additives (JECFA) established a tolerable intake for total mercury of 5 μg/kg body weight. The PTWI was considered applicable to dietary exposure to total mercury from foods other than fish and shellfish. As to fish and shellfish, the PTWI for methylmercury is 1.6 μg/kg body weight [68]. The national food safety standards of China (GB2762-2012) specify the allowable limit for total mercury in food: the maximum concentrations is 0.02 mg/kg in cereals, 0.01 mg/kg in fresh vegetables, 0.05 mg/kg in meat, 0.01 mg/kg in edible mushrooms, and 0.05 mg/mL in eggs and egg products. The limitation for methylmercury in aquatic animals is 0.1 mg/kg.

#### 5.2.3.3.2 Uptake and Metabolism in Humans

Liquid metallic mercury is poorly absorbed from the gastrointestinal tract. However, some cases of mercury poisoning have been found that mercury was deposited in diverticula, fistulas, or abscesses in the gastrointestinal tract or after aspiration in the lungs. Moreover, methylmercury is the main chemical form absorbed into the food chain with high toxicity. Appropriately 95% of methylmercury ingested is absorbed from the intestinal tract after consumption. After that, it is transported to all tissues and target organs via the bloodstream [69].

It takes a very long time to remove mercury from the body from the fecal route. Usually, methylmercury is secreted in the bile from where a fraction is reabsorbed in the gallbladder and gastrointestinal tract [70]. However, methylmercury may also be secreted across the intestinal membrane due to the functions of intestinal flora in the gastrointestinal tract, which include breaking the carbon-Hg bond, converting MeHg to poorly absorbed iHg [71]. Finally, methylmercury is mostly excreted in the feces. So, mercury intake should be controlled under a certain standard and monitored carefully in case of accumulating toxic Hg metals in body, which occurs when the amount of Hg absorbed exceeds that being excreted.

#### 5.2.3.3.3 Toxicity and Clinical Effects

Methylmercury is one of the most toxic nerve agents, which causes severe damages on the developing human brain. It can easily pass the placental barrier and the blood-brain barrier to cause concerns of mercury exposure during pregnancy [72]. Methylmercury is proved as a possible carcinogenic agent by the International Agency for Research on Cancer and defined as a high risk to humans (group 2B).

If a child's parents contain a high level of mercury in the body (congenital cases), it is very likely that the child will be exposed by mercury inherently and show a higher level of symptoms than the parents. Symptoms are quite severe showing disorders of nervous functions and highly delayed developmental skills. Common symptoms of mercury toxicity include sensory action disorders, ataxia, narrowing field of vision, hearing impairment, balance impairment, speech impediment, trembling in hands and feet, and cognitive impairment. In terms of the statistics of the Environmental Protection Agency and National Academy of Sciences, it showed that 8–10% of American women have mercury levels that would expose their own children to neurological disorders [73].

Moreover, the main target organ for mercury is the brain. And, the impairment of brain can result in malfunctioning of the nerves, kidneys, and muscles through damaging the membrane potential and interrupting with intracellular calcium homeostasis. Due to the high stability constants of mercury, it can bind to freely available thiols [74].

#### 5.2.3.4 Arsenic (As)

Arsenic is a metalloid in group VA and period 4 of the periodic table that associates a wide variety of minerals. Arsenic has the physical properties of atomic number 33, atomic mass 75, density 5.72 g/cm$^3$, melting point 817 °C, and boiling point 613 °C and exhibits fairly complex chemistry and can be present in several oxidation states (−III, 0, III, V). The normal inorganic forms of arsenic are the trivalent arsenate and the pentavalent arsenate. The organic forms mainly refer to the methylated metabolites, including monomethylarsonic acid (MMA), dimethylarsinic acid (DMA), and trimethylarsine oxide. The exposure to inorganic arsenic is the dominant reason that makes human poisoned. In general, arsenic in nature is mainly from volcanic eruptions and soil erosion, while arsenic is found in industrial products (insecticides, herbicides, fungicides, algicides) in human society [75].

##### 5.2.3.4.1 Distribution in Food

Foods usually account for most of the daily intake of arsenic if the environment (like soil) doesn't contain high arsenic levels. The dietary intake of arsenic could be very likely from fish, shellfish, meat, poultry, dairy products, and cereals. However, the organic compounds (e.g., arsenobetaine) that have relatively low toxicity are also present in those fish and shellfish containing a certain amount of arsenic. On the contrary, if arsenic levels are especially high in the environment, foods (e.g., rice) made of water and food crops that are highly polluted with arsenic conduce to total daily intake. The presence of arsenic in fish has been found in several species such as sardine, chub mackerel, horse mackerel, blue fish, carp, mullet, tuna, and salmon [76].

Based on the risk assessment report from the 72nd meeting of the Food and Agriculture Organization (FAO) and the Joint FAO/WHO Expert Committee on Food Additives (JECFA), the provisional tolerable weekly intake (PTWI) for inorganic arsenic should be re-evaluated and proposed combined with the toxicity, epidemiology, and exposure assessment of it. It should be taken into consideration that the lower limit on the benchmark dose for a 0.5% increased incidence of cancer (BMDL0.5) was determined to be 3.0 μg/kg body weight per day (2–7 μg/kg body weight per day based on the range of estimated total dietary exposure) [77]. The national food safety standards of China (GB2762-2012) specify the allowable limit for total arsenic in food: the maximum concentration is 0.5 mg/kg in cereals and cereal products; 0.5 mg/kg in fresh vegetable, meat, and edible mushrooms; and 0.01 mg/mL in milk and dairy products. The limitation for inorganic arsenic in aquatic animals is 0.5 mg/kg but 0.1 mg/kg in fish.

#### 5.2.3.4.2 Uptake and Metabolism in Humans

When arsenic is present in the body, it can accumulate in soft tissue organs such as the liver, spleen, kidneys, and lungs. However, the best long-term storage site for arsenic is keratin-rich tissues, such as the skin, hair, and nails. Therefore, it is suggested that a better way to determine the daily intake of arsenic is to measure the arsenic levels in these biological specimens. In order to confirm arsenic poisoning, testing biological samples for arsenic needs at least 48 h after arsenic from seafood is absorbed by the body. Otherwise, the test may be affected by the presence of arsenobetaine, which is a relatively harmless form of arsenic as arsenobetaine is usually accumulated in fish at high levels of concentration.

Methylation is the main metabolic pathway for inorganic arsenic in humans. Arsenic trioxide is methylated to two major metabolites via a nonenzymatic process, firstly to MMA, which is further methylated enzymatically to DMA before excretion in the urine [78]. Previously, this methylation process was thought as a pathway of arsenic detoxification; however, recent studies have showed that some methylated metabolites of arsenic may be more toxic if they contain trivalent forms. In this biotransformation process, these inorganic arsenic species (iAs) are converted enzymatically to methylated arsenicals, which are the end metabolites and the biomarker of chronic arsenic exposure.

#### 5.2.3.4.3 Toxicity and Clinical Effects

The toxicity of arsenic in food relates to its existence form and the valence state. The element arsenic does not dissolve in water and shows no toxicity, while arsenic compounds such as oxides, salts, and organic compounds are toxic.

Acute arsenic poisoning is infamous for its lethality, which stems from arsenic's acute arsenic poisoning that is especially fatal to human due to its damages to the integrity of the blood vessels, gastrointestinal tissue, heart, and brain. The high lethality of acute arsenic poisoning is owing to the misjudgment of symptoms similar to other diseases including vomiting, abdominal pain, and diarrhea. If the patients are not treated in time, much more severe symptoms are followed such as numbness and tingling of the extremities, muscle cramping, and death, in extreme cases [78].

Chronic exposure to lower levels of arsenic can cause some abnormal patterns of skin hyperpigmentation. Then, skin lesions and hard patches on the palms of the hands and soles of the feet are followed. What's more, chronic arsenic exposure also causes a markedly elevated risk for developing a number of cancers, such as, most notably, cancers of the liver (angiosarcoma), skin, lung, bladder, and possibly the kidney and colon [79]. Recently, the dose that could increase the risk for cancer has become the focus of particularly intense scrutiny in the United States and many other countries due to some points that the standards for exposures to arsenic should be lowered [80, 81].

### 5.2.4 Conclusion

Nowadays, many researches on the content of heavy metals in food are conducted in the majority of countries worldwide. There are studies that indicate that heavy metals such as arsenic cadmium, lead, and mercury occur naturally. But, with developments of human industries, the environmental contaminations are becoming more and more serious due to anthropogenic activities. It is sure that these heavy metals are common systemic toxicants, which can cause people severe diseases including cardiovascular diseases, developmental abnormalities, neurological and neurobehavioral disorders, diabetes, hearing loss, hematological and immunologic disorders, and various types of cancer. Importantly, the influences of heavy metals to health are associated with the following factors: type of heavy metals, its own chemical forms, period of exposure, and dose. A lot of studies have shown that exposure of toxic heavy metals would cause long-term health problems in human populations. Although the acute and chronic symptoms are known for some heavy metals, there are lots of unknown aspects of mixtures of toxic elements: whether they are also fatal to human. Therefore, a large number of researches in heavy metals should be conducted to protect human from suffering from diseases caused by heavy metals. Although solving the contaminations of heavy metals is a great challenge for all mankind, it is still very likely that the problems concerned with heavy metals will be solved in the future.

## 5.3 Persistent Organic Pollutants

Although the laws, which forbid using most chlorinated persistent organic pollutants (POPs) such as polychlorinated biphenyls and organochlorine pesticides and dioxins, have been formulated several years ago in most developed countries, the risks of being exposed to these chemicals in the general population still exist as the food chains have already been contaminated by them. Also, POPs accumulated in human adipose tissue are a dominant source of internal exposure because POPs are slowly but continuously released from adipose tissue to the circulation. However, with the knowledge of toxicology and analytical capabilities developing, both potential toxicities and health hazards caused by contaminants of POPs have been found. Some notorious environmental pollutants such as polycyclic aromatic hydrocarbons (PAHs), dibenzodioxins, dibenzofurans, and polychlorinated biphenyls will be mentioned in this chapter.

### 5.3.1 Polycyclic Aromatic Hydrocarbons (PAHs)

Polycyclic aromatic hydrocarbons (PAHs) generally can be found in mixed aromatics, usually accompanying with an enormous amount of diverse organic compounds containing two or more aromatic rings [82]. They can be distributed into two varieties: low-molecular-weight compounds composed of fewer than four rings and high-molecular-weight compounds of four or more rings. Some structures of common PAHs are shown in Fig. 5.2 [83, 84]. They may be generated and released during incomplete combustion or pyrolysis of organic matter, industrial processes, and other human activities.

The physical properties of PAHs have some differences in their molecular weight and structure. PAHs are easily miscible in organic solvents due to its high lipotropy. Normally, each additional ring added to the PAHs could decrease the aqueous solubility of PAHs. In addition, PAHs also show other properties such as light

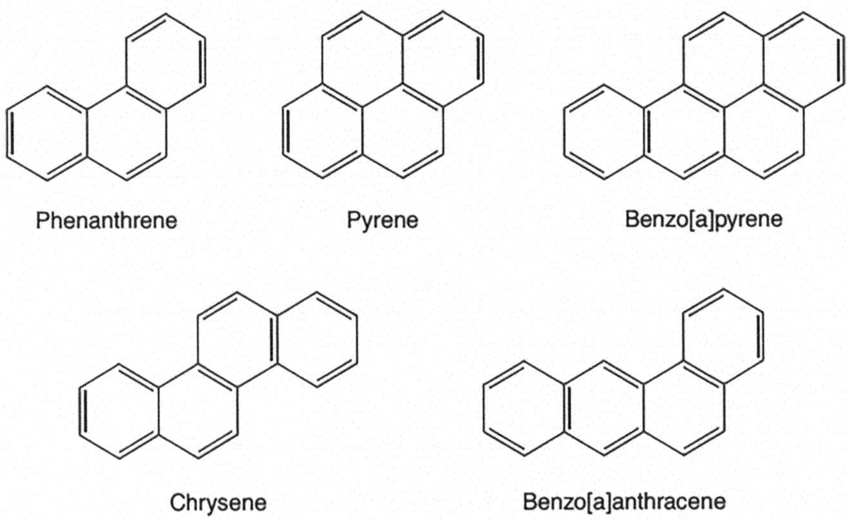


**Fig. 5.2** Structures of some common PAHs

sensitivity, heat resistance, conductivity, emissivity, corrosion resistance, and physiological action [84].

PAHs are composed of many different chemicals including derivatives of PAHs, such as nitro-PAHs, hydroxy-PAHs, and heterocyclic PAHs. In 1970s, EPA named 16 PAHs, which can be usually found in environmental monitoring samples, as follows: *acenaphthene*, *acenaphthylene*, *anthracene*, *fluoranthene*, *fluorene*, *naphthalene*, *phenanthrene*, *pyrene*, *benz[a]anthracene*, *benzo[b]fluoranthene*, *benzo[k] fluoranthene*, *benzo[ghi]perylene*, *benzo[a]pyrene*, *chrysene*, *dibenz[a,h]anthracene*, and *indeno[1,2,3-cd]pyrene* [85]. Some researches show that most of the PAHs could be the critical factors for the hazard and risk characterizations due to its carcinogenic and genotoxic potential. Therefore, it suggested to use benzo[*a*] pyrene as a marker of occurrence of carcinogenic PAHs in food as a result of the examinations of PAH profiles in food and the evaluation of a carcinogenicity study of coal tar mixtures in mice [86].

#### 5.3.1.1 Contamination of Food by PAHs

It has been proven that food consumption is an essential pathway for human exposure to environmental contaminants, which was the main source of PAH intake. The food products, like beef, pork, fish, and marine products; dairy products; vegetables; beverages; grain; and animal fats and oils, usually contain a bit higher concentration of PAHs [87]. Food can be contaminated with PAHs in the following ways: (a) PAHs can pass through the food chain by contaminated crops and marine life, and (b) PAHs in food are also generated due to food preparations and cooking methods such as smoking, roasting, barbecuing, frying, drying, and steaming [87].

Sometimes, soils are contaminated by atmospheric fine particulate matter containing PAHs through atmospheric sedimentations. The PAHs will be absorbed in the crops like cereals, fruits, and vegetables due to the polluted soils they grew in. When plants grew in contaminated soils used to feed grazing herds, PAHs can inevitably fetch up in dairy products [87]. Additionally, rivers and seas can be contaminated with crude or engine oil. For instance, industrial waste will generate deposit sediments containing both petrogenic and pyrogenic PAHs, which could cause more serious contaminations. The first potential targets near contaminated sediments for exposure to PAHs are filter-feeding bivalves due to their huge dependence on water and biological accumulation of PAHs. Moreover, predation provides a portal for PAHs to enter into the food chain [87].

The PAHs in food are usually formed through cooking methods especially like smoking, roasting, and grilling. And, smoking is a main influencing factor to the PAH levels found in smoked foods [88]. Smoked fish and meat usually contain less than 1 ppb PAHs, and benzo[*a*]pyrene is most often found at levels of 0.1–0.5 ppb [89]. In exceptional cases, the outer of heavily smoked products always can be found with higher levels of PAHs. Except for smoking, grilling, or broiling of meat and fish on fire, these can also result in PAHs. Firstly, the PAHs can be endogenously formulated on the surface of the food due to the high temperatures. The burning of the fuel in the grilling can also generate PAHs. Last but not least, PAHs are generated when melted fat drips down on the fire [90]. In addition, foods prepared by other means can also generate PAHs. For instance, benzo[*a*]pyrene (15–28 ppb) is enriched in the soot and skin of coffee beans, which had been roasted by direct contact with the combustion gases [91].

On account of the high consumption, cereals were considered as a primary factor to the intake of PAHs. The other main reason that could make people exposed of PAHs is vegetable fats and oils containing a higher concentration of PAH levels in this food group [92]. Although the concentration of PAHs in those food prepared by smoking and grilling can be extremely high, these foods do not play a vital role in the exposure of PAHs because they only account for a fraction of the daily diet. Nevertheless, they do play a chief part in higher PAH intakes if people eat plenty of them by daily diet.

#### 5.3.1.2 Health Effects

Fifteen PAHs have been proven to be carcinogenic in animals through various routes of exposure [93]. Several of them can cause skin tumors in the two-stage mouse skin carcinogenesis system. In addition, the local sarcomas can also be caused at the injection site. Others are able to generate lung tumors through intravenous injection or intratracheal instillation or inhalation [94]. According to previous studies, benzo [*a*]pyrene seemed to be the most potent carcinogen. The carcinogenic PAHs tested in foods have all been proven to be genotoxic and may be reasonably considered to be human carcinogens [95].

##### 5.3.1.2.1 Acute or Short-Term Health Effects

The acute toxicity of PAHs on human health will be decided by the exposure concentration, exposure duration, and route of exposure [96]. It has been reported that short-term exposure to PAHs can lead to dysfunction of lung in asthmatics and thrombotic influences on people suffering from coronary heart disease [96]. So far, the ability of PAHs with different concentrations to cause human fall ill in the short term is still absolutely figured out. On the contrary, it is sure that occupational exposures to high levels of pollutant mixtures containing PAHs can lead to symptoms such as eye irritation, nausea, vomiting, and diarrhea [97]. In addition, mixtures of PAHs certainly showed that it could bring about skin irritation and inflammation. Anthracene, benzo(*a*)pyrene, and naphthalene are the direct sources of skin irritants, while anthracene and benzo(*a*)pyrene are validated to be skin sensitizer sowing to the results of skin allergy tests in animals and humans [98].

5.3.1.2.2 Chronic or Long-Term Health Effects

Being exposed to low levels of some PAHs for a long period has been considered as the cause of cancer in in vivo study in experimental animals. Animal studies have also indicated the exposure of PAHs is adverse to the reproduction and development, while such effects have not commonly been detected in humans [99]. Moreover, exposure to PAHs may induce cataracts and increase the risk of kidney and liver damage and jaundice. If a large amount of naphthalene is inhaled or swallowed by human, it can lead to the breakdown of red blood cells [100]. According to evidences from a number of epidemiological studies, all cases concerned with the health effects of PAH mixtures have been validated that their carcinogenic potentials are deeply associated with lung cancer, some skin cancers, and bladder cancers [101]. What's more, due to PAHs' potential interference with hormone systems, they can disrupt reproduction and immune function [92]. Since DNA damage induced by PAH exposure has been validated by multitudinous researchers, it seems that long-term exposure to PAHs can increase the risks of cell damage via gene mutation and cardiopulmonary mortality [92, 102].

#### 5.3.1.3 Regulatory Limits in Food

A bunch of mandatory and suggested standards, which are composed of China national standard or Guobiao (GB) standard, have been regulated by the Standardization Administration of China. In 2012, the ministry of health issued the food safety national standard GB 2762-2012 (maximum levels of contaminants in foods), which regulates maximum levels of certain contaminants in foods including benzo[*a*]pyrene. Maximum contamination levels of four food categories for benzo[*a*] pyrene are set for (Table 5.1).

### *5.3.2 Dioxins and Dioxin-Like Compounds*

Dioxins and dioxin-like compounds (DLCs) consist of polychlorinated dibenzo-*p*-dioxins (PCDDs), polychlorinated dibenzofurans (PCDFs), and part of polychlorinated biphenyls (PCBs). PCDDs, PCDFs, and PCBs are persistent organic pollutants (POPs) with similar chemical structure and are chemically considered as

**Table 5.1** The limits of maximum level of benzo[*a*]pyrene in different types of food in China

| Food type/name | Maximum levels (μg/kg) |
|---|---|
| Paddy, wheat | 5 |
| Smoked or baked meats | 5 |
| Smoked or baked aquatic products | 5 |
| Fats and oils and fat emulsions | 10 |

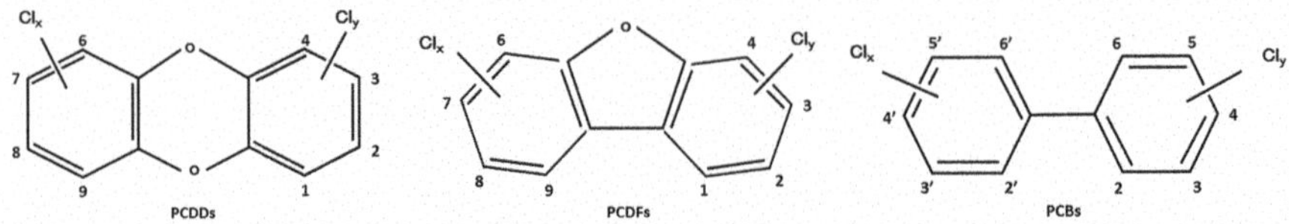


**Fig. 5.3** General chemical structure of PCDDs, PCDFs, and PCBs

halogenated aromatic hydrocarbons (Fig. 5.3). Although there are 75, 135, and 209 different CDDs, CDFs, and PCBs, respectively, only 7, 10, and 12 of them are referred to have dioxin-like activity [103, 104]. For example, dioxin-like activity of PCDDs and PCDFs demands chlorination of the parent compounds at the 2, 3, 7, and 8 positions; the activity of PCBs calls for chlorination at 4 or more positions. The most toxic and harmful of the dioxins is tetrachlorodibenzo-*p*-dioxin (TCDD) [103]. Although TCDD has 22 different isomers that usually are generated in mixtures in the environment, the 2,3,7,8-tetrachloro isomer has been focused significantly due to its unusual degree of toxicity [105].

During the industrial and natural combustion processes, PCDD/Fs can be released substantially. The massive chlorine and a large production of chlorine-containing materials are used in industrial manufactures, whereas the main sources of PCBs are related to accidental events and environmental reservoirs (soil, sediments, and clay) after illegal discharges of contaminated material [105]. So, both PCDD/Fs and PCBs ubiquitously exist in the environment. All these dioxin congeners have high lipophilicity/low water solubility, high stability, and low vapor pressure, which make them have the access to accumulate in the food chain. Although the ways of being exposed by dioxin are diverse including inhalation of air, dermal absorption, consumption of drinking water, and consumption of food, it seems no doubt indeed that food is by far the dominant pathway for most people who don't experience occupational exposure. Recently, some studies showed that dietary intake of 2,3,7,8-TCDD occupies nearly 98% of human exposure [106]. Public attentions to the safety of the food supply and potential adverse outcomes to DLC exposure are increasing rapidly, especially focusing on sensitive and highly exposed population groups.

#### 5.3.2.1 Contamination of Food for PCDDs, PCDFs, and PCBs

All humans are at risk of a certain "background" exposure to dioxins. PCDD/Fs and PCBs are greatly difficult to degrade naturally and chemically. Therefore, they can penetrate the food chain and bioaccumulate in the human body finally. Researches indicated that over 90% of human exposure to dioxins conventionally happens through the food supply. Accordingly, food intake is the dominant approach for human exposure to PCDD/Fs and PCBs, particularly high-surface vegetables and animal fatty foods [106]. However, levels of exposure through food supply may be quite different due to different regions, hinging on the consumptions of local food.

Most foods, except milk, are manufactured in bulk and distributed widely, so that foods purchased by consumers are less likely to reflect different levels of PCDD/Fs and PCBs [103]. Thus, the exposure of the general population through the food supply seems to be consistent to a certain degree. On the other side, foods that are caught or harvested in the wild can show levels of dioxins in the local environment. So, those who depend on local wild creatures for food may be more easily to approach higher exposure levels [103].

#### 5.3.2.1.1 Vegetables and Grain

PCDD/Fs and PCBs can contaminate plants and crops through the soil, groundwater, and air. Particularly, the soil, considered as one of the main sinks for airborne POPs, provides a potential container, like a significant reservoir, for dioxin contamination under grazing conditions. Moreover, dioxins in the soil can contaminate waterways through runoff and sediments [103]. The methods applied to measure the levels of PCDD/Fs and PCBs in soils and vegetation can be used to monitor trends in their abundance and their consequences as consequence of natural and anthropogenic changes [107].

A study conducted by Hülster et al. indicated that some plant species from the family Cucurbitaceae that grow on the ground usually contain a bit higher levels of PCDD/Fs. The principal reason could be owing to the uptake of atmospheric deposition of PCDD/Fs though roots and leaves [108]. The retention of airborne PCDD/Fs and PCBs, including absorption from the vapor phase by the waxy cuticle and retention of particulate-bound contaminants, is the key factor of contaminating those plants growing on the ground [106]. Studies also prove that cooking processes, such as washing and boiling, could diminish the amount of dioxin from the leafy vegetables [109].

In addition, a few dioxin levels could be remained on the surface of grain by-products, bran or middling, recycled grains, and vegetable by-products owing to concentrated surface contamination caused by local incinerators or persistent soil contamination from past herbicide application [103]. These products may be fed to some livestock, which accounts for a substantial part of the rations. And importantly, it increases exposure risk, while larger animal populations are unlikely affected.

#### 5.3.2.1.2 Animal Fatty Foods

The major way of food-producing animals to be exposed by PCDD/Fs and PCBs is by feeding. Recently, a bunch of cases showed that high levels of PCDD/F and PCB contamination were found in manufactured fodder, which carried over into foods [106]. The regulatory authorities of the EU have set different PCDD/F and PCB contamination threshold levels for the compliance of both marketed feeds and foods [104].

Normally, milk fat contains PCDD/Fs and PCBs. Since the first reports of the detection of PCDD/Fs in cow's milk in 1987, many studies on milk have been conducted, which probably provided more data about milk than any other foodstuff. When expressed on a fat basis, the concentration levels in milk products are the same as in the milk from which they were produced. So, some researches have indicated that the differences among milk, butter, and cheese may be partly caused by issues of accuracy and representativeness. In the meantime, it may also happen because milk products are exported and imported to a greater extent and over greater rangers than milk itself [106].

Compared with milk and dairy products, meat is a more heterogeneous and complicated group. The situation is that different ethnic groups and individuals have extremely different tastes or customs, which could generate considerable variable kinds of meat consumption. Usually, the levels of both PCDD/Fs and PCBs in pork are much lower than beef relatively [106]. However, as for other species, the levels in them are still unknown. In terms of offal products, more and more evidences show that the liver produced from farm animals like cows, pigs, and sheep may contain a high level of dioxins that have exceeded the limitations, although the livestock are given compliant animal feed and are exposed only to normal background levels of dioxin contamination in the environment [106]. In 2010, a study from Fernandes et al. investigated the PCDD/F and PCB levels in commonly consumed offals and offal products [110]. The results gained in this preliminary research proved the fact that liver samples indeed contained the highest levels of PCDD/Fs, especially in liver samples from the deer and lamb liver. In terms of lamb liver, there were nine samples that showed PCDD/Fs levels are beyond the maximum limit of EU [104, 110]. Fish or aquatic organisms may not be exposed by POPs, such as dioxins, directly if they live near the bottom [103]. Generally, the fat contents in fish can partly determine the levels of PCDD/Fs and PCBs in them. Other factors such as the extent to which the fish migrate, the number of times they spawn, and their ages, size, and feeding habits can also determine the levels of PCDD/Fs and PCBs in fish [103]. The aquatic food chain is the prominent pathway for fish exposed by DLCs. The DLCs are fat-soluble and easy to concentrate through food chains, which accumulate in the larger fish. The vital factor in exposure process is desorption of the DLCs from particulate matter and subsequent entry into the food chain. After DLCs are ingested, they slowly accumulate in different tissues in fish according to the tissue lipid content and nonspecific binding proteins to a certain extent. In conclusion, the older, larger, and oilier the fish is, the higher DLCs levels can be detected in it [103, 106].

#### 5.3.2.2 Health Effects

The negative health effects of dioxins are associated with their metabolic resistance and lipophilic nature. Dioxin-like compounds are known endocrine disruptors [106]. Both animal researches and short-term human exposure have already validated that the dioxins could have adverse impressions on human health. PCDD,

PCDF, and PCB technical mixtures are usually applied in toxicity studies. In short, some serious toxicological effects such as effects on the thyroid, immune function, the reproductive system, behavior, and carcinogenicity can be induced by POP mixtures.

#### 5.3.2.2.1 Carcinogenicity

Researches bore out that TCDD is considered as a human carcinogen by EPA while other DLCs are likely to be human carcinogens [111]. A lot of studies have already showed that TCDD is one of the most toxic and carcinogenic compounds in different kinds of animal species. For example, a rat experiment showed that the number of liver tumors increased significantly if rats were treated with 10 or 100 ng of TCDD/kg body weight. As the doses increased, a mass of tumors including the mouth, nose, lungs, and liver occurred generally in both male and female rats [105]. The TCDD levels detected in female rats are three times as potent a carcinogen as aflatoxin B1, which is referred to as one of the most potent hepatocarcinogens known [105].

Even though massive results of experiments with experiment animals clearly show the carcinogenicity of TCDD, it is still limited whether TCDD could induce cancer in humans according to the current epidemiological evidence. A few well-controlled studies have proved that TCDD exposure plays a role in formation of cancers. The study of a Swedish group showed the use of TCDD-contaminated phenoxy herbicides could cause sarcomas of muscles, nerves, and fat tissue in people. On the basis of two studies, people exposed to phenoxy herbicide were five to six times more likely to develop a tumor compared to an unexposed control group [105].

Many TCDD-induced effects are extremely similar to effects caused by PCBs and other structurally related compounds. Hardell's group has discovered a significant relation between PCB levels in fatty tissue and non-Hodgkin's lymphoma [112]. According to epidemiological study's findings, capacitor manufacturing workers and electrical utility workers, which are exposed to various PCB technical mixtures frequently in their work, had higher cancer occurrence rates compared with the general population [113–117]. Further studies included a large number of subjects, but relatively, a few of them worked more than 10 years, which limited the follow-up period [118].

#### 5.3.2.2.2 Noncancer Effects

It is reported that the general exposure to dioxins and DLCs could cause several potential noncancer health effects in the light of several evaluations of DLC toxicity [103]. Epidemiological studies on nervous system developmental disorders were conducted in different countries and regions. Rogan et al. carried out this type of study in North Carolina where it was observed that hyporeflexia, hypotonicity, and delayed motor development up to 24 months postnatally were linked to the prenatal

PCB body burden of the mothers as indexed by total PCBs in early milk samples. Other similar studies suggested that visual recognition memory at 7 months was negatively associated with total PCB levels [119, 120]. According to this type of research in Europe, Huisman et al. recruited a cohort of healthy mother-infant pairs in both Groningen and Rotterdam. Negative relations were confirmed between the PCDD/F and PCB levels in milk and neurological status and psychomotor development [19]. In general, these results indicate that the neurobehavioral impairment concerned with PCB exposure is likely to be permanent.

Adverse effects of PCDD/F and PCB mixtures on human reproduction have also been reported. The exposure pathway included the consumption of contaminated rice oil and fish and occupational exposures. It was reported in some studies that increased PCB exposure was associated with menstrual irregularities, miscarriages, shorter length in menstrual cycle, and spontaneous fetal death [121, 122]. According to male reproductive functions, a PCB-associated risk of infertility has been found in some studies. Another study conducted by Mocarelli et al. even revealed that male births decreased relative to all births exposed to TCDD in Seveso (Italy) [123].

Another important worry with respect to the toxicity of dioxins and DLCs may be the effects on immune functions and thyroid hormone status. One evaluation shows that immune response decreases and changes in T-lymphocyte differentiation occurred in people exposed to TCDD from the Times Beach contamination. Moreover, people eating relatively large amounts of Baltic Sea fish showed changes in T-cell lymphocytes [103]. Data from several research studies performed by Weisglas-Kuperus et al. indicate that PCB exposure affects the immune functions of infants and children. In infants, researches even suggested that PCB and PCDD/F exposures were linked with reduced T3 and T4 in the infants at age 2 weeks and 3 months while TSH was increased [124, 125]. Besides, a population-based study suggested that enlarged thyroid gland volume was linked with technical PCB mixture exposure in workers at a PCB production facility and nearby residents. [126]

#### 5.3.2.3 Regulatory Limits in Food

The EU was the first union to regulate extensive and comprehensive threshold for PCDD/Fs, while there were almost no standards for dioxins around world at that time. The provisions were applied in July 2002, which set the maximum limits for PCDD/Fs in food and animal feed. Later, dioxin-like PCBs were inclusive in the regulation in 2006 [106, 127–129]. Further, nondioxin-like PCBs were also later included in 2011. Maximum limits for these contaminants are set for several food categories (Table 5.2) [26]. However, China national standard (GB2762-2012) only sets a maximum level (0.5 mg/kg) of the sum of certain PCBs in aquatic products including PCB28, PCB52, PCB101, PCB118, PCB138, PCB153, and PCB180.

**Table 5.2** Maximum levels for dioxins and PCBs as laid down in Regulation (EU) No. 1259/2011 [127]

| Foodstuffs | | Maximum levels | | |
|---|---|---|---|---|
| | | Sum of dioxins (WHO-PCDD/F-TEQ) | Sum of dioxins and dioxin-like PCBs (WHO-PCDD/F-PCB-TEQ) | Sum of PCB28, PCB52, PCB101, PCB138, PCB153, and PCB180 (ICES—6) |
| 6.1 | Meat and meat products (excluding edible offal) of the following animals: | | | |
| | Bovine animals and sheep | 2.5 pg/g fat | 4.0 pg/g fat | 40 ng/g fat |
| | Poultry | 1.75 pg/g fat | 3.0 pg/g fat | 40 ng/g fat |
| | Pigs | 1.0 pg/g fat | 1.25 pg/g fat | 40 ng/g fat |
| 6.2 | The liver of terrestrial animals referred to in 5.1 and derived products thereof | 4.5 pg/g fat | 10.0 pg/g fat | 40 ng/g fat |
| 6.3 | Muscle meat of fish and fishery products and products thereof, with the exemption of the following: | 3.5 pg/g wet weight | 6.5 pg/g wet weight | 75 ng/g wet weight |
| | — Wild-caught eel | | | |
| | — Wild-caught freshwater fish, with the exception of diadromous fish species caught in fresh water | | | |
| | — Fish liver and derived products | | | |
| | — Marine oils | | | |
| | The maximum level for crustaceans applies to muscle meat from the appendages and abdomen. In case of crabs and crab-like crustaceans (*Brachyura and Anomura*), it applies to muscle meat from appendages. | | | |
| 6.4 | Muscle meat of wild-caught freshwater fish, with the exception of diadromous fish species caught in fresh water and products thereof | 3.5 pg/g wet weight | 6.5 pg/g wet weight | 125 ng/g wet weight |
| 6.5 | Muscle meat of wild-caught eel (*Anguilla anguilla*) and products thereof | 3.5 pg/g wet weight | 10.0 pg/g wet weight | 300 ng/g wet weight |

(continued)

**Table 5.2** (continued)

| Foodstuffs | | Maximum levels: Sum of dioxins (WHO-PCDD/F-TEQ) | Sum of dioxins and dioxin-like PCBs (WHO-PCDD/F-PCB-TEQ) | Sum of PCB28, PCB52, PCB101, PCB138, PCB153, and PCB180 (ICES—6) |
|---|---|---|---|---|
| 6.6 | Fish liver and derived products thereof with the exception of marine oils referred to in point 5.7 | – | 20.0 pg/g wet weight | 200 ng/g wet weight |
| 6.7 | Marine oils (fish body oil, fish liver oil, and oils of other marine organisms intended for human consumption) | 1.75 pg/g fat | 6.0 pg/g fat | 200 ng/g fat |
| 6.8 | Raw milk and dairy products, including butter fat | 2.5 pg/g fat | 5.5 pg/g fat | 40 ng/g fat |
| 6.9 | Hen eggs and egg products | 2.5 pg/g fat | 5.0 pg/g fat | 40 ng/g fat |
| 6.10 | Fat of the following animals: | | | |
| | Bovine animals and sheep | 2.5 pg/g fat | 4.0 pg/g fat | 40 ng/g fat |
| | Poultry | 1.75 pg/g fat | 3.0 pg/g fat | 40 ng/g fat |
| | Pigs | 1.0 pg/g fat | 1.25 pg/g fat | 40 ng/g fat |
| 6.11 | Mixed animal fats | 1.5 pg/g fat | 2.50 pg/g fat | 40 ng/g fat |
| 6.12 | Vegetable oils and fats | 0.75 pg/g fat | 1.25 pg/g fat | 40 ng/g fat |
| 6.13 | Foods for infants and young children | 0.1 pg/g wet weight | 0.2 pg/g wet weight | 1.0 ng/g wet weight |

## 5.4 Process-Induced Toxicants

Process-induced toxicants are a series of contaminants produced during food manufacture process in food factory or at home. These chemical contaminants are usually related to food materials, storage conditions, and processing methods. They are common in human life because they can be produced in daily cooking. They are reported to have genotoxicity and carcinogenicity. But their epidemiological results are still to be verified for deficient detection methods and different dietary tradition. In this section, four of process-induced toxicants, heterocyclic aromatic amine, acrylamide, 3-monochloropropanol-1,2-diol, and *N*-nitrosamine, are described here.

## 5.4.1 *Heterocyclic Aromatic Amine*

### 5.4.1.1 Brief Introduction

Heterocyclic aromatic amines (HAAs) are a series of potent carcinogenic and mutagenic chemicals, which were first reported by Lijinsky and Shubick in 1964 [130]. They found the mutagenic activity of these compounds on the surface of the cooked fish and beef. These mutagenic chemicals are grouped and named as HAAs.

To date, at least 25 chemicals are isolated and identified at ng/g levels in cooked foods as HAAs [131]. And some interesting study results show that HAAs are almost everywhere in human life. They not only exist in well-done or grilled meat but also are found in drinks (coffee and alcohol beverages), the environment (breath of air, cigarette smoke, cooking fumes, rainwater, and river), human tissue (hair), and even biological fluids (plasma, urine, bile, and milk from health women). These HAAs are featured by two to five condensed aromatic cycles with at least one nitrogen atom in the aromatic ring. Furthermore, one exocyclic amino group is usually necessary except harman, norhaman and Lys-P-1 [132]. The representative structures of HAAs are shown in Fig. 5.4.

### 5.4.1.2 Hazard Identification and Characterization

The discovery of HAAs lasts for a long time in human history. Widmark found in 1939 that the substance in roasted food will increase the risk of cancer in mice test [133]. However, Lijinsky and Shubick in 1964 reported that mutagenic compounds formed during fish and meat cooking process can be grouped into aromatic hydrocarbons [130]. Then, Sugimura found these HAAs with mutagenicity in smoked condensates and charred surface of fish and meat [134]. From then on, more and more HAA types are recognized and paid attention to by the public. The main focus is on the health effect of HAAs in ordinary human life because the carcinogenicity and mutagenicity of HAAs are constantly confirmed by bacterial, mammalian cell line, and long-time animal tests.

Knize et al. in 1995 showed mutagenic activity of foods in daily diet using Ames *Salmonella* assay [135]. Furthermore, the mutagenicity of HAAs has also been demonstrated on mammalian cells with a number of tests, including gene mutation, chromosome aberrations, sister chromatid exchange, DNA strand breaks, DNA repair, and oncogene activation.

Animal tests in rodents and macaques showed that HAAs will cause liver, breast, and colon cancers. Metabolism analysis revealed that cytochrome P450 plays a key role in the carcinogenicity of HAAs [136]. Cytochrome P450 mediates a key activation step that N-hydroxylation of HAAs will form adducts to nucleic acids and lead to mutations. Other important researches indicate that the mutagenic activity of HAAs is much higher than traditional mutagens/carcinogens, e.g., 100-fold higher than aflatoxin B1 and 2000-fold higher than benzo[*a*]pyrene. So

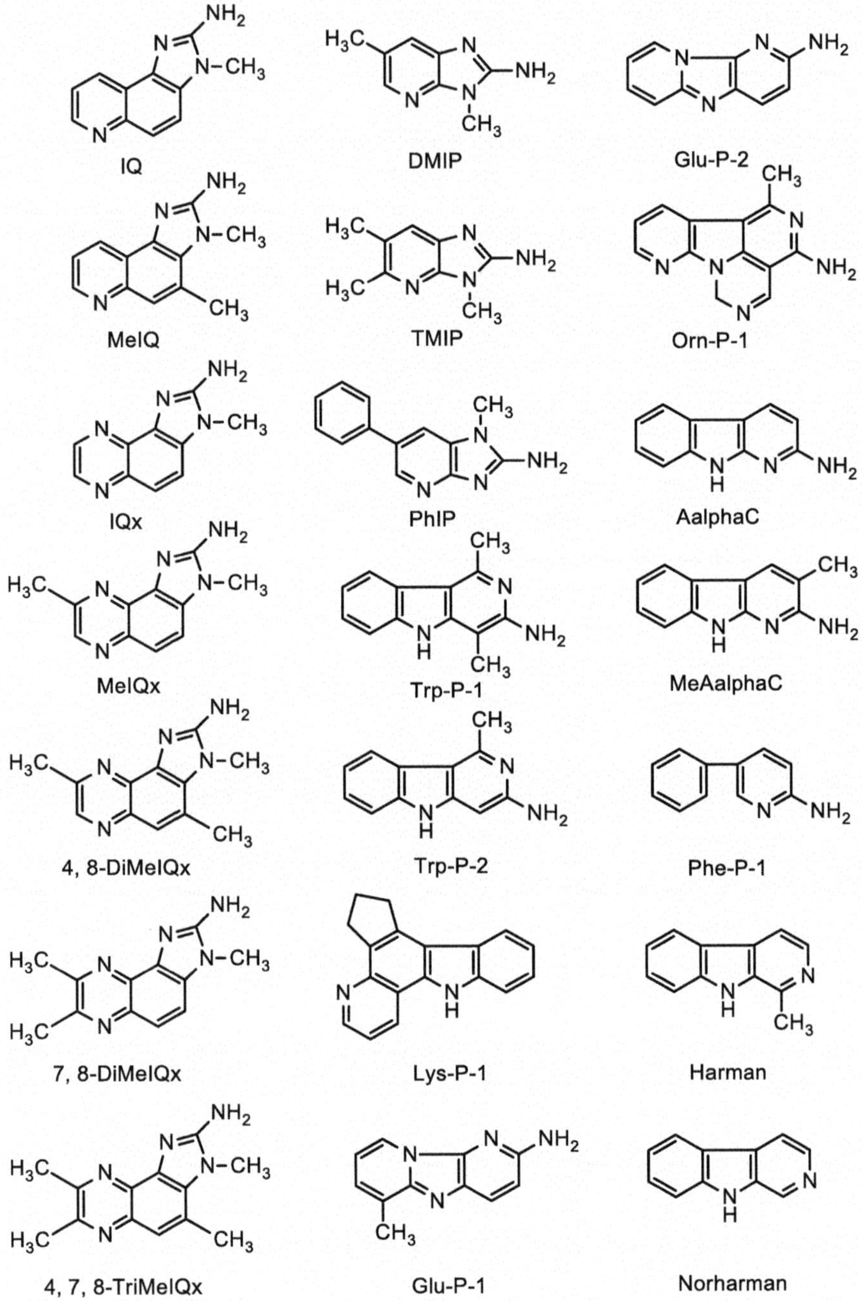


**Fig. 5.4** Representative structures of HAAs

in 1993, eight HAAs (MeIQ, 8-MeIQx, PhIP, AαC, MeAαC, Trp-P-1, Trp-P-2, and Glu-P-1) were listed as possible human carcinogens (class 2B) and one HAA (IQ) as a probable human carcinogen (class 2A) by the International Agency for Research on Cancer (IARC). A reduced exposure to these compounds was recommended. And in 2004, four HAAs (IQ, MeIQ, 8-MeIQx, and PhIP) were classified as reasonably anticipated to be human carcinogens in the National Toxicology Program.

Previous lab data have provided solid evidence that most of HAAs in human diets have mutagenicity and carcinogenicity, which will increase the risk of various types of human cancers. However, statistics from epidemiological research are not sufficient to support the lab results. For example, some epidemiological studies from Sinha in 2000 and 2001 show that HAAs in dietary food are positive risk factors for breast and colorectal cancers [137, 138]. But other studies consider that these HAAs cannot be associated with the incidence of human cancers in the colon, rectum, bladder, or kidney. Augustsson et al. consider that daily intake level of HAAs in the study population is lower than carcinogenic dose [139]. Gunter et al. indicate benzo[*a*]pyrene in red meat contributes to colorectal carcinogenesis [140]. In addition, the conclusion from experts of the World Cancer Research Fund (WCRF) and the American Institute for Cancer Research in 1997 indicated that the risk of developing colorectal cancer probably increases and the risk of pancreas, breast, prostate, and kidney cancers possibly increases with high intake of red meat, not HAAs [136]. The difference between lab data and epidemiological statistics maybe due to several factors, such as distinguishable cooking and eating habits among different countries, lack of dose response of possible relationship between HAAs and human cancer, inaccurate estimation of HAA exposure in epidemiological studies, deficient analytical method to determine the trace of HAAs in food preparation and metabolic process, difficulty in analysis of synergistic or antagonistic effect from complex gradients in food, incomparable metabolic fate of HAAs between human beings and other species, and the genetic factors like diverse acetylation status between populations. Therefore, a further research system should be established to clarify the relationship between the mechanism of carcinogenesis and the intake level of HAAs in daily diet.

#### 5.4.1.3 Formation Mechanism of HAA

Generally, HAAs are classified into two groups according to the synthesis temperature during the cooking process [141]. The first group is known as amino-carbolines or pyrolytic HAAs, including Phe-P-1, AαC, MeAαC, harman, norharman, and Trp-P-2. These HAAs are formed through pyrolytic reaction between amino acids and proteins at temperatures higher than 300 °C. This process of pyrolytic reaction is accompanied by production of many chemical intermediates, which will condense to form potent mutagens in the end. The second group is called aminoimidazole-azaarenes or thermic HAAs, such as DMIP, IQ, MeIQ, and IQx. This group of HAAs is usually formed through the reaction among free amino acids, creatin(in)e and hexoses, at temperatures between 100 and 300 °C. The products can be classified as imidazopyridine, imidazoquinoline, and imidazoquinoxaline derivatives. It should be noted that human is more susceptible to thermic HAAs because these highly mutagenic chemicals are often formed under the family cooking condition (<300 °C).

#### 5.4.1.4 Analysis Method of HAA

The difficult analysis of HAAs lies in complex matrices and diverse substances with low concentration. A series of analytical techniques have been adopted to solve these problems during the process of sample extraction, chromatography analysis, and mass spectrometry detection [131]. Solid-phase extraction (SPE) is often used to collect and concentrate substances from the complex matrices for HAA analysis. A combination of strong cation exchange cartridge and a reverse phase (C-18) cartridge is also necessary for SPE method. During the process of analysis and detection, chromatography technique and mass spectrometry are usually coupled to determine HAAs. Using techniques like high-performance liquid chromatography-mass spectrometry and gas chromatography-mass spectrometry, HAAs can be detected at a concentration of ng/g level.

#### 5.4.1.5 Modulation and Mitigation Strategies of HAA

Many factors influence the formation of HAAs in foods. It is generally accepted that the formation of HAAs during dietary process is influenced by factors like cooking temperature, time, way of cooking, and food material [142–144]. For example, the amounts and varieties of HAAs increase with the increase of cooking temperature and time. HAAs are almost not detected in fried food at less than 150 °C. And at a cooking temperature higher than 190 °C, the levels of HAAs increase sharply. Long cooking time also has positive correlation with the formation of HAAs.

Food material is a complex factor related to HAAs. Firstly, the fat content causes the difference in the formation of HAAs between white meat (chicken meat) and red meat (beef and pork). Fat is an effective heat-transfer agent, which needs a short time to reach a fixed meat surface temperature. It means that fat reaches the cooking temperature more rapidly than other parts of meat, which leads to a shorter heat exposure time and less HAA formation. Secondly, the total free amino acids, creatine, and other related nitrogen compounds in different meats influence formation of HAAs. PHIP and IFP are more easily detected in chicken meat, while the content of 8-MeIQx is much higher in beef and pork than that in chicken meat. Thirdly, the formation of HAAs is correlated to glycogen content in meat. Researches show that high glycogen content regulated by RN-allele in certain pigs significantly decreases the yield of HAAs (about 50% lower).

Cooking method also plays a key role in affecting the formation of HAAs. It usually includes three high-temperature cooking methods (panfrying, grilling/

barbecuing, and oven broiling) and some milder ways of cooking like microwaving, stewing, boiling, oven roasting, and deep-frying. It is common sense that the three high-temperature cooking methods produce high content of HAAs during the cooking process. It seems to be caused by three reasons: (1) the direct contact with hot surface of cookers, (2) high temperature in the course of grilling/barbecuing, and (3) rapid loss of water content from food leading to a relatively dry surface. And lower content of HAAs is produced using those milder cooking methods. Especially, stewing and boiling of meat will not produce detectable HAAs for the cooking temperature less than 100 °C. Compared to the cooking methods with direct contact between food and cookers, roasting is a special cooking method that mainly transfers heat to food by air. This indirect way of cooking will cause less HAA formation.

Cooking techniques may help reduce the formation of HAAs. It is reported that keeping the turn of meat at a frequency of 1 min leads to a lower average amount of total HAAs than turning meat just once.

Besides the above factors, the formation of HAAs also has correlation with other factors, such as degree of doneness; meat drippings and pan residues; use of sugar, oil, and antioxidants; content of free amino acids; and divalent cations in food. Food with high doneness level contains high content of HAAs. Meat drippings and pan residues often contain high level of HAAs. Excessive use of sugar while cooking food is helpful to reduce the formation of HAA. Free amino acids in food have a high correlation with HAA formation because many amino acids like proline, tryptophan, phenylalanine, serine, and threonine are involved in the reaction with precursors of HAAs. The addition of divalent cations like iron can enhance formation of 8-MeIQx. The divalent iron ion participates in the synthesis of HAAs as free radicals through a free radical–mediated reaction mechanism.

## 5.4.2 *Acrylamide*

### 5.4.2.1 Brief Introduction

Acrylamide (AA) is an organic compound with physicochemical properties defined as odorless, white, and crystalline solid. The molecular structure is shown in Fig. 5.5, and a series of physicochemical properties are listed in Table 5.3. AA is widely applied in the field of chemical engineering, construction, papermaking, environmental engineering, and biological research because its monomeric form can be easily cross-linked into polymeric form [145]. For example, AA-containing sodium dodecyl sulfate-polyacrylamide gel electrophoresis (SDS-PAGE) is a good tool in

**Fig. 5.5** Structure of AA

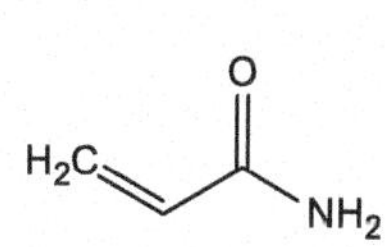

**Table 5.3** Physicochemical properties of AA

| | |
|---|---|
| CAS | 79-06-1 |
| Molecular formular | $C_3H_5NO$ |
| Molecular weight | 71.08 g/mol |
| Boiling point | 125 °C at 25 mmHg |
| Melting point | 84.5 °C |
| Solubility | Soluble in water, alcohols, acetone, and acetonitrile |
| Reactivity | Reacts with acids, bases, and oxidizing agents |
| Stability | Polymerization at temperature above the melting point |

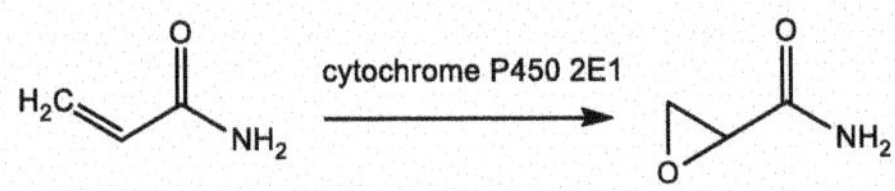


**Fig. 5.6** Conversion of AA to GA by cytochrome P450 2E1

biochemistry to separate charged molecules in mixtures by molecular masses in an electrical field.

#### 5.4.2.2 Hazard Identification and Characterization

Hazard identification and characterization are essential to risk assessment for AA. Studies show that AA is easily absorbed by the human gastrointestinal tract and rapidly distributed to different tissues. AA can be converted by cytochrome P450 2E1 in the liver to a reactive metabolite, glycidamide (GA), which is shown in Fig. 5.6 [146]. GA is able to react with nucleic acids to form adducts, which will lead to gene mutation and chromosomal aberration. This reactive metabolite is also ready to become adducts with many proteins. Furthermore, GA is found to form hemoglobin adduct in the blood system, which is important to assess the professional exposure to AA.

As early as 1995, rodent experiments showed that AA had carcinogenic activity [147]. It can induce tumors in several organs like the breast, thyroid, and testis. Long-term animal experiments showed that AA had reproductive, developmental, and nervous toxicities. Therefore, AA was grouped as a probable human carcinogen (class 2A) by the International Agency for Research on Cancer (IARC) in 1994 [148].

Epidemiological studies are carried out to assess the risk of AA [149, 150]. The hemoglobin adduct has be found in the blood system from Swedish workers with several years' exposure of AA. As for risk assessment of AA from dietary source, great data gaps still exist between epidemiological studies and lab statistics. Epidemiology studies have shown that AA is only likely to cause cancers of the ovaries, endometrium, breast, and oral cavity, except the kidney. These data are not sensitive enough to show the significant association between AA exposure and cancer risk. The extremely inconsistent reproducibility of epidemiology data is mainly caused by low-level exposure, inaccurate food consumption questionnaires, insensitive detection method for internal exposure, and possible internal hormonal mechanisms involved in the absorption and metabolism.

#### 5.4.2.3 Formation Mechanism of AA

AA is not a natural ingredient in dietary food. It is a by-product of Maillard reaction occurring during the process of heating treatment like baking, frying, and roasting [151]. Generally, a list of cooked foods with high carbohydrate is involved in the formation of AA, including potato chips, fries, breads, cereals, and coffee. The AA content in these foods is listed in Table 5.4 [149]. This cooking procedure to produce AA is featured by color browning of food and high content of AA. The reducing sugars (glucosc or fructosc) and thc amino acids (mainly asparagine) in foods are responsible for the Maillard reaction. In brief, a carbonyl group of a reducing sugar at first reacts with the amino acid asparagines to form an *N*-glycosyl conjugation. Then, this conjugation is prone to forming a decarboxylated Schiff base accompanied by dehydration at high temperature more than 120 °C. In the end, two reaction routes are involved in the formation of AA. One is that the Schiff base decomposes into AA and an imine directly. The other is that Schiff base hydrolyzes into aminopropionamide and a carbonyl compound. This aminopropionamide will form AA with the elimination of an ammonia group. In principle, asparagine alone is able to form AA directly with low yields in an inefficient way. Reducing sugars are able to improve the efficiency of AA formation. In addition, studies also show AA can be formed by other routes without asparagines. They are not important enough to be described here.

**Table 5.4** Content of AA in some foods (European Food Safety Authority, 2008 report)

| Food item | Acrylamide (μg/kg) | | |
|---|---|---|---|
| | Mean | Median | Maximum |
| *Potatoes* | | | |
| Chips (French fries) | 278 | 211 | 2466 |
| Crisps | 614 | 416 | 4382 |
| Home-cooked products | 180 | 65 | 3025 |
| *Bread/cereals* | | | |
| Biscuits | 205 | 120 | 1940 |
| Bread, soft | 38 | 15 | 528 |
| Bread, crisp | 234 | 109 | 1538 |
| Breakfast cereals | 160 | 74 | 2072 |
| *Others* | | | |
| Coffee, roasted | 206 | 167 | 1524 |

#### 5.4.2.4 Analysis Method of AA

The content of AA in foods is closely associated with the food itself, especially the components of food and the ways to treat food. Accurate determination of AA in foods is the precondition for mechanism study on AA formation, food production technique improvement, AA content control, and health risk assessment. To date, many analysis methods are published for AA determination, such as capillary electrophoresis-mass spectrometry (CE-MS), liquid chromatography-mass spectrometry (LC-MS), nonaqueous capillary electrophoresis (NACE), high-performance liquid chromatography-mass spectrometry (HPLC-MS), pressurized fluid extraction (PFE), matrix solid-phase dispersion (MSPD), gas chromatography-mass spectrometry (GC-MS), solid-phase microextraction-gas chromatography (SPME-GC), enzyme-linked immunosorbent assay (ELISA), and microemulsion electrokinetic chromatography (MEEKC) [151–154]. The main differences among these analysis methods lie in sample cleanup and AA extraction, which have a great impact on the result of AA determination. And due to the high solubility of AA in water, LS-MS and GC-MS are considered as fast, accurate, and reproducible methods and most widely used in the world. The advantage of liquid chromatography (LC) is that AA can be easily solved in aqueous solvent. This aqueous extraction directly from foods is compatible with the LC method. Derivatization is avoided in this method with the advantage of sensitivity and stability. And gas chromatography (GC) is also a matured method to detect brominated AA with high sensitivity. In addition, it is reported that LC-MS may be superior to GC-MS method for AA determination.

#### 5.4.2.5 Modulation and Mitigation Strategies of AA

There are many factors that influence the formation of AA, such as the content of AA precursors, pH condition, heating condition (heat and time), and water activity. Firstly, AA precursors including reducing sugars and amino acids (mainly asparagine) have a great effect on AA formation. The two precursors are involved in Maillard reaction, which is the main mechanism of AA formation. And it is generally considered that high content of reducing sugars and amino acids increases the formation of AA. The content of the two precursors can be influenced by the type of raw materials, storage condition, and fertilizers [145]. For example, the content of the two precursors is different among potatoes, cereals, and coffee. It is also the same to different flours (rye flour, 0.41–0.44 g asparagine $kg^{-1}$; spelt flour, 0.06–0.12 g asparagine $kg^{-1}$; wheat flour, 0.05–0.25 g asparagine $kg^{-1}$). Potatoes stored at 8 °C have lower level of glucose and fructose than those at 4 °C because reducing sugars in potatoes markedly increase at 4 °C. Sulfur and nitrogen in fertilizers are reported to play a key role in the content of precursors. Decrease of sulfur or increase of nitrogen will elevated the level of asparagines. Secondly, pH condition is an important factor, which influences the activity of amino acids and reducing sugars in Maillard reaction. A low-pH condition is able to inhibit the formation of Schiff base, which is the chemical intermediate to form AA. Therefore, using some edible acids can reduce the content of AA in cooked foods. Thirdly, heating condition can directly influence the formation of AA. Scientific researches show that AA formation begins at 120 °C. The content of AA increases with the increase of cooking temperature and achieves the highest level at 180 °C. Long time of cooking also increases AA formation. But after AA content reaches the peak level, prolonged cooking time and elevated temperature induce the decrease of AA content in foods. The elimination of AA by evaporation or degradation takes place at the same time of AA formation. Last but not least, water activity in foods has a great effect on AA formation. High water content induces low content of AA. And 11–21% water content causes highest AA level in foods.

#### 5.4.2.6 Biomarker Strategy to Monitor AA

The monitoring and risk assessment on AA have become an inevitable task since it is demonstrated to be a genotoxic chemical. Some studies adopt the food consumption questionnaire strategy, and others detect the biomarkers of AA in blood such as AA hemoglobin adducts and AA glycidamide adduct to evaluate the level of dietary AA exposure [148]. However, as mentioned above, epidemiological studies have shown that no significant association is found between AA exposure and cancer risk. It may be attributed to the different dietary habits in populations and imprecise food consumption questionnaires. To solve these problems, biomarker strategy is selected as a relative accurate route to measure dietary exposure of AA. This strategy aims to determine a steady-state biomarker level under the long-time AA exposure because AA or its derivative, glycidamide, is able to form adduct with hemoglobin. And this biomarker can steadily circulate in blood system over 120 days, equal to the life span of red blood cells. Using this strategy, scientists find that long-time AA exposure probably increases the incidence of estrogen receptor-positive breast cancer.

### *5.4.3 3-Monochloropropane-1,2-Diol*

#### 5.4.3.1 Brief Introduction

Chloropropanol is a family of food contaminants produced in the process of edible oil refinement. It can be considered to be the derivatives of triacylglycerol (TAG) [155]. According to the substitution number of hydroxyl group by chlorine atom, chloropropanol can be divided into two groups, monochloropropanol (MCPD) and dichloropropanol (DCP). As for the substitution site of hydroxyl group by chlorine atom, the molecular structure of MCPD includes 2-monochloropropane-1,3-diol (2-MCPD) and 3-monochloropropane-1,2-diol (3-MCPD). The chemical structure of DCP has 1,3-dichloropropane-2-ol (1,3-DCP) and 2,3-dichloropropane-1-ol

| | | |
|---|---|---|
| $R_1$ and $R_2$ and $R_3$ = acyl group | TAG | Glycidol |
| $R_1$ and $R_2$ and $R_3$ = OH | Glcerol | |
| $R_2$ = Cl, $R_1$ and $R_3$ = OH | 2-MCPD | |
| $R_3$ = Cl, $R_1$ and $R_2$ = OH | 3-MCPD | |
| $R_1$ and $R_3$ = Cl, $R_2$ = OH | 1, 3-DCP | |
| $R_2$ and $R_3$ = Cl, $R_1$ = OH | 2, 3-DCP | |

**Fig. 5.7** Analogue structures of TAG and structure of glycidol

(2,3-DCP). The structures of these molecules are shown in Fig. 5.7. It is generally considered that the content of MCPD is much higher than that of DCP produced during the food processing. And 3-MCPD is the main component of MCPD. So 3-MCPD is a representative substance for food contaminant monitoring.

#### 5.4.3.2 Hazard Identification and Characterization

3-MCPD can be found not only in refined oil but also in a list of dietary foods (Table 5.5). In 2004, Svejkovská et al. reported that 3-MCPD in foods is mainly in an esterified form [156]. However, to date, the toxicological evaluation of 3-MCPD esters is mainly based on free 3-MCPD. In vitro studies have shown that 3-MCPD esters will be degraded by pancreatic lipases, and the released free 3-MCPD can be rapidly absorbed in the gastrointestinal tract. The efficiency of this enzymatic reaction for monoesters and diesters of 3-MCPD is quite different. Almost 100% monoesters can be degraded into 3-MCPD in 1 min; 45%, 65%, and 90% diesters can be enzymatically hydrolyzed to 3-MCPD in 1, 5, and 90 min, respectively [157]. According to the hydrolyzation of 3-MCPD esters by lipases, the German Federal Institute for Risk Assessment (BfR) assumes a 100% release of 3-MCPD from its esterified form in the gut to evaluate the potent hazard of 3-MCPD esters.

Microbiological and cellular experiments indicate that 3-MCPD has mutagenic activity in vitro, but no mutagenic activity is observed in vivo [158].Animal experiments indicate that 3-MCPD has obvious renal toxicity in rats and mice. 3-MCPD can induce acute glomerular nephritis in male SD rats and severe proteinuria and glucosuria in male Wister rats [157]. It can increase the relative kidney weights. The toxicological mechanism is considered to be caused by the inhibition of glycolysis by metabolites of 3-MCPD associated with β-chlorolactate pathway. Studies using primates show that 3-MCPD can cause damage to bone marrow with clinical signs of hemorrhage, depression, weakness, and hematological symptoms like anemia, leucopenia, and severe thrombocytopenia. Furthermore, 3-MCPD is reported to have reproductive toxicity, neurotoxicity, and immunotoxicity. For example, 3-MCPD is able to impair male fertility by decreasing sperm motility, changing sperm morphology, and causing epididymal lesions [157]. The mechanism is suggested to be the inhibition of spermatozoa glycolysis enzymes by metabolites of 3-MCPD. Some studies report that 3-MCPD can cause neurotoxic changes in astrocytes and induce severe forelimb and hind limb paralysis in mice at single doses

**Table 5.5** Content of 3-MCPD in different foods

| Food | Range (mg $kg^{-1}$) |
|---|---|
| Biscuits | 0.3–0.7 |
| Bread | <0.01–0.04 |
| Bread, toasted | 0.06–0.16 |
| Cereal | <0.01–1.40 |
| Crackers | 0.1–1.14 |
| Crispbread | 0.42–0.58 |
| Crisps | 0.05–1.19 |
| French fries | 0.04–0.40 |
| Chicken, grilled | 0.26–0.74 |
| Ham | n.d.-2.64 |
| Salami | 0.88–6.41 |
| Coffee | <0.1–0.39 |
| Malt and beer | 0.01–0.65 |
| Frying oils (fresh and used) | <0.15–16.2 |
| Margarine (fat portion) | <0.15–7.7 |
| Refined coconut oil | 1.42–1.69 |
| Refined palm kernel oil | 0.85–1.40 |
| Refined palm oil | 1.39–4.17 |
| Refined vegetable oils | <0.15–18.8 |
| Unrefined vegetable oils | <0.15–0.31 |
| Coffee creamer | 0.13–0.73 |
| Cream (aerosol) | 0.05–0.73 |
| Infant formula | <0.08–0.59 |
| Infant formula (fat portion) | 0.57–4.10 |
| Milk, growing up | 0.06–0.29 |
| Human breast milk | <0.011–0.076 |

Source: Adapted from reports published by the International Life Sciences Institute (www.ilsi.org/)

of 90 mg/kg body weight. 3-MCDP is also demonstrated to decrease many parameters such as thymus weight, cellularities of the spleen and thymus, the antibody-forming cell response to SRBC, NK cell activity, and peritoneal macrophage activity. Given the toxicological data mentioned above, in 2001, a provisional maximum tolerable daily intake (TDI) of 2 μg/kg body weight for 3-MCPD was set by the Joint FAO/WHO Expert Committee on Food Additives. And in 2011, 3-MCPD was classified as possible human carcinogens (class 2B) by IARC. In addition, it should be noted that glycidol, the essential metabolite of 3-MCPD esters, was classified as probably carcinogenic to humans (class 2A) by IARC in 2000 and considered as "reasonably anticipated to be a human carcinogen" by the US National Toxicology Program (NTP) in 2007.

#### 5.4.3.3 Formation Mechanism of 3-MCPD

There are three possible mechanisms for the formation of 3-MCPD [159]. The first is acyloxonium ion intermediate theory, which is briefly described below (Fig. 5.8). TAG is hydrolyzed into diacylglycerols (DAG) during the oil processing, and DAG is further hydrolyzed into monoacylglycerols (MAG). Under acid condition, the hydroxyl group on DAG and MAG is prone to be protonated, which results in the elimination of one water molecule to form cyclic DAG- and MAG-derived acyloxonium ion intermediate. The next step is that chlorine anion opens the acyloxonium ion cycle to form 3-MCPD by a nucleophilic addition. The acyloxonium ion intermediate has been demonstrated by scientists using infrared spectroscopy and isotope labelling technique. The second mechanism to form 3-MCPD is glycidol intermediate theory. Glycidol is an important intermediate existing in the process of oil refinement. The epoxy structure of glycidol can be directly nucleophilic-attacked by chlorine anion to form 3-MCPD. The third mechanism is that under the condition of acid catalytic hydrolysis, the acyl group of TAG or DAG is nucleophilic-substituted by chlorine anion directly to form 3-MCPD.

#### 5.4.3.4 Analysis Method of 3-MCPD

Indirect and direct strategies are often used to determine the content of 3-MCPD in foods [160, 161]. It is noteworthy that 3-MCPD in foods is mainly in an esterified form. The principle of indirect strategy is to convert 3-MCPD ester to 3-MCDP. It aims to measure the total content of 3-MCPD in samples. This strategy is briefly divided into several steps: extraction of samples from raw materials, addition of an internal standard to samples, transesterification to obtain 3-MCPD under acid or alkaline condition, neutralization of the reaction mixture, salting out to purify 3-MCPD, derivatization of 3-MCPD, and determination using GC-MS method. Direct strategy does not require complex procedures of sample processing and is able to directly determine 3-MCPD esters in samples. LC-MS/MS and LC-TOFMS methods are often used in this strategy. Comparatively, the two methods have much difference in sample preparation, chemical standard, and chemical structure verification. In terms of sample preparation, there are several steps involved in the preparation of test samples using indirect method. During this process, the quality of samples can be influenced by many factors. For example, the addition of acid or alkaline is possible to change the concentration of 3-MCPD or form by-product in samples to influence measured values. In terms of direct method, very easy steps like

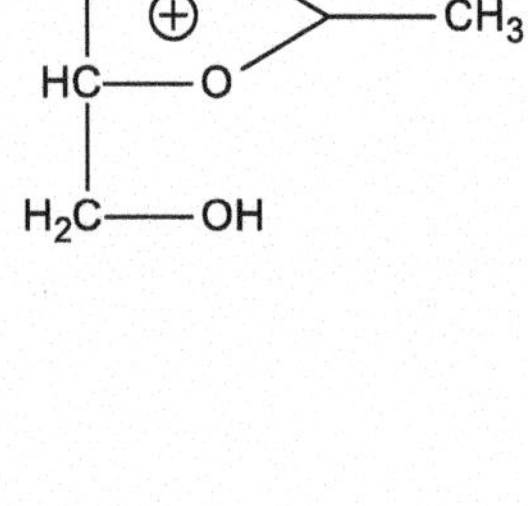


**Fig. 5.8** Structure of cyclic acyloxonium ion intermediate

dissolution and homogenization are needed for sample preparation, which can avoid interference in determination. In terms of reference preparation, indirect method can be easily applied to almost all types of vegetable oils because few chemical standards are needed. But a large number of reference materials have to be prepared using direct method because substances with any structures can be included into the assays. In terms of structure verification, indirect method cannot provide any detailed information on the chemical structure of the different esters. But the definite chemical structure of 3-MCPD esters can be recognized using direct method. It is important for toxicological study on 3-MCPD because varieties of 3-MCPD esters may have different potent toxicity and biological fate. To some extent, the two methods can be an alternative for each other to evaluate the precision and trueness of data.

#### 5.4.3.5 Modulation and Mitigation Strategies of 3-MCPD

The occurrence of 3-MCPD esters is closely associated with the type of raw materials and way of processing foods [159]. For example, the components of crude vegetable oils are important to the formation of 3-MCPD esters. The content of 3-MCPD ester precursors like chloride ions, TAG, and DAG in crude palm oils is much higher than that in seed oils. So, the refined palm oil contains more 3-MCPD esters than seed oils like maize, rapeseed, soybean, and sunflower. Furthermore, it is reported that the highest content of 3-MCPD esters is found in the refined oils, but it almost cannot be detected in the oils without refinement. Many studies show that during the process of oil refinement, degumming, bleaching, and deodorization stage play a crucial role in the increase of 3-MCPD esters. Water degumming can remarkably reduce 3-MCPD esters by 84%. Bleaching with synthetic magnesium silicate is able to reduce a further 10% of 3-MCPD esters by removal of related precursors. The heating temperature and time in deodorization stage are the main factors involved in the formation of 3-MCPD esters in oil refinement. Elevated temperature and extended time lead to an increase in 3-MCPD esters. In addition, 3-MCPD esters can be found in thermally treated cereals, retailed foods, infant formula, and human breast and goat milk. The generation of 3-MCPD esters in roasted cereals (barley, rye, and wheat) is dependent on time and temperature of food processing. Almost all the retail foods like potato crisps, French fries, biscuits, and coffee have some common characteristics of ingredient fats and oils, heating processing, and high surface area to volume ratios. It should be noticed that the significance of 3-MCPD esters in human breast and goat milk lies in the probable relationship between dietary intake and potent metabolism of 3-MCPD esters.

Many efforts are made to mitigate 3-MCPD esters in various foods. This aim is achieved mainly in two aspects. One is controlling the content of 3-MCPD ester precursors. For example, methods like reducing precursors in raw materials and avoiding introduction of precursors during the food processing are often used. The other is optimizing process of food manufacture. Heating temperature, time, water content, pH condition, and production techniques are the key points to decrease the content of 3-MCPD esters in foods.

### 5.4.4 N-*Nitrosamine*

#### 5.4.4.1 Brief Introduction

*N*-Nitrosamine (NA) is a large group of chemical compounds with the structural similarity of a nitroso group. NA compounds can be divided into two groups, volatile NA (VNA) and nonvolatile NA (NVNA), which is associated with the molecular weight of NAs [162]. VNA with lower molecular weight is characterized by alkyl or monocyclic secondary amines. It can be isolated from foods by solvent extraction or by distillation. NVNA usually has long alky chain substitution with higher molecular weight. It is usually extracted from foods using polar organic solvents. Several representative structures of NAs are shown in Fig. 5.9.

#### 5.4.4.2 Exposure to NA

Generally, the exposure to NAs can be divided into exogenous and endogenous route [163]. On one hand, the exogenous NAs are perhaps the most widespread food contaminants in the world. It is well-known that treating foods with nitrites is a traditional method to enhance the coloring, flavoring, and preservation of foods. Meanwhile, proteins abundant in human common foods (fish, meat, vegetable, and even beer) can be degraded into amino compounds. The two parts of precursors, nitrites and amino compounds, easily interact with each other to form NAs, and then, these NAs are consumed by people in everyday diet. The contaminated foods include cured meats, cheeses, smoked fish, meat, malt, milk products, and spices (Table 5.6) [164, 165]. The estimated total exposure to NAs is approximately 1 μmol $day^{-1}$. Dietary intake contributes over 70% of the NA exposure. Asian people are more easily exposed to NAs because of high content of NAs in seafood and barbeque food in their dietary intake. On the other hand, endogenous NAs are formed in the human stomach, in which acid condition and abundant precursors can promote the formation of NAs. The precursors may come from food ingested outside or digested inside. The contribution of endogenous NAs cannot be neglected because some reports have shown that they play a key role in human cancer. Free iron, released from haem in red or processed meat, is suggested to affect the formation of endogenous NAs. Some studies show it is prone to be absorbed by the digestive tract and promote NA formation. In addition, industrial pollution will result in the exposure to NAs. For example, NAs widely exist in the rubber product for daily use. It is reported that every 5 g rubber nipple contains NAs of 42–617 μg/kg. And migration exposure from water happens to ion exchange resin too.

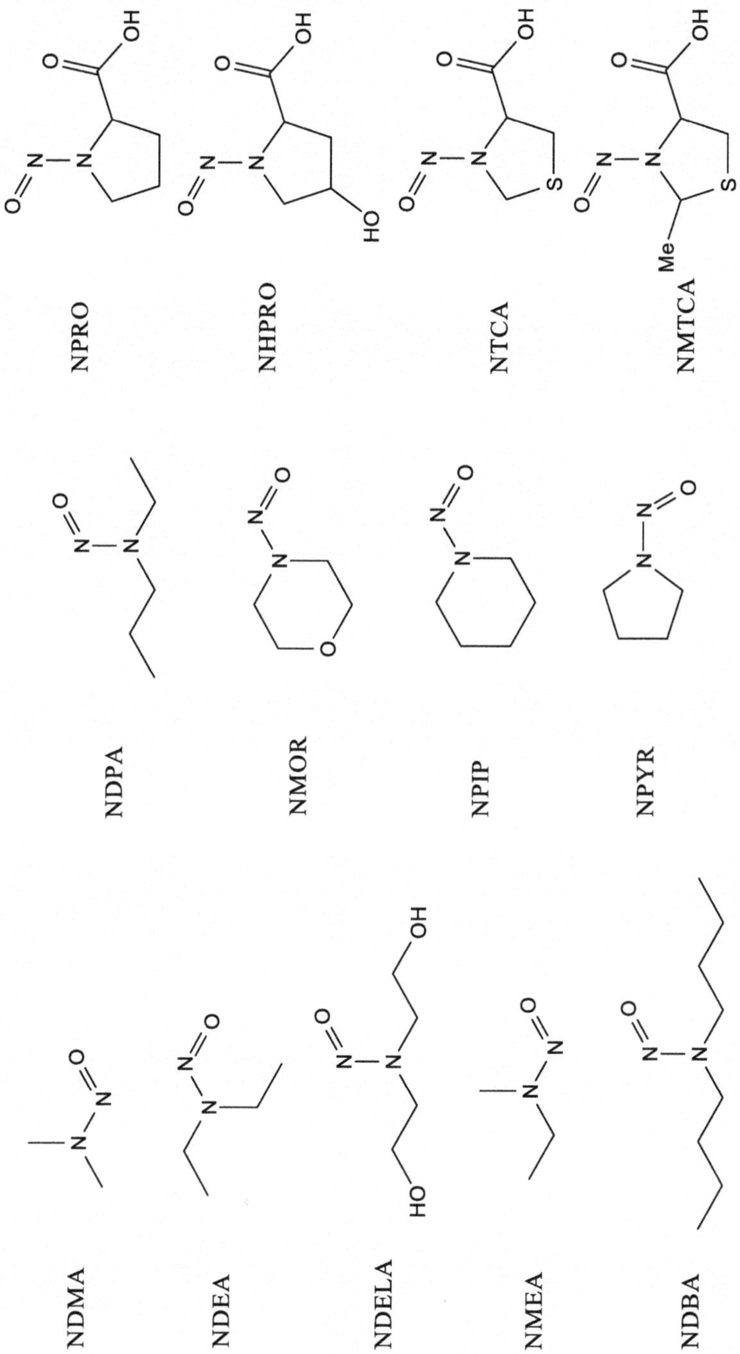


**Fig. 5.9** Representative structures of NAs

**Table 5.6** Range of volatile *N*-nitrosamine measured in selected foods (1973–2001)

| Food | *N*-nitrosamine (μg $kg^{-1}$) |
|---|---|
| Cheese | Not detected–0.5 |
| Milk | Not detected–0.45 |
| Bacon (cooked) | Not detected–6.5 |
| Ham | Not detected–0.1 |
| Salami | Not detected–0.33 |
| Fish (cooked) | Not detected–13.1 |
| Oil | Not detected–0.38 |
| Beer | Not detected–6.8 |

#### 5.4.4.3 Hazard Identification and Characterization

Most of the VNAs are reported to be carcinogenic [162]. *N*-Nitrosodimethylamine (NDMA) and *N*-nitrosodiethylamine (NDEA) were classified as probable human carcinogens (class 2A) by IARC in 1978. *N*-Nitrosodiethanolamine (NDELA), *N*-nitrosomethylethylamine (NMEA), *N*-nitrosodibutylamine (NDBA), *N*-nitrosodipropylamine (NDPA), *N*-nitrosomorpholine (NMOR), *N*-nitrosopiperidine (NPIP), and *N*-nitrosopyrrolidine (NPYR) were classified as possible human carcinogens (class 2B). However, majority of NVNAs are assumed to be weak carcinogens except *N*-nitrosoproline (NPRO), *N*-nitrosohydroxyproline (NHPRO), *N*-nitrosothiazolidine-4-carboxylic acid (NTCA), and *N*-nitroso-2-methyl-thiazolidine-4-carboxylic acid (NMTCA). Carcinogenicity of the four NVNAs remains to be verified by toxicological studies in vivo.

To date, the carcinogenicity of more than 300 NA compounds has been verified by scientific tests [165]. And about 90% of the tested NAs are demonstrated to be carcinogenic to 40 animal species, even to primates. The carcinogenic NAs can be targeted to a list of organs like the liver, lung, kidney, bladder, pancreas, esophagus, and tongue, which is dependent on the species. The probable carcinogenic mechanism of NAs to promote the cancer incidence can be concluded as follows: (1) NAs interact with DNA to form *N*-nitroso DNA adduct, (2) cells are not capable of repairing DNA damage, and (3) there is base exchange of the K-ras oncogene. The first carcinogenic route of NAs is the most important because the reactive structure of NAs is ready to induce the alkylation of DNA.

Many epidemiological investigations have been conducted to establish the connection between NAs and human cancers, e.g., the correlation between pregnant exposure to cured meats and brain tumors in children [162, 164, 165]. Unfortunately, most of these epidemiological studies do not have a definitive conclusion, except for the study on the correlation between esophageal cancer in certain parts of China and native foods with substantial concentration of NAs. The failure of these epidemiological studies may be results from two aspects. One is lack of reliable database in these studies to distinguish one or several special kinds of NAs, which are crucial to tumorigenesis. NA is a group of chemical compounds with different structures. The obtained data in these studies are the total exposure to NAs, which leads to the failure to find VIP chemical. The other is the source of NAs. NAs not only come from exogenous food exposure but also can be generated in the stomach of human. This endogenous carcinogenic exposure cannot be clearly predicted up to now.

#### 5.4.4.4 Formation Mechanism of NA

The formation of NA in foods is a complicated process (Fig. 5.10) [166]. Two groups of precursors are involved in the chemical reaction. One group is nitrite or nitrogen oxides. The other group is amino compounds (usually secondary or tertiary amines). The two groups of precursors interact with each other and form NA under proper condition. The reaction rate is dependent on pH condition (usually acid), precursor concentration, and type of amino compound. Secondary amines have strongest activity compared to primary and tertiary amines. Quaternary amines are also able to form NA but with low yield.

#### 5.4.4.5 Modulation and Mitigation Strategies of NA

NAs can be formed in a series of food manufacturing procedures [167, 168]. Many factors are likely to affect NA formation, such as meat quality, fat content, way of processing, and maturation time. They are usually helpful to prevent the pollution of precursors or inhibit the formation of NAs. Besides these approaches, antioxidants are effective alternative to remove NAs in foods, which can be used as additives in food processing. It is reported that adding different forms of vitamin C like ascorbate/ascorbic acid/erythorbic acid is successfully able to decrease the content of NAs. The strong affinity between vitamin C and nitrite reduces the precursors and inhibits the formation of NAs. Moreover, some studies also show that polar antioxidants like vitamin C are not good inhibitors because most of hazardous NAs are formed in lipid phase like fat. It is difficult for polar antioxidants in aqueous phase to diffuse into lipid phase where NAs generate. Therefore, it is recommended that a combination of polar (erythorbic acid) and nonpolar (ascorbyl palmitate) antioxidant more efficiently inhibits formation of NAs.

## 5.5 Food Irradiation and Other Emerging Technologies

Food irradiation is an application that can eliminate nonsporing pathogens and spoilage microorganisms by exposing food, such as X-rays and grays, to ionize radiation. It could be applied by randomly damaging DNA to make the insects and

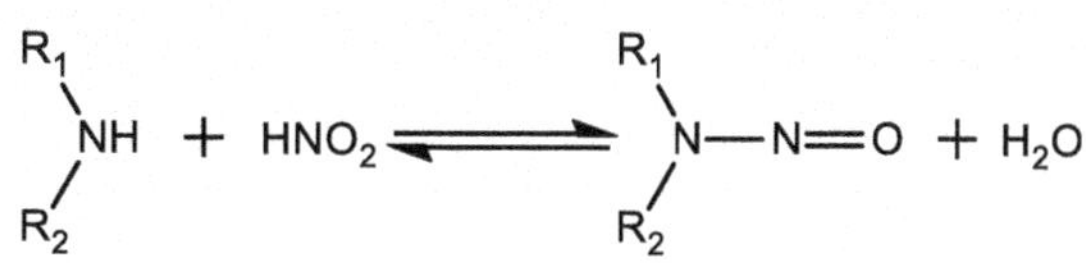


**Fig. 5.10** Formation mechanism of NA

parasites die or sterile. Moreover, the root/tuber crops such as potatoes and onions are also inhibited from sprouting. Generally, fruits and vegetables can extend their shelf lives after exposure [169]. Other applications of food irradiation include delaying ripening, increasing fruit moisture, and improving the rehydration capacity. Food can also be sterilized by being exposed under high-dose irradiation.

### *5.5.1 History of Food Irradiation*

Scientists started the food irradiation researches in 1895 by using the newly discovered X-rays. However, due to the development of the uses of atomic energy, radiation and pasteurization started to be applied in food preservation, medical supply sterilization, and the structural changes of building materials until 1960s [170].

Much research on food irradiation was carried out in the United Kingdom and the United States in the 1930s and 1940s, which led to the effectiveness and technology of food irradiation being perfected. The International Food Irradiation Project (IFIP) was launched in 1970; this was a 19-nation fund to finance research into studying and developing food irradiation procedures. The IFIP was completed in 1982 and was replaced by collaboration between the WHO, the Food and Agriculture Organization (FAO), and the International Atomic Energy Agency (IAEA) called the International Consultative Group on Food Irradiation (ICGFI). Joint FAO/IAE/WHO Expert Committees were set up, including one on the Wholesomeness of Irradiated Foods and the Joint Expert Committee on Food Irradiation (JECFI) to explore irradiation procedures and safety. Their findings were adopted by the Codex Alimentarius Commission (CAC) and became the general standard for irradiated foods and later the Recommended International Code of Practice for the operation of radiation facilities for the food treatment [171–176].

### *5.5.2 Types of Radiation [170]*

The adequate wavelength of electromagnetic radiation to treat foods is between 103 and 101 nM (also called ionizing radiations). Under the circumstance, the radiation can convert atoms and molecules to ions by removing electrons. Energetic charged particles and high-energy photons such as electrons, X-rays, and gamma rays are generally considered as the ionizing radiation. However, not all types of ionizing radiation are inappropriate for food irradiation. Some types of ionizing radiation do not penetrate thoroughly into the irradiated material, and the others make the material radioactive. Thus, the Codex General Standard for Irradiated Foods has provided the following comment in 1984:

1. The γ-rays, which is applied for irradiating foods, should choose the radio nuclides 60Co and 137Cs.
2. The machine, which used to generate the X-rays, should operate less than 5 MeV as an energy level.
3. The machine, which used to generate the electrons, should operate less than 10 MeV as an energy level.

### *5.5.3 The Irradiation By-Products [177]*

In previous research, there were almost 65 products that had been characterized as unique irradiation by-products, and only six in the research were confirmed as naturally occurring or found in other situations. Also, it is noteworthy that the free radical compounds are not unique to food irradiation. One compound used to sterilize hooks was proved to be the reason that the hooks in slaughterhouses, which were used to hang carcasses, were contaminated.

With the researches in the recent decades and more than a thousands studies, the side effect of irradiation of foods has not been found.

As most of preservation and cooking methods, some chemical changes happen in irradiation treatment. The atoms have lost their electrons and then formed ions when high-energy particles strike the matter. The new compounds, which didn't exist in the food before treatment, may be created by the interaction with the newly generated radiolytic products. And, some new compounds could produce off-flavors. In addition, meat can regulate this interaction during the irradiation process by maintaining a low temperature. And the conversion of water to hydrogen peroxide is also the most common chemical reaction during treatment. All of these reactions occurred in food preservation, and the unique reactions from irradiation are almost not detrimental.

The FDA suggested that "very few of these radiolytic products are unique to irradiated foods and about 90% of the radiolytic products are known as natural components of food." Some of these radiolytic products are fatty acids, amino acids, and hydrocarbon, which are commonly found in the waxy coating on various fruits including grapes, pears, and apples. Other products, such as fatty compounds, are the same to the compounds found from common cooking methods of meat. And the other 10% is similar to natural ingredients in food. The chemistry of irradiation is easy to predict, and the type of food or other food components present cannot affect the products of an individual component such as proteins. There is no obvious evidence of hazards that have been found by critically testing the toxicity of radiolytic products. The total nutrient retention in irradiated foods is similar to other methods, and it also does not impair the activity of some nutrients. In terms of macronutrients, the proteins are split or aggregated after methionine degraded. Similarly, the double bonds of polyunsaturated fatty acids also produce off-flavors after split. Furthermore, part of micronutrients, such as vitamins C, E, K, B6, and riboflavin, disappeared. The reduction of vitamin C (also known as ascorbic acid) has been reported. Although the ascorbic acid becomes the dehydroascorbic acid, it is a minor change in nutritional standpoint. When the tocopherol interacts with oxygen, it can be very sensitive to irradiation. Vitamin K is relatively stable compared with other macronutrients. To reduce the damage of irradiation on vitamins, some methods exclude oxygen and light and then keep the food at a low-temperature environment and also use the lowest dose, which was needed to the treatment. However, there are minimal losses and changes in the process of improving techniques, such as keeping a vacuum or low-temperature environment during the radiation milieu.

### 5.5.4 *Dose Considerations*

Dose is the amount of radiation received by an irradiated object. The basic unit is the gray (Gy), which means 1 Gy = 1 J (unit to energy)/kg of irradiated material. In the irradiation of food, workers commonly refer to the dose as being "high" (more than 10,000 Gy) or "low" (less than 10,000).

The dosage requirements for beneficial effects vary widely, depending on the desired effect as well as the particular food being treated. Sterilization obviously requires the highest doses. Research has also shown that it is at these high levels that serious side effects begin to show up, especially in foods of animal origin. Off-flavors, off-colors, and textural defects are some of the side effects caused by high-dose irradiation of such products as meat, poultry, eggs, and dairy products. Sometimes, the side effects can be minimized; sometimes, they cannot.

#### 5.5.4.1 Potential Applications: Low-Dose Levels

It is certain that properly packaged meats and fish, if irradiated with low doses and kept under refrigeration, have a significantly longer shelf life than the similar products that are unirradiated. Food-borne pathogens (e.g., salmonellae) can be substantially reduced or eliminated by low-dose irradiation. In addition, parasites (e.g., trichina) are readily destroyed by low-level irradiation. Unprocessed fruits and vegetables irradiated at low-dose levels can be rendered free of insects; a substantial reduction in spoilage microorganisms can also be accomplished. In addition, senescence (overripening) in fruit and sprouting in tubers and bulbs can be delayed substantially. These effects provide for a significant increase in the shelf life of many produce items. Cereal products can be freed of insect pests by low-dose irradiation. In all of these low-dose applications, there are substantially no adverse side effects from irradiation.

#### 5.5.4.2 Potential Applications: High-Dose Levels

High-dose irradiation can be used to sterilize the metal cans in which meat and fish are hermetically sealed. These products are kept indefinitely at room temperatures and are superior in sensory quality to similar heat-sterilized products. However, to obtain the optimal quality, it is necessary to irradiate these products while frozen, and certain additives must be used to protect texture. Naturally, these additional requirements increase production costs.

Spices and condiments can be sterilized by high-dose irradiation. This greatly enhances their value for use in producing formulated foods because of the high microbial loads usually found in commercial spices and condiments. Hospital diets for patients highly susceptible to infections can be sterilized by high-dose irradiation. Although not generally as acceptable as regular hospital food, these irradiated diets seem to be acceptable to such patients and, of course, do protect them from infections due to food-borne pathogens.

## References


1. Sagratini G, Mañes J, Giardiná D, et al. Analysis of carbamate and phenylurea pesticide residues in fruit juices by solid-phase microextraction and liquid chromatography–mass spectrometry. J Chromatogr A. 2007;1147:135–43.
2. Food and Agriculture Organization of the United Nations. Submission and evaluation of pesticide residues data for the estimation of maximum residue levels in food and feed; 2002, Rome, First Edition.
3. Jakubowski M, Trzcinka-Ochocka M. Biological monitoring of exposure: trends and key developments. J Occup Health. 2005;47:22–48.
4. Barr DB, Needham LL. Analytical methods for biological monitoring of exposure to pesticides: a review. J Chromatogr B Analyt Technol Biomed Life Sci. 2002;778:5–29.
5. Aitio A. Guidance values for the biomonitoring of occupational exposure. State of the art. Med Lav. 2006;97:324–31.
6. Maroni M, Colosio C, Ferioli A, et al. Introduction. Toxicology. 2000;143:5–118.
7. Angerer J, Aylward LL, Hays SM, et al. Human biomonitoring assessment values: approaches and data requirements. Int J Hyg Environ Health. 2011;214:348–60.
8. Göen T, Schaller K-H, Drexler H. Biological reference values for chemical compounds in the work area (BARs): an approach for evaluating biomonitoring data. Int Arch Occup Environ Health. 2011;85(5):571–8.
9. Casida JE. Pyrethrum flowers and pyrethroid insecticides. Environ Health Perspect. 1980;34: 189–202.
10. Verschlyle RD, Aldridge WN. Structure-activity relationship of some pyrethroids in rats. Arch Toxicol. 1980;45:325–9.
11. Clark JM, Symington SB. Advances in the mode of action of pyrethroids. Top Curr Chem. 2012;314:49–72.
12. Casida JE, Quistad GB. Golden age of insecticide research: past, present, or future? Annu Rev Entomol. 1998;43:1–16.
13. California Department of Pesticide Regulation (CDPR). Pesticide use status. Government of California. http://www.cdpr.ca.gov/docs/pur/purmain.htm. Accessed June 2007.

14. DEFRA, Department for Environment, Food, and Rural Affairs, Pesticide usage statistics. Central Science Laboratory and Scottish Agricultural Science Agency, UK. 2006. http://pusstats.csl.gov.uk/index.cfm.
15. Soderlund DM. Molecular mechanisms of pyrethroid insecticide neurotoxicity: recent advances. Arch Toxicol. 2012;86(2):165–81.
16. Anand SS, Kim KB, Padilla S, et al. Ontogeny of hepatic and plasma metabolism of deltamethrin in vitro: role in age-dependent acute neurotoxicity. Drug Metab Dispos. 2006;34(3):389–97.
17. Anand SS, Bruckner JV, Haines WT, et al. Characterization of deltamethrin metabolism by rat plasma and liver microsomes. Toxicol Appl Pharmacol. 2006;212(2):156–66.
18. Kim KB, Anand SS, Kim HJ, et al. Age, dose, and time-dependency of plasma and tissue distribution of deltamethrin in immature rats. Toxicol Sci. 2010;115(2):354–68.
19. Elliot M, Janes NF. Synthetic pyrethroids—a new class of insecticide. Chem Soc Rev. 1978;7: 473–505.
20. Soderlund DM, Clark JM, Sheets LP. Mechanisms of pyrethroid neurotoxicity implications for cumulative risk assessment. Toxicology. 2002;171:3–59.
21. Glomot R. Toxicity of deltamethrin to higher vertebrates, deltamethrin (monograph). Paris: Roussel-Uclaf Research Centre; 1982. p. 109–36. Chapter 4
22. Gray AJ, Soderlund DM. Mammalian toxicology of pyrethroids. In: Hutson DH, Roberts TR, editors. Insecticides. New York: Wiley; 1985. p. 193–248.
23. CDC, Center for Disease Control and Prevention. Toxicological profile for pyrethrins and pyrethroids. Agency for Toxic Substances and Disease Registry, US Department of Health and Human Services, Atlanta, Sept 2003.
24. Crofton KM, Kehn LS, Gilbert ME. Vehicle and route dependent effects of a pyrethroid insecticide, deltamethrin, on motor function in the rat. Neurotoxicol Teratol. 1995;17(4): 489–95.
25. Pham HC, Navarro-Delmasure C, Pham HC, et al. Toxicological studies of deltamethrin. Int J Tissue React. 1984;6(2):127–33.
26. Wolansky MJ, McDaniel KL, Moser VC, et al. Influence of dosing volume on the neurotoxicity of bifenthrin. Neurotoxicol Teratol. 2007;29(3):377–84.
27. Williamson EG, Long SF, Kallman MJ, et al. A comparative analysis of the acute toxicity of technical-grade pyrethroid insecticides and their commercial formulations. Ecotoxicol Environ Saf. 1989;8(1):27–34.
28. Shafer TJ, Meyer DA, Crofton KM. Developmental neurotoxicity of pyrethroid insecticides: critical review and future research needs. Environ Health Perspect. 2005;113:123–36.
29. Narahashi T. Neuroreceptors and ion channels as the basis for drug action: past, present, and future. J Pharmacol Exp Ther. 2000;294(1):1–26.
30. Crofton KM, Howard JL, Moser VC. Inter-laboratory comparisons of motor activity experiments: implications for neurotoxicological risk assessments. Neurotoxicol Teratol. 1991;13 (6):599–609.
31. Metker L, Angerhofer RA, Pope CR, Swentzel KC. Toxicological evaluation of 3-(phenoxyphenyl) methyl (±)-cis, trans-3-(2,2-dichlor-oethenyl)-2,2-dimethylciclopropanecarboxylate (permethrin). Sudy #51-0831-78. Department of the Army, US Environmental Hygiene Agency, Aberdeen Proving Ground, Maryland, December 1 1977.
32. Ray DE. Function in neurotoxicity: index of effect and also determinant of vulnerability. Clin Exp Pharmacol Physiol. 1997;24(11):857–60.
33. Weinert D, Waterhouse J. Diurnally changing effects of locomotor activity in body temperature on laboratory mice. Physiol Behav. 1998;63(5):837–43.
34. Gordon CJ. Effect of cage bedding on temperature regulation and metabolism of group-housed female mice. Comp Med. 2004;54(1):50–5.
35. Gordon CJ. Temperature and toxicology: an integrative, comparative, and environmental approach. Boca Raton: CRC Press; 2005.

36. Soderlund DM. Toxicology and mode of action of pyrethroid insecticides. In: Krieger R, editor. Hay's handbook of pesticide toxicology. 3rd ed. Amsterdam: Elsevier; 2010.
37. Dorman DC, Buck WB, Trammel HL, et al. Fenvalerate/N,N-diethyl-m-toluamide (Deet) toxicosis in two cats. J Am Vet Med Assoc. 1990;196(1):100–2.
38. Dorman DC, Beasley VR. Neurotoxicology of pyrethrin and pyrethroid insecticides. Vet Hum Toxicol. 1991;33(3):238–43.
39. Lawrence LJ, Casida JE. Pyrethroid toxicology: mouse intracerebral structure-toxicity relationships. Pest Biochem Physiol. 1982;18:9–14.
40. Tchounwou PB, Yedjou CG, Patlolla AK, et al. Heavy metals toxicity and the environment. EXS. 2012;101:133–64.
41. Nagajyoti PC, Lee KD, Sreekanth TVM. Heavy metals, occurrence and toxicity for plants: a review. Environ Chem Lett. 2010;8:199–216.
42. Usharani B, Vasudevan N. Impact of heavy metal toxicity and constructed wetland system as a tool in remediation. Arch Environ Occup Health. 2016;71(2):102–10.
43. Shiowatana J, McLaren RG, Chanmekha N, et al. Fractionation of arsenic in soil by a continuous flow sequential extraction method. J Environ Qual. 2001;30(6):1940–9.
44. Ross SM. Toxic metals in soil–plant systems. Chichester: Wiley; 1994. p. 469.
45. Silveira MLA, Alleoni LRF, Guilherme LRG. Biosolids and heavy metals in soils. Sci Agric. 2003;60(4):64–111.
46. Verkleji JAS. The effects of heavy metals stress on higher plants and their use as bio monitors. In: Markert B, editor. Plant as bioindicators: indicators of heavy metals in the terrestrial environment. New York: VCH; 1993. p. 415–24.
47. D'Amore JJ, Al-Abed SR, Scheckel KG, et al. Methods for speciation of metals in soils: a review. J Environ Qual. 2005;34(5):1707–45.
48. Lacerda LD. Global mercury emissions from gold and silver mining. Water Air Soil Pollut. 1997;97:209–21.
49. Yanqun Z, Yuan L, Jianjun C, et al. Hyper accumulation of Pb, Zn and Cd in herbaceous grown on lead-zinc mining area in Yunnan, China. Environ Int. 2005;31:755–62.
50. Bradford WI. Urban storm water pollutant loadings a statistical summary through. JWPCF. 1997;49:610–3.
51. Wuana RA, Okieimen FE. Heavy metals in contaminated soils: a review of sources, chemistry, risks and best available strategies for remediation. Int Scholar Res Netw ISRN Ecol. 2011;2011:402647.
52. Lasat MM. Phytoextraction of metals from contaminated soil: a review of plant/soil/metal interaction and assessment of pertinent agronomic issues. J Hazard Subst Res. 2000;2:1–25.
53. Campbell PGC. Cadmium—a priority pollutant. Environ Chem. 2006;3(6):387–8.
54. Flora SJS, Flora G, Saxena G. Environmental occurrence, health effects and management of lead poisoning. In: Casas JS, Sordo J, editors. Lead. Amsterdam: Elsevier B.V.; 2006.
55. WHO. Preventing disease through healthy environments: exposure to lead: a major public health concern. Geneva: World Health Organization; 2010.
56. WHO. Lead exposure. In: Comparative quantification of health risks. Geneva: World Health Organization; 2004. p. 1495–542.
57. WHO Food Additives Series 64. 73rd meeting of the Joint FAO/WHO Expert Committee on Food Additives (JECFA). Geneva: World Health Organisation; 2011.
58. Duruibe JO, Ogwuegbu MOC, Egwurugwu JN. Heavy metal pollution and human biotoxic effects. Int J Phys Sci. 2007;2:112–8.
59. Baldwin DR, Marshall WJ. Heavy metal poisoning and its laboratory investigation. Ann Clin Biochem. 1999;36(3):267–300.
60. Chunhabundit R. Cadmium exposure and potential health risk from foods in contaminated area, Thailand. Toxicol Res. 2016;32(1):65–72.
61. WHO. Exposure to cadmium: a major public health concern. In: Preventing disease through healthy environments. Geneva: World Health Organization; 2010.

62. WHO. Cadmium. In: Guidelines for drinking-water quality, 3rd edition incorporating 1st and 2nd addenda. Vol. 1. Recommendations. Geneva: World Health Organization; 2008. p. 317–9.
63. Bosch AC, O'Neill B, Sigge GO, et al. Heavy metals in marine fish meat and consumer health: a review. J Sci Food Agric. 2016;96:32–48.
64. Kima K-H, Kabir E, Jahan SA. A review on the distribution of Hg in the environment and its human health impacts. J Hazard Mater. 2016;306:376–85.
65. Pirrone N, Cinnirella S, Feng X, et al. Global mercury emissions to the atmosphere from anthropogenic and natural sources. Atmos Chem Phys. 2010;1010:5951–64.
66. Storelli MM, Marcotrigiano GO. Fish for human consumption: risk of contamination by mercury. Food Addit Contam. 2000;17:1007–11.
67. Castro-Gonzalez MI, Mendez-Armenta M. Heavy metals: implications associated to fish consumption. Environ Toxicol Pharmacol. 2008;26:263–71.
68. WHO. Exposure to mercury: a major public health concern. In: Preventing disease through healthy environments. Geneva: World Health Organization; 2010.
69. Diez S. Human health effects of methylmercury exposure. Rev Environ Contam Toxicol. 2009;198:111–32.
70. Matta G, Gjyli L. Mercury, lead and arsenic: impact on environment and human health. J Chem Pharm Sci. 2016;9(2):718–25.
71. Clarkson T. The toxicology of mercury and its chemical compounds. Crit Rev Toxicol. 2006;36:609–62.
72. Clarkson TW, Magos L, Myers GJ. The toxicology of mercury—current exposures and clinical manifestations. N Engl J Med. 2003;349:1731–7.
73. Ely JTA. Mercury induced Alzheimer's disease: accelerating incidence? Bull Environ Contam Toxicol. 2001;67:800–6.
74. Haley BE. Mercury toxicity: genetic susceptibility and synergistic effects. Med Veritas. 2005;2(2):535–42.
75. WHO. Exposure to arsenic: a major public health concern. In: Preventing disease through healthy environments. Geneva: World Health Organization; 2010.
76. Vieira C, Morais S, Ramos S, et al. Mercury, cadmium, lead and arsenic levels in three pelagic fish species from the Atlantic Ocean: intra- and inter-specific variability and human health risks for consumption. Food Chem Toxicol. 2011;49:923–32.
77. Joint FAO/WHO Expert Committee on Food Additives (JECFA). Evaluation of certain contaminants in food (Seventy-second report of the Joint FAO/WHO Expert Committee on Food Additives). WHO Food Additives Series, No 959; 2012.
78. Mudhoo A, Sharma SK, Garg VK, et al. Arsenic: an overview of applications, health, and environmental concerns and removal processes. Crit Rev Environ Sci Technol. 2011;41:435–519.
79. Tchounwou PB, Patlolla AK, Centeno JA. Carcinogenic and systemic health effects associated with arsenic exposure—a critical review. Toxicol Pathol. 2003;31(6):575–88.
80. Col M, Col C, Soran A, et al. Arsenic-related Bowen's disease, palmar keratosis, and skin cancer. Environ Health Perspect. 1999;107:687–9.
81. IPCS. Arsine: human health aspects. Geneva, World Health Organization, International Programme on Chemical Safety (Concise international chemical assessment document no. 47); 2002.
82. Zelinkova Z, Wenzl T. The occurrence of 16 EPA PAHs in food—a review. Polycycl Aromat Compd. 2015;35(2–4):248–84.
83. Masih J, Masih A, Kulshrestha A, Singhvi R, Taneja A. Characteristics of polycyclic aromatic hydrocarbons in indoor and outdoor atmosphere in the north central part of India. J Hazard Mater. 2010;177:190–8.
84. Masih J, Singhvi R, Kumar K, Jain VK, Taneja A. Seasonal variation and sources of polycyclic aromatic hydrocarbons (PAHs) in indoor and outdoor air in a semi arid tract of northern India. Aerosol Air Qual Res. 2012;12:515–25.

85. Mumtaz M, George J. Toxicological profile for polycyclic aromatic hydrocarbons (PAHs). In: Public Health Service, US Department of Health and Human Services, editors. Atlanta: Agency for Toxic Substances and Disease Registry; 1995, p. 487.
86. Culp SJ, Gaylor DW, Sheldon WG, Goldstein LS, Beland FA. A comparison of the tumors induced by coal tar and benzo[a]pyrene in a 2-year bioassay. Carcinogenesis. 1998;19:117–24.
87. Huang M, Prossing TMP. Contaminants: polycyclic aromatic hydrocarbons (PAHs). Amsterdam: Elsevier BV; 2014.
88. Larsen JC, Poulsen E. Mutagens and carcinogens in heat-processed food. In: Miller K, editor. Toxicological aspects of food. London: Elsevier; 1987. p. 205–52.
89. Adrian J, Billaud C, Rabache M. Part of technological processes in the occurrence of benzo (a)pyrene in foods. World Rev Nutr Diet. 1984;44:155–84.
90. Deshpande SS. Handbook of food toxicology. New York: Marcel Dekker; 2002.
91. Yannai S. Toxic factors induced by processing. In: Liener IE, editor. Toxic constituents of plant food stuffs. New York: Academic Press; 1980. p. 371–427.
92. Kim K-H, Jahan SA, et al. A review of airborne polycyclic aromatic hydrocarbons (PAHs) and their human health effects. Environ Int. 2013;60:71–80.
93. USDHHS. Polycyclic aromatic hydrocarbons. In: Sixth annual report on carcinogens-summary. Washington, DC: U.S. Dept. Health Human Services; 1991.
94. IARC. Monographs on the evaluation of the carcinogenic risk of chemicals to humans. Vol. 32: polynuclear aromatic compounds, part 1: chemical, environmental and experimental data. Lyon: International Agency for Research on Cancer; 1983.
95. USDHHS. Toxicological profile for polycyclic aromatic hydrocarbons (PAHs). Washington, DC: U.S. Dept. Health Human Services; 1993.
96. ACGIH (American conference of governmental industrial hygienists). Polycyclic aromatic hydrocarbons (PAHs) biologic exposure indices (BEI). In: American conference of governmental industrial hygienists, Cincinnati, 2005.
97. Unwin J, Cocker J, Scobbie E, Chambers H. An assessment of occupational exposure to polycyclic aromatic hydrocarbons in the UK. Ann Occup Hyg. 2006;50(4):395–403.
98. IPCS (International Programme on Chemical Safety). Polycyclic aromatic hydrocarbons, selected non-heterocyclic. 2010. http://www.inchem.org/documents/ehc/ehc/ehc202.htm.
99. Wells PG, McCallum GP, Lam KC, Henderson JT, Ondovcik SL. Oxidative DNA damage and repair in teratogenesis and neurodevelopmental deficits. Birth Defects Res C Embryo Today. 2010;90(2):103–9.
100. Srogi K. Monitoring of environmental exposure to polycyclic aromatic hydrocarbons: a review. Environ Chem Lett. 2007;5(4):169–95.
101. Armstrong BG, Hutchinson E, Fletcher T. Cancer risk following exposure to polycyclic aromatic hydrocarbons (PAHs): a meta-analysis. Rep No 068. Sudbury, UK: this health and safety executive. 2002. http://www.hse.gov.uk/research/rrhtm/rr068.htm.
102. Kuo CY, Hsu YW, Lee HS. Study of human exposure to particulate PAHs using personal air samplers. Arch Environ Contam Toxicol. 2003;44:454–9.
103. Institute of Medicine (US) Committee on the implications of dioxin in the food supply. Dioxins and dioxin-like compounds in the food supply: strategies to decrease exposure. Washington, DC: The National Academies Press; 2003.
104. Fernández-González R, Yebra-Pimentel I, Martínez-Carballo E, et al. A critical review about human exposure to polychlorinated dibenzo-p-dioxins (PCDDs), polychlorinated dibenzofurans (PCDFs) and polychlorinated biphenyls (PCBs) through foods. Crit Rev Food Sci Nutr. 2015;55(11):1590–617.
105. Shibamoto T, Bjeldanes LF. Introduction to food toxicology. 2nd ed. Maryland Heights: Elsevier; 2009.
106. Rose M. Environmental contaminants: dioxins, furans, and dioxin-like polychlorinated biphenyls. Amsterdam: Elsevier BV; 2014.

107. Schuhmacher M, Nadal M, Domingo JL. Levels of PCDD/Fs, PCBs, and PCNs in soils and vegetation in an area with chemical and petrochemical industries. Environ Sci Technol. 2004;38(7):1960–9.
108. Hülster A, Muller JF, Marschner H. Soil-plant transfer of polychlorinated dibenzo-p-dioxins and dibenzofurans to vegetables of the Cucumber family (Cucurbitaceae). Environ Sci Technol. 1994;28:1110–5.
109. Tsutsumi T, Iida T, Hori T, et al. Recent survey and effects of cooking processes on levels of PCDDs, PCDFs and CoPCBs in leafy vegetables in Japan. Chemosphere. 2002;46:1443–9.
110. Fernandes A, Mortimer D, Rose M, Gem M. Dioxins(PCDD6 Fs) and PCBs in offal: occurrence and dietary exposure. Chemosphere. 2010;81(4):536–40.
111. EPA (U.S. Environmental Protection Agency). Exposure and human health reassessment of 2,3,7,8-tetrachlorodibenzo-p-dioxin (TCDD) and related compounds. Draft final report. Washington, DC: EPA; 2000.
112. Hardell L, Van Bavel B, Lindstrom G, et al. Higher concentrations of specific polychlorinated biphenyl congeners in adipose tissue from nonHodgkin's lymphoma patients compared with controls without a malignant disease. Int J Oncol. 1996;9:603–8.
113. Brown DP. Mortality of workers exposed to polychlorinated biphenyls—an update. Arch Environ Health. 1987;42:333–9.
114. Loomis D, Browning SR, Schenck AP, Gregory E, Savitz DA. Cancer mortality among electric utility workers exposed to polychlorinated biphenyls. Occup Environ Med. 1997;54 (10):720–8.
115. Tironi A, Pesatori A, Consonni D, Zocchetti C, Bertazzi PA. Mortality among women workers exposed to PCBs. Epidemiol Prev. 1996;20:200–2.
116. Sinks T, Steele G, Smith AB, Watkins K, Shults RA. Mortality among workers exposed to polychlorinated biphenyls. Am J Epidemiol. 1992;136:389–98.
117. Gustavsson P, Hogstedt C. A cohort study of Swedish capacitor manufacturing workers exposed to polychlorinated biphenyls (PCBs). Am J Ind Med. 1997;32:234–9.
118. Moysich KB, Menezes RJ, Baker JA, Falkner KL. Environmental exposure to polychlorinated biphenyls and breast cancer risk. Rev Environ Health. 2002;17(4):263–77.
119. Rogan WJ, Gladen BC, McKinney JD, et al. Neonatal effects of transplacental exposure to PCBs and DDE. J Pediatr. 1986;109:335–41.
120. Jacobson JL, Jacobson SW. Intellectual impairment in children exposed to polychlorinated biphenyls in utero. N Engl J Med. 1996;335:783–9.
121. Kusuda M. A study on the sexual functions of women suffering from rice-bran oil poisoning. Sanka to Fujinka. 1971;38:1062–72.
122. Gerhard I, Daniel B, Link S, et al. Chlorinated hydrocarbons in women with repeated miscarriages. Environ Health Perspect. 1998;106:675–81.
123. Mocarelli P, Gerthoux PM, Ferrari E, et al. Paternal concentrations of dioxin and sex ratio of offspring. Lancet. 2000;355:1858–63.
124. Koopman-Esseboom C, Morse DC, Weisglas-Kuperus N, et al. Effects of dioxins and polychlorinated biphenyls on thyroid hormone status of pregnant women and their infants. Pediatr Res. 1994;36:468–73.
125. Weisglas-Kuperus N, Vreugdenhil HJ, Mulder PG. Immunological effects of environmental exposure to polychlorinated biphenyls and dioxins in Dutch school children. Toxicol Lett. 2004;149:281–5.
126. Longnecker MP, Klebanoff MA, Brock JW, et al. Maternal levels of polychlorinated biphenyls in relation to preterm and small-for gestational-age birth. Epidemiology. 2005;16(5):641–7.
127. European Union. Commission regulation (EU) no 1259/2011, Official Journal of European Union, L320/23 1259/2011; 2011.
128. World Health Organization (WHO). Fact sheet no 25: dioxins and their effect on human health. http://www.who.int/mediacentre/factsheets/fs225/en/
129. http://ec.europa.eu/food/food/chemicalsafety/contaminants/dioxins_en.htm European Commission, DG SANCO.

130. Lijinsky W, Shubick P. Benzo(a)pyrene and other polynuclear hydrocarbons in charcoal-broiled meat. Science. 1964;145:53.
131. Oz F, Kaya M. Heterocyclic aromatic amines in meat. J Food Process Preserv. 2011;35:739.
132. Shabbir MA, Raza A, Anjum FM, Khan MR, Suleria HAR. Effect of thermal treatment on meat proteins with special reference to heterocyclic aromatic amines (HAAs). Crit Rev Food Sci Nutr. 2015;55:82.
133. Widmark EMP. Nature. 1939;143:972.
134. Sugimura T, Nagao M, Kawachi T, Honda M, Yahagi T, Seino Y, Sato S, Matsukara N, Shirai A, Sawamura M, Matsumoto H. Mutagens-carcinogens in food, with special reference to highly mutagenic pyrolytic products in broiled foods. In: Hiatt HH, Watson JD, Winsten JA, editors. Origins of human cancer. New York: Cold Spring Harbour Laboratory; 1977. p. 1561–77.
135. Knize MG, Sinha R, Rothman N, Brown ED, Salmon CP, Levander OA, Cunningham PL, Felton JS. Food Chem Toxicol. 1995;33:545.
136. Alaejos MS, Afonso AM. Compr Rev Food Sci Food Saf. 2011;10:52.
137. Sinha R, Gustafson DR, Kulldorff M, Wen WQ, Cerhan JR, Zheng W. J Natl Cancer Inst. 2000;92:1352.
138. Sinha R, Kulldorff M, Chow WH, Denobile J, Rothman N. Cancer Epidemiol Biomark Prev. 2001;10:559.
139. Augustsson K, Skog K, Jägerstad M, Dickman PW, Steineck G. Lancet. 1999;353:703.
140. Gunter MJ, Probst-Hensch NM, Cortessis VK, Kulldorff M, Haile RW, Sinha R. Carcinogenesis. 2005;26:637.
141. Muhammad AS, Ali R, Faqir MA, Moazzam RK, Hafiz ARS. Crit Rev Food Sci Nutr. 2017;55:82.
142. Jinap S, Mohd-Mokhtar M, Farhadian A, Hasnol N, Jaafar S, Hajeb P. Meat Sci. 2013;94:202.
143. Viegas O, Novo P, Pinto E, Pinho O, Ferreira I. Food Chem Toxicol. 2012;50:2128.
144. Liao G, Wang G, Xu X, Zhou G. Meat Sci. 2010;85:149.
145. Pedreschi F, Mariotti MS, Granby K. J Sci Food Agric. 2014;94:9.
146. Sumner SC, Fennell TR, Moore TA, Chanas B, Gonzalez F. Chem Res Toxicol. 1999;12: 1110.
147. Friedman MA, Dulak LH, Stedham MA. Fundam Appl Toxicol. 1995;27:95.
148. Riboldi BP, Vinhas ÁM, Moreira JD. Food Chem. 2014;157:310.
149. Kahkeshani N, Saeidnia S, Abdollahi M. J Food Sci Technol. 2015;52:3169.
150. Wenzl T, De La Calle MB, Anklam E. Food Additiv Contam. 2003;20:885.
151. Zyzak DV, Sanders RA, Stojanovic M, Tallmadge DH, Eberhart BL, Ewald DK, Gruber DC, Morsch TR, Strothers MA, Rizzi GP, Villagran MDJ. Agric Food Chem. 2003;51:4782.
152. Vikström AC, Wilson KM, Paulsson B. Food Chem Toxicol. 2010;48:820.
153. Arvanitoyannis IS, Dionisopoulou N. Crit Rev Food Sci Nutr. 2014;54:708.
154. Wenzl T, De La Calle MB, Anklam E. Food Addit Contam. 2003;20:885.
155. Zelinková Z, Svejkovská B, Velíšek J, Doležal M. Food Addit Contam. 2006;23:1290.
156. Divinova V, Svejkovská B, Dolezal M, Velíšek J, Czech J. Food Sci. 2004;22:182.
157. Bakhiya N, Abraham K, Gürtler R, Appel KE, Lampen A. Mol Nutr Food Res. 2011;55:509.
158. Hamlet C, Sadd P, Crews C, Velíšek J, Baxter D. Food Addit Contam. 2002;19:619.
159. Hamlet CG, Asuncion L, Velíšek J, Doležal M, Zelinková Z, Crews C. Eur J Lipid Sci Technol. 2011;113:279.
160. Razak RAA, Kuntom A, Siew WL, Ibrahim NA, Ramli MR, Hussein R, Nesaretnam K. Food Control. 2012;25:355–60.
161. Lampen A, Scholz G, Weisshaar R, Wenzl T. Food Addit Contam Part A. 2013;30:11.
162. Herrmann SS, Duedahl-Olesen L, Granby K. Food Control. 2015;48:163.
163. Tricker AR, Kubacki S. J Food Addit Contam. 1992;9:39.
164. Lijinsky W. Mutat Res/Genet Toxicol Environ Mutag. 1999;443:129.
165. Stuff JE, Goh ET, Barrera SL, Bondy ML, Forman MR. J Food Compost Anal. 2009;1:22.
166. Herrmann SS, Granby K, Duedahl-Olesen L. Food Chem. 2015;174:516.

167. Stuff JE, Goh ET, Barrera SL, Bondy ML, Forman MR. J Food Compos Anal. 2009;22:S42.
168. Herrmann SS, Duedahl-Olesen L, Christensen T, Olesen PT, Granby K. Food Chem Toxicol. 2015;80:137.
169. Todd ECD. Foodborne diseases: overview of emerging food technology, Encyclopedia of food safety. Amsterdam: Elsevier BV; 2014.
170. Deshpande SS. Handbook of food toxicology. Marcel Dekker, Inc.; 2002.
171. Venugopal V, Doke SN, Thomas P, et al. Radiation processing to improve the quality of fishery products. Crit Rev Food Sci Nutr. 1999;39(5):391–440.
172. Cope S, Frewer LJ, Renn O, Dreyer M. Potential methods and approaches to assess social impacts associated with food safety issues. Food Control. 2010;21(12):1629–37.
173. Fischhoff B, Slovic P, Lichtenstein S, Read S, Combs B. How safe is safe enough? A psychometric study of attitudes towards technological risks and benefits. Policy Sci. 1978;9:127–52.
174. Frewer L, Lassen J, Kettlitz B, Scholderer J, Beekman V, Berdal KG. Societal aspects of genetically modified foods. Food Chem Toxicol. 2004;42(7):1181–93.
175. Hackwood S. The irradiation processing of foods. In: Thorne S, editor. Food irradiation. London: Elsevier Applied Science; 1991. p. 7.
176. World Health Organization. Safety and nutritional adequacy of irradiated food, WHO report, Geneva, Switzerland; 1994.
177. Stanley, et al. Food irradiation. Food Nutr Toxicol. 2004.

# Chapter 6
# Food Additives

**Jiang Liang**

**Abstract** Food additives are the nonnutritive substances added intentionally to food, generally in small quantities, to improve its appearance, flavor, texture, freshness, or storage properties. Food additives play an important role in the food industry and perform a variety of functions in foods. According to the different technological functions, food additives can be divided into several categories, including antioxidants, acids, food colors, preservatives, sweeteners, and flavoring agents. The scope and amount of food additives used in food are based on the strict scientific toxicological evaluations and safety assessments. Safety assessments are required before new food additives ae permitted for use in food manufacture. Also, if an approved food additive has new data on the toxicity or exposure, the need to re-evaluate its risk to human health will be triggered. Risk assessment is a scientifically based process consisting of the four steps: (1) hazard identification, (2) hazard characterization, (3) exposure assessment, and (4) risk characterization. Many countries now have formulated principles and methods for risk assessment of food additives. Supervision and administration are necessary for standardizing the production, business operation, and use of food additives. The regulation systems of food additive of Codex Alimentarius Commission (CAC), Europe, the United States, and China are described, respectively.

**Keywords** Food additive · Function · Classification · Safety assessment · Regulation

## 6.1 Introduction and Definition

In ancient times, our ancestors could extract natural pigment from plants, use nitrite in meat preservation and color protection, and add herbs and spices to improve the flavor of foods and pickle vegetables in vinegar. Today, consumers demand and

J. Liang (✉)
China National Center for Food Safety Risk Assessment, Beijing, People's Republic of China
e-mail: liangjiang@cfsa.net.cn



L. Zhang (ed.), *Nutritional Toxicology*,
https://doi.org/10.1007/978-981-19-0872-9_6

enjoy a food supply that is flavorful, nutritious, safe, convenient, colorful, and affordable. So, the advances in food additives make that possible. Early in 1955, the FAO/WHO Joint Expert Committee for Food Additives (JECFA) originally defined additives as "nonnutritive substances added intentionally to food, generally in small quantities, to improve its appearance, flavor, texture, or storage properties." While the scope of this definition was somewhat narrow and did not include flavorings and nutrients, with the development of the food industry, many new additives with more functions have been introduced. Food additives are defined in different ways in different countries or regions [1].

The *Codex Alimentarius* expands the scope of the food additive definition: "any substance not normally consumed as a food by itself and not normally used as a typical ingredient of the food, whether or not it has nutritive value, the intentional addition of which to food for a technological (including organoleptic) purpose in the manufacture, processing, preparation, treatment, packing, packaging, transport or holding of such food results, or may be reasonably expected to result, (directly or indirectly) in it or its by-products becoming a component of or otherwise affecting the characteristics of such foods." The term does not include contaminants or substances for maintaining or improving food nutritional qualities [2–4].

In European Union legislation (Regulation (EC) 1333/2008), food additives are described as substances "that not normally consumed as a food in itself and not normally used as a characteristic ingredient of food, whether or not it has nutritive value, the intentional addition of which to food for a technological purpose in the manufacture, processing, preparation, treatment, packaging, transport or storage of food results, or may be reasonably expected to result, in it or its by-products becoming directly or indirectly a component of such foods." Preparations from foods and other natural material are intended to have a technological effect in the final food, or those obtained by selectively extracting of components (e.g., pigments) relative to the nutritive or aromatic components should be regarded as food additives within the meaning of the regulation. However, substances should not be considered as food additives when they are used to impart flavor and/or taste or for nutritional purposes, such as salt replacers, vitamins, and minerals [5].

The US Food and Drug Administration (FDA) defines a food additive as "any substance the intended use of which results or may reasonably be expected to result—directly or indirectly—in its becoming a component or otherwise affecting the characteristics of any food." This definition includes any substance that is directly or indirectly used in the process of food production, processing, treatment, packaging, transportation, or storage. Direct food additives are those added to a food for a specific purpose. Indirect food additives are those that become components of the food at trace levels due to its packaging, storage, or other processing [6].

In the food safety law of the People's Republic of China and national food safety standards of using food additives (GB2760–2014), food additives are defined as synthetic or natural substances added to food to improve food quality, color, smell, and taste and meet the need for corrosion protection, preservation, and processing technology. Nutrition fortifier, food spices, basic substances in gum candy, and food processing aids used in the industry of material are also included [7].

## 6.2 Function of Food Additives

Today, food additives play an important role in the food industry and perform a variety of functions in foods. Most consumers rely on the technological, aesthetic, and convenient benefits provided by additives.

### *6.2.1 Improve Taste, Texture, and Appearance of Food*

The appropriate use of food additives can significantly improve the sensory quality of food and meet the consumers' needs of food flavor and taste. Spices and sweeteners are added to improve the food taste. Food colors can maintain or enhance food appearance. Emulsifiers, stabilizers, and thickeners give foods better texture and consistency. Some additives are used to adjust the acidity and alkalinity of foods, while other ingredients help maintain the taste and appeal of foods with reduced fat content [8].

### *6.2.2 To Maintain or Improve Safety and Freshness of Food*

Fresh foods without preservatives may deteriorate easily. In order to maintain food quality within the shelf life, preservatives like sorbic and benzoic acids are used to delay food spoilage caused by mold, air, bacteria, fungi, or yeast. Antioxidants, such as butylhydroxyanisole (BHA) and butylhydroxytoluene (BHT), can inhibit fats and oils from becoming rancid or developing an off-flavor [6].

### *6.2.3 To Maintain or Enhance Nutritive Value of Food*

Food preservatives and antioxidants are used to prevent oxidative deterioration of food, which play an important role in maintaining food nutritive value. For example, antioxidants are important to ensure oxidative stability of vegetable oils. Moreover, food nutrient fortifier like vitamins, minerals, and fiber may improve the nutritional quality of food, which is important to prevent malnutrition [6].

### 6.2.4 *Required in Food Processing*

Lubrication, antifoaming, leaching, stability, and solidification are required in food processing. The appropriate food additives are added for the need of food processing, formulation, and sensory properties [6].

## 6.3 Food Additive Classification

According to various sources, food additives can be divided into three main groups: (1) substances isolated from edible plants or from other living materials, for example, alginates (E 401), agar (E 406), and carrageenan (E 407) isolated from seaweeds; lecithin (E 322) from soybeans; and pectin (E 440) from fruits; (2) substances contained in foodstuffs but the production of which by chemical synthesis is cheaper, such as antioxidant ascorbic acid or vitamin C (E 300) and yellow dye $\beta$-carotene (E 160a); and (3) substances not found in nature and obtained only synthetically, such as sweetener saccharin (E 954) or antioxidant *t*-butylhydroxyanisol (E 320) [9].

As shown in Table 6.1, based on the different technological functions, food additives can be divided into several categories, including antioxidants, acids, food colors, preservatives, sweeteners, and flavoring agents [10].

### 6.3.1 *Antioxidants*

Oxidative deterioration, one of the common types of food deterioration, is characterized with the undesirable change in color or flavor caused by oxygen in air. This

**Table 6.1** Food Additive Classification by Their Functionality in Foods

| Additive | Function | Examples |
|---|---|---|
| Antioxidants | Prevent rancidity and enzymatic browning | Ascorbic acid, BHT, BHA, EDTA, ethoxyquin |
| Food colors | Impart desired colors | Carotenes and other natural pigments, synthetic food colors |
| Preservatives | Prevent the spoilage of foods caused by the microorganisms or oxidation | Sodium benzoate, calcium propionate, potassium sorbate, sodium nitrite |
| Sweeteners | Impart a sweet taste in foodstuffs or as tabletop sweeteners | Sucrose, lactose, glucose, fructose, corn syrup, maple syrup, molasses, honey, saccharin, cyclamate, aspartame, xylitol |
| Flavoring agents | Supplement, enhance, or modify original flavor without enhancing its own characteristic flavor | MSG, disodium inosinate, disodium guanylate |

process involves the oxidation of food ingredients. Oxidation can cause changes in food color or flavor as well as decrease the nutritive value and even sometimes produce toxic metabolites. Foods rich in lipids and polyunsaturated fatty acids are very sensitive to oxidation. Oxidative deterioration of fat not only can destroy vitamins A, D, E, K, and C but also destroy essential fatty acids with a pungent and offensive off-flavor. In some cases, toxic by-products can be produced from oxidative reactions. The most effective method of preventing oxidative degradation is the use of antioxidants with the mechanisms of scavenging free radicals or oxygen or by inhibiting enzymes that facilitate oxidation process. The antioxidants that can prevent oxidation to retain the color and flavor of food are widely used to extend the shelf life of food. The antioxidants may be classified as natural and synthetic according to their origin. The most common natural antioxidant additives are ascorbic acid (vitamin C) and ascorbates. For the relatively weak antioxidant properties of natural antioxidants, synthetic antioxidants including the phenol derivatives Butyl hydroxy anisd (BHA), Butylated hydroxytoluene (BHT), tert-Butylhydroquinone (TBHQ), and propyl gallate are more widely used in food processing [11].

### *6.3.2 Food Colors*

Many natural food colors are unstable in heat or oxidation. The perception and acceptability of food are influenced by its color, taste, and flavor. Thus, food coloring used in commercial food production as well as domestic cooking can make food more attractive, appealing, and appetizing: to compensate color loss caused by exposure to light, air, extreme temperature, moisture, and storage conditions; to correct natural changes in color; to enhance natural colors; and to allow consumers to identify products, like candy flavors or medicine dosages.

According to the origins, food colors are classified into two types. Many natural colorants extracted from plants, animals, or microorganisms can be used as food color. Plant-derived pigments including carotenoids, anthocyanins, chlorophyllin, and betanin are the main categories used in food products.

Other sources of natural colorants or specialized derivatives include carmine from the cochineal insect, monascus color produced by fermentation of genus *Monascus*, and caramel by heating sugar at high temperature.

While the variety of natural colorants are very limited, synthetic food colors began to replace the natural colors in the late nineteenth century. Compared with natural colors, many synthetic food colorants are easy to produce, low in cost, and good in coloring properties. According to the chemical structure, synthetic colorants can be classified as diazo, oxanthracene, and diphenylmethane. Synthetic coloring agents including cochineal, amaranth red, sunset yellow, lemon yellow, bright blue, indigo, new red, and red moss are approved to be used in some food and drinks [7, 9].

### 6.3.3 Preservatives

Preservatives are used to prevent the spoilage of foods caused by microorganisms or oxidation. Many microorganisms, including bacteria, fungi, and yeast, can have adverse effects on the appearance, taste, or nutritional value of foods. Moreover, some toxins produced by the organisms can pose high risks to human health. Antimicrobial preservatives prevent degradation by bacteria. Nitrates and nitrites, as their sodium or potassium salts, are often used as preservatives in processed meat products. They help to prevent growth and toxin production of pathogenic bacteria like *Clostridium botulinum* and to improve the color of the meat products by keeping them in red or pink. The role of preservatives is not to kill bacteria (bactericidal) but to retard their action by inhibiting their activity (bacteriostatic action). Preservative can reduce the risk of foodborne infections, decrease microbial spoilage, and maintain fresh properties and nutritional quality. In a broader sense, the term preservative includes not only compounds that inhibit microorganisms but also compounds that prevent chemical and biochemical deterioration [12].

### 6.3.4 Sweeteners

Sweeteners are important to food flavors. Sweeteners, or sugar substitute, are added to foods to substitute the sweetness normally provided by sugars without contributing significantly to available energy. Moreover, their addition to food can have beneficial effects as they are beneficial to prevent and control diseases such as obesity, diabetes, and tooth decay.

Sugar substitutes are produced by plant extracts or chemical synthesis. Steviol glycosides, a natural plant extracts from the leaves of a South American shrub with the scientific name as *Stevia rebaudiana*, are found to contain zero calories and be approximately 300 times sweeter than sugar. Some sugar substitutes including sorbitol and xylitol are sugar alcohols with similar or less sweetness and fewer calories as compared with sugar. They are found to naturally occur in many fruits and vegetable and were produced by catalytic hydrogenation of the appropriate reducing sugar. Those that are not produced by nature are generally called artificial sweeteners. Artificial sweeteners including saccharin, aspartame, acesulfame-K, and sucralose are widely used in processed foods and beverages. Artificial sweeteners including saccharin have been scrutinized intensely for decades for the controversy over their safety toward human health. Studies dating to the 1970s suggested the link of saccharin to bladder cancer in laboratory rats. While no consistent evidence was indicated in human epidemiology studies, the artificial sweetener, aspartame, was reported to induce cancer in the livers and lungs of male Swiss mice in a long-term carcinogenicity study. Moreover, artificially sweetened soft drinks were reported to be correlated with preterm delivery in a cohort of pregnant women. After evaluating the two studies in 2011, European Food Safety Authority (EFSA) concluded that the information available in these publications could not provide sufficient evidence or support clear causal relationship to reconsider previous evaluations. Thus, there's no reliable scientific evidence that artificial sweeteners has caused cancer or other serious health problems. Many studies showed that artificial sweeteners consumed in limited quantities are confirmed to be generally safe [13–15].

### *6.3.5 Flavoring Agents*

Flavoring agents are substances used to impart taste and/or smell to food and/or to intensify the existing flavors of products. Natural or synthetic flavor enhancers have been used to activate receptors for the umami or savory taste, enhance the palatability, and thus increase the acceptability of foods.

Flavorings have a long history of safe use in a wide variety of foods, including ice cream, jam, soft drinks, cookies, cereals, cakes, and yogurts. Flavoring substances are generally extremely low in consumption, but there are several thousand individual aromatic substances in commercial use worldwide. Since the middle of the nineteenth century, numerous flavor chemicals have been synthesized. At present, more than 3000 synthetic chemicals are used as flavor agents. Flavoring agents used in food industry can be divided into three groups: natural flavors extracted from fruits or vegetables; natural identical flavors, which are chemically synthesized molecules identical to their natural counterparts; artificial flavors that impart a flavor characteristic of a particular food, but do not resemble in molecular terms the natural flavor molecules [16].

## 6.4 Safety Evaluation of Food Additives

The General Standard for Food Additives (GSFA) claims that a food additive is authorized to use only when such use has technological necessity or advantage and does not pose an appreciable health risk to consumers. Risk analysis has been widely applied in the regulation of food additive uses. Regarding to protecting the health of the consumers, principles for risk analysis have been formulated by the Codex Committee on Food Additives (CCFA) and Codex Committee on Contaminants in Foods (CCCF). Risk analysis defined by the Codex Alimentarius Commission (CAC) is a process consisting of three closely linked parts: risk assessment, risk management, and risk communication. As the international agency taking charge of the risk assessment of food additives, JECFA has conducted risk assessment for over 1500 kinds of food additives [17–19].

Many countries now have formulated principles and methods for risk assessment of food additives. Moreover, safety assessments are required before new food additives are permitted for use in food manufacture. Also, if an approved food additive has new data on toxicity or exposure, the need to re-evaluate its risk to human health will be triggered. The purpose of risk assessment on food additive is to determine whether the food additive can be allowed for use in food and in what amounts, according to the basic principle of the good manufacturing practice (GMP). Risk assessment is a scientifically based process consisting of four steps: (1) hazard identification, (2) hazard characterization, (3) exposure assessment, and (4) risk characterization [20].

### 6.4.1 Hazard Identification and Characterization

Hazard identification of food additives is to determine the potential adverse effects of food additives on human health. The data or information needed in hazard identification include epidemiological studies, animal toxicology studies, and in vitro tests. Animal toxicological studies are mostly commonly used in hazard identification of food additives to identify potential adverse effects, the dose–effect relationship, and target organs of toxicity.

For chemicals with non-genotoxic effects, the threshold level, such as the no-observed-adverse-effect level (NOAEL) or lowest-observed-adverse-effect level (LOAEL) for the critical effect, can be determined from the animal toxicological data and be used to derive the acceptable daily intake (ADI). The default 100-fold safety/uncertainty factor is applied to the NOAEL to derive the ADI. Invariably, this 100-fold default factor reflects a tenfold factor for experimental animal-to-human extrapolation and a tenfold variation in sensitivity within the human population. ADI is the maximum amount of any substance in food or drinking water that can be ingested (orally) on a daily basis over a lifetime without an appreciable health risk. The ADI is usually expressed as milligrams of the chemical ingested per kilogram of bodyweight [21].

For genotoxic compounds with no-threshold toxic effects, the daily dose that would induce extra cancer incidence in humans can be estimated by using information on the carcinogenic dose in animal experiments. Currently, 1 chance in 100 million of extra cancer risk per lifetime was considered as safe [22, 23].

### 6.4.2 Exposure Assessment

The exposure assessment of food additives is a combined qualitative and quantitative evaluation according to the estimated dietary exposures of food additives. The exposure assessment of food additives is performed based on the food consumption data and the concentration of the food additive in food. Evaluating the dietary exposure of food additives requires three elements: (1) food additive concentration in food, (2) the consumption level of the food consumed, and (3) the average body weight of the target population (kg). The general equation for dietary exposure is expressed as follows:

$$\text{Dietary exposure} = \sum \frac{\text{Concentration of food additive in food} \times \text{Food consumption}}{\text{Body weight}}$$

The estimation of the food additive dietary exposure can be compared with the relevant ADI, if available, in the step of the risk characterization. At present, a stepwise or tiered intake assessment framework is recommended by JECFA for additives (including flavoring), contaminants, and nutrients. Generally, the initial step of the assessment framework is often to choose substances for evaluation by conservative screening methods. If no safety concerns are identified, there is need for additional exposure assessment. Where potential safety concerns are identified, more specific or refined data is required in the subsequent steps of the framework.

According to the *Guideline for the Simple Evaluation of Dietary Exposure to Food Additives* (CAC/GL 3–1989, Revision 2014), the screening methods should overestimate dietary exposure of high consumers by using conservative assumptions for food consumption and food additive concentration.

A screening method named as the "budget method" has been used to estimate the theoretical maximum daily dietary exposure levels of some food additives. The result is compared with the ADI for the substance to assess the risk. The budget method has been used as a tool for screening at an early stage to establish monitoring priorities in assessing additives by JECFA . The method relies on assumptions that (1) the levels of consumption of foods and beverages considered are maximum physiological levels of consumption; (2) the levels contained in foods and beverages are to be the highest maximum levels of the additive reported in any category for foods and for beverages, respectively; and (3) a default proportion of foods and of non-milk beverages that may contain it is 12.5% for solid foods and 25% for beverages. The budget method is a simple, rapid, and conservative screening method that can easily be performed.

The theoretical maximum daily intake (TMDI) is another screening method using more refined consumption data of the target population. It is calculated by multiplying the average daily food consumption level for each food category or foodstuff by the authorized maximum use level of the additive established by Codex standards or by national regulations.

If the conservatively estimated intake amount of the substance exceeds its ADI, the subsequent steps of assessment incorporating more specific and refined concentration data are required. In the guideline of the WHO, the refined assessment is defined as the estimated daily intake (EDI), in which the concentration level of the additive can be based on the actual use level of the additive by industry or an approximation as close as possible to the actual use levels according to good manufacturing practice (GMP) [24, 25].

The assessment for food additive intake can generally be divided into pre-market and post-market assessments. For the former case, the concentration data are mainly from the maximum usage levels provided by applicants or food producers. For the latter case, the concentration data can be obtained from the food producers' reports, food industry surveys, surveillance and monitoring, total diet studies, and available scientific literatures. The food consumption data mainly comes from the national food consumption surveys [26].

### 6.4.3 Risk Characterization

In the step of risk characterization, the ADI values derived from the hazard characterization are compared with the estimated daily intake (EDI) of certain food additives. If the EDI of an additive is below or close to the ADI, the food additive generally is considered safe for its intended use, and no further information on risk characterization need to be provided, while a small or occasional dietary exposure in excess of the ADI value based on a subchronic or chronic study does not necessarily imply that adverse health effects will occur in humans.

If the ADI may be significantly exceeded, further advice is required on the potential health consequences; i.e., severity and target population and uncertainties should be taken into consideration. For a new food additive, the approval is recommended if the estimated daily intake does not exceed the ADI. However, if the ADI is exceeded, the approval of the new additive is not recommended.

In this case, the applicants may be required to submit additional new toxicology data to support an increase in the ADI value or to lower the use levels in food and even eliminate some uses completely.

For an approved food additive, it should be continuously observed for potential harmful effects. When a safety concern or any harmful effect is newly identified as a result of progress in toxicological research or other factors, reassessment of the additive that has previously been assessed should be carried out promptly.

In cases where the data are not sufficient to derive a health-based guidance value for a substance producing threshold effects, the margin of exposure (MOE) between the doses at which effects are observed in animals and the estimated human dietary exposure may be applied [18].

### 6.4.4 The Management of Food Additives

With the rapid development of the processed foods, food additives are more widely used. Supervision, administration, and safety evaluation have been established to standardize the production, business operation, and use of food additives.

#### 6.4.4.1 CAC

The main role of Codex Committee on Food Additives and Contaminants (CCFAC) is to recommend appropriate food additive standards to CAC. CCFAC specifically considers the technological justification and need for proposed use levels of food additive. CCFAC also assigns priorities for food additive evaluation to JECFA.

Estimated daily intakes and their compartments with the ADI need to be taken into account when approving or establishing permitted maximum levels of additives in food. Different approaches are required in the process of risk management of food additives. The scope and use level of the additive in individual foods can be specifically controlled by particular risk management options.

Previously, CCFAC incorporated the use of food additive into individual Codex food standards. With the development of the draft of General Standard for Food Additives (GSFA), CCFAC is changing its approach of risk management and is establishing general risk analysis approaches, which can be applied to all foods or food categories. The GSFA covers the use of additives in all foodstuffs. The approach recommended to be used in the GSFA provides the framework that needs to be considered in exposure assessments [2].

#### 6.4.4.2 European Union

According to EU legislation, food additives must be authorized before they can be used in foods.

A formal request including an application files on the substance and scientific data on its intended uses and use levels were required to be submitted to the European Commission. Then, the application files were sent to EFSA to evaluate the safety of the substance for its intended uses on request. Then according to the EFSA's safety assessment, the European Commission decides whether to authorize the substance. The authorized food additives applying for new uses follow the same procedure.

EFSA is mainly responsible for the safety assessment of food additives, with three main tasks: (1) assessment on the safety of new food additives or proposed new uses of approved food additives before they get the authorization for use in the EU, (2) reassessing the intake risk of all food additives already authorized for use, and (3) responding to ad hoc requests from the European Commission to reassess some food additives on the basis of new scientific data and/or changing the use conditions. As part of the food additive safety assessment, EFSA seeks to determine an acceptable daily intake (ADI) for each substance as possible (e.g., when there is sufficient data or information). When reassessing previously authorized additives, EFSA may confirm or modify existing ADIs after reviewing of all available evidence. When there are not sufficient data to establish an ADI, the value of a margin of safety can be calculated to evaluate if the estimated intake of the substance might be of potential concern. Once authorized, these substances are included in the EU list of permitted food additives laid down in Regulation (EC) 1333/2008. Authorized food additives must also comply with approved purity criteria in Regulation (EU) 231/2012 [5, 27].

#### 6.4.4.3 The United States

Chemical substances added to food regulated by the US FDA are classified into the following types:

1. Food additives, colors, and flavoring substances, which are authorized to be produced and used in food products after the evaluation with no potential health risk. The application materials for the safety evaluation by FDA include the composition and properties of the substance, the typical consumed amount, short-term and long-term health effects, and other safety factors. With the accumulated new evidences, as well as the development of analytical techniques, the safety of food additives is required to be re-evaluated every few years. According to the Federal Food, Drug, and Cosmetic Act (FFDCA), food additives are regulated by the US FDA, while food additives for meat and poultry products are certified by both FDA and USDA. The legislation of food flavoring substances is undertaken by the FEMA. The pre-market pigments used in food and other fields must be approved by the FDA.
2. Generally recognized as safe ("GRAS") substances, which are generally recognized by qualified experts to adequately show to be safe under the conditions of its intended use and so is exempted from the food additive tolerance requirements of the usual Federal Food, Drug, and Cosmetic Act (FFDCA). The approval of the substance in the GRAS list either are based on general recognition of safety scientific application procedures with the generally available and accepted scientific data or are through experience based on a substantial consumption history for food use by a significant number of consumers. The substances will be removed from the GRAS list if new evidence of adverse effects is found.
3. Prior-sanctioned substances, which are approved for specific uses in foods prior to September 6, 1958.
4. Formerly used substances, which include the prohibited substances labeled as "PROHIBITED" or "PROHIBITED WITH EXCEPTIONS"; the delisted color additives labeled as "DELISTED"; and some substances recognized as "no longer FEMA GRAS" [4, 6, 12].

#### 6.4.4.4 China

According to the food safety laws and regulations in China, National Health Commission of PRC (the former National Health and Family Planning Commission) is responsible for the development of national food safety standards and the safety evaluation of food additives. Food additives should be technically necessary and evaluated to be safe and reliable based on a risk assessment before the approval for use. "Regulations on the Administration of New Varieties of Food Additives" defines the scope, the basic requirements, and application procedures for new varieties of food additives. Raw materials for production, chemical structures and physical properties, production process, toxicological safety evaluation data or inspection report, and quality specification inspection report should be included in the safety assessment materials for application. "Standards for Use of Food Additives" (GB2760), "Standard for Use of Food Nutrition Fortifiers" (GB 14880), and "Procedures and Methods on Food Toxicology Safety Evaluation" (GB 15193) are developed for safety evaluation and standards [7, 27].

## 6.5 Conclusion

Food additives are substances added intentionally to foodstuffs to perform certain functions, for example, to improve taste, texture, and appearance of food; to preserve foods better; or to improve nutrition quality. Food additives are closely related to the development of the food industry. The scope and amount of the food additives added to food are based on the strict scientific toxicological evaluations and safety assessments. Supervision and administration are necessary for standardizing the production, business operation, and use of food additives.

## References


1. Board E. JOINT FAO/WHO conference on food additives. World Health Organization Technical Report, vol. 55; 1956. p. 1–14.
2. Laganà P, Avventuroso E, Romano G, et al. The Codex Alimentarius and the European legislation on food additives: chemistry and hygiene of food additives. Springer International Publishing; 2017. p. 23–31.
3. CODEX STAN 192–1995 (Revision 2016). Codex General Standard for Food Additives (GSFA).
4. World Health Organization (WHO). Codex alimentarius commission procedural manual. 24th ed. Food and Agricultural Organisation of the United Nations, World Health Organisation; 2015. p. 1–231.
5. European Commission. Regulation (EC) No 1333/2008 of the European Parliament and the council of 16 December 2008 on food additives. Off J Eur Union. 2008;L354:16–33.
6. US Food and Drug Administration (FDA), Overview of Food Ingredients, Additives & Colors. https://www.fda.gov/Food/IngredientsPackagingLabeling/FoodAdditivesIngredients/ucm094211.htm#foodadd
7. National Health and Family Planning Commission of PRC. GB2760-2014 China and National food safety standards of using food additives.
8. Wang J, Sun B. Food safety chemistry: chemistry and safety of food additives. CRC Press; 2014. p. 254–71.
9. Deshpande SS. Handbook of food toxicology: food additives. CRC Press; 2002. p. 222–76.
10. Omaye ST, Watson DH. Introduction to food toxicology: food additives. Amsterdam: Elsevier Academic Press; 2009. p. 116–32.
11. Schyvens C. Food additives: antioxidants. In: Encyclopedia of Food Safety. Amsterdam: Elsevier Academic Press; 2014. p. 449–84.
12. Davidson PM, Taylor TM. Chemical preservatives and natural antimicrobial compounds. In: Food Microbiology: Fundamentals And Frontiers; 2013. p. 713–45.

13. EFSA. Statement of EFSA on the scientific evaluation of two studies related to the safety of artificial sweeteners. EFSA J. 2011;9:2089.
14. Weihrauch MR. Artificial sweeteners--do they bear a carcinogenic risk? Ann Oncol. 2004;15: 1460–5.
15. Ditschun TL, Winter CK. Food toxicology: food additives. CRC Press; 2001. p. 194.
16. Kaitano TE. Encyclopedia of food safety food additives: flavors and flavor enhancers. In: Encyclopedia of food safety. Amsterdam: Elsevier, Academic Press; 2014. p. 466–70.
17. FAO, WHO. Codex Alimentarius: general standard for food additives. In: Codex Alimentarius General Standard for Food Additives; 2011.
18. Hathaway SC. Risk assessment procedures used by the Codex Alimentarius Commission and its subsidiary and advisory bodies. Food Control. 1993;4:189–201.
19. World Health Organization. Safety evaluation of certain food additives and contaminants: seventy-third meeting of the Joint FAO/WHO expert committee on food additives (JECFA). World Health Organization; 2011.
20. EFSA. Statement on a conceptual framework for the risk assessment of certain food additives re-evaluated under Commission Regulation (EU) No 257/2010. EFSA Journal. 2014;12:3697.
21. Renwick A. The use of an additional safety or uncertainty factor for nature of toxicity in the estimation of acceptable daily intake and tolerable daily intake values. Regul Toxicol Pharmacol. 1995;22:250–61.
22. FAO and WHO. Principles and methods for the risk assessment of chemicals in food. World Health Organization; 2009. 6-2-92
23. Dybing E, O'Brien J, Renwick AG, et al. Risk assessment of dietary exposures to compounds that are genotoxic and carcinogenic-an overview. Toxicol Lett. 2008;180:110–7.
24. Douglass J, Barraj L, Tennant D, et al. Evaluation of the budget method for screening food additive intakes. Food Addit Contam. 1997;14:12.
25. FAO and WHO. Guidelines for the simple evaluation of dietary exposure to food additives CAC/GL 3–1989 Adopted 1989. Revision 2014.
26. Gürtler R. Risk assessment of food additives. Berlin: Springer; 2014.
27. Wang J, Liu JJ, Law SO. Comparative research on food additive regulations and standards between European Union and China. J Food Safety Qual. 2015;6:3753–7.

# Chapter 7
# Food Contact Materials

Haixia Sui

**Abstract** Food contact materials (FCM) are all materials and articles intended to come into direct or indirect contact with food. The definition of food contact materials and articles is essentially the same in different countries but with slight differences. Food contact materials are made of base materials, together with additives, adjuvants, and polymerization aids based on different purposes, to glue, protect, and impress base materials. Food contact materials must be sufficiently designed or controlled to prevent substances from being transferred to food in amounts that could endanger human health, result in an unacceptable change in the composition of food, or result in deterioration of its organoleptic properties. Risk assessment needs to be done for all migrants, including intentionally added substances (IASs) and non-intentionally added substances (NIAS). Bisphenol A (BPA) is a chemical produced in large quantities for use primarily in the production of polycarbonate plastics used in various products and epoxy resins used as protective layers for food and beverage cans and as coatings for drinking water storage tanks. Regulatory bodies and expert groups worldwide have conducted extensive risk assessment on BPA in the past 10 years. The Consortium Linking Academic and Regulatory Insights on BPA Toxicity (CLARITY-BPA) program was developed to study the full range of potential health effects from exposure to BPA. The European Food Safety Authority's (EFSA) Panel on Food Contact Materials, Enzymes and Processing Aids (CEP) has started to reassess the potential hazards of BPA in food and review again the temporary safe level set in EFSA's previous 2015 full-risk assessment.

**Keywords** Food contact material · IAS · NIAS · Migration · BPA

---

H. Sui (✉)
Division III of Risk Assessment, China National Center for Food Safety Risk Assessment, Beijing, People's Republic of China
e-mail: suihaixia@cfsa.net.cn



L. Zhang (ed.), *Nutritional Toxicology*,
https://doi.org/10.1007/978-981-19-0872-9_7

## 7.1 Introduction

Food comes into contact with many kinds of materials and articles during its production, processing, transportation, and storage, before its final consumption. Such materials and articles are called food contact materials (FCMs). FCMs are either intended to be brought into contact with food, already in contact with food, or can be reasonably assumed to contact food under normal or foreseeable use. The definition of food contact materials is essentially the same in different countries, with slight differences on authorization scope or target of authorization. In America, as defined in the 1997 Food and Drug Administration Modernization Act (FDAMA), food contact substance refers to any substance intended for use as a component of materials in manufacturing, packing, packaging, transporting, or holding food, and such use is not intended to have a technical effect in food. Food contact substance covers indirect food additives, including polymers, monomers, additives, polymerization aids, adjuvants, equipment components, packaging compounds subject to irradiation, and some secondary direct food additives, such as boiler water additives, sugar processing additives, ion exchange resins, and antimicrobials used in the processing of meats, vegetables, and poultry [1]. In the EU, FCM covers packaging materials, kitchenware and tableware, containers for transporting food, and machinery for processing food. Fixed public or private water supply equipment is not considered as food contact materials and articles [2]. In China, the definition of food contact materials and articles is similar to the EU definition [3].

The general principles of safety and inertness for all FCM is the same; that is, materials should not release their constituents into food at levels harmful to human health and change food composition, taste, or odor in an unacceptable way [4]. To preclude substances from being transferred to food in large quantities, any material or article intended to come into direct or indirect contact with food must be sufficiently inert.

In addition to traditional food contact materials, there are two kinds of new food contact materials, active food contact materials and articles and intelligent food contact materials and articles. Active food contact materials and articles are intentionally designed to actively maintain or improve the status of the food. Therefore, they are not inert by design. Intelligent food contact materials and articles are designed to monitor the status of the food. Both of these types of materials and articles may be in contact with food. However, active and intelligent food contact materials and articles should not change the composition or the organoleptic properties of food or give information about the condition of the food that could mislead consumers.

Food contact materials are made of base materials, such as plastics, rubber, and paper together with additives, coatings, and printing inks for different purposes. Food contact articles are objects, equipment, containers, packaging, and various utensils that are used for manufacturing, preparing, conserving, flowing, transporting, or handling food or that are presented as such [5]. To enhance the ability or mechanical properties of the food contact materials and articles or increase the packaged food's shelf life, intentionally added substances (IASs) are intentionally added in the manufacturing and production processes. IASs are specifically added during the production process of FCMs and have a function either in the manufacturing process or in the final product. Starting substances and monomers used to build the polymer, the main structural component, for example, plastic and coating, are considered to be IAS. Except these substances of known origin, there are also some non-intentionally added substances (NIAS) into food contact materials and articles with unknown origin. In addition, intentionally adding an IAS that contains known impurities does not cause the impurities to become IAS, for example, impurities present in the IAS or degradations created during the synthetic process. Therefore, food contact materials and articles can also be considered to be mixture of substances with known or unknown identity/origin.

The primary functions of food packaging is to protect food products from outside influences and damage, contain food, and provide consumers with traceable information conveniently and concisely [6, 7]. For example, packaging labels include product identification, nutritional value, ingredient table, net weight, price tag, and manufacturer information. Traceability is defined as "the ability to follow the movement of a food through specified stage(s) of production, processing and distribution" by the Codex Alimentarius Commission [8]. The objectives of traceability include improving supply management, facilitating trace back for food safety and quality, and differentiating and marketing foods with subtle or undetectable equality attributes. Convenience also has a great effect in package innovation such as ease of access, handling, and disposal; product visibility; resealing ability; and microwave ability. Thus, packaging plays a vital role in reducing the effort required to prepare and deliver foods.

Packaging may not always provide benefits to food. Despite protecting foods from contamination and preserving the safety and quality of food, packaging is also a potential source of contamination through the migration of material components into foodstuffs. Depending on the physical/chemical parameters, chemical composition of the FCM, and nature of the food, FCMs may transfer their ingredients (both IAS and NIAS) to foods. Migration of chemicals from FCMs can impact food quality and safety. Migration may lead to excessive exposure to certain chemicals, which could be harmful to human health and bring an unacceptable change to the composition of the food; thus, it must be evaluated and controlled.

The extent of migration is affected by various factors such as the physicochemical properties of the packaging material, the food itself (e.g., fat content), temperature, storage time, and the ratio of packaging to food volume (smaller-size packaging has a larger surface-to-volume ratio). Human health risk assessment is only relevant for migration with molecular weight (MW) below 1000 daltons (Da), because it is conventionally assumed by the European Food Safety Authority (EFSA) that above this molecular weight, substances are not absorbed by the body and therefore may be excluded from any calculations of migration and exposure. However, it is important to note that NIAS with MW above 1000 Da can be a major portion of the migrants.

There are two main types of migrants, intentionally added substances and non-intentionally added substances. IASs are deliberately added during the production of food contact materials and have a function in the production process or in the final product. NIAS are chemicals present in a food contact material that are not intentionally added during the production process. NIAS include impurities of raw materials, undesired by-products, and various contaminants from recycling processes. In general, NIAS are not known by consumers and are a challenge for the FCM manufacturer to detect and remove. Both process contaminants and environmental contaminants are considered as NIAS. The former is mostly known chemicals with unknown levels such as lubricant contaminants from storage/transport, while the latter is unknown and unpredictable contaminants from the FCM. As a special class of migrants, oligomers might be known or unknown to the manufacturer but mostly unknown to consumers.

When analyzing NIAS from FCM, usually, not all interested compounds can be detected with state-of-the-art analytical techniques. Due to the lack of chemical properties of unidentified substances, it has been complicated to analyze NIAS. Extracted samples can be separated by chromatography and analyzed by mass spectrometry. Sample extraction is one of the most critical steps because incomplete transfer may hinder comprehensive analysis. Furthermore, mass spectra may not assign a known chemical structure. Using direct thermal desorption techniques can eliminate the extraction step, but it may be even more difficult to interpret the results due to complicated fragmentation patterns. Finally, risk assessment is needed for certain concentration of NIAS. However, the lack of analytical standards and the use of internal standards may lead to high uncertainty [9].

Depending on the type of material used, different kinds of chemicals can migrate from packaging into food. For inert materials, such as stainless steel, ceramic, and glass, migration occurs only when chemicals from the inner surface contact food directly, and these chemicals transfer from the inner surface to food by surface exchange. It is impossible to chemically diffuse through the packaging material (printing inks, adhesives) because of the small pore sizes of packaging materials. However, plasticizers (like epoxidized soybean oil or phthalates) from the closure can migrate into and contaminate glass-packaged oily food. However, migrations can be reduced by careful manufacturing or using specially developed low migrations closures. For non-inert materials, such as paper, board, or plastics, migration occurs easily and directly. In addition, chemicals may also migrate from the outside through the packaging. For example, printing inks have been found to migrate through paperboard into dry food. Smaller molecules can migrate through the paper-based materials due to its large pore size. However, migration can be significantly reduced by using barrier materials. A carton with an inner bag made of aluminum foil or plastic is an example of packaging with barrier properties.

When the outer and inner layers of food-packaging materials are in direct contact with each other during manufacutre or storage, the components of the printed outer layer may transfer to the inner layers. This kind of migration is called “off-set migration,” which occurs when paper cups are stacked into each other or beverage carton sheets are stored in rolls.

In order to determine the extent of chemical transfer from packaging into food, migrants are measured in food simulants or actual food. Food simulants are liquids intended to simulate different types of foodstuff. To simplify the chemical analysis, food is substituted by food simulants. To better simulate, it's necessary to identify the differences of chemical migration into simulants and actual foodstuff. Migration can also be modeled based on diffusion theory.

Based on chemical properties and food properties, there are three kinds of food simulants: hydrophilic (water-based), lipophilic (fatty foods), and amphiphilic (foods with both watery and fatty properties). Chemical detection and quantification require specific analytical methods toward different types of food simulants. For example, vegetable oil can simulate the migration of chemicals into oily food: 10% ethanol (aq.) or 4% acetic acid (3% acetic acid in the United States) as food simulants. Tenax, a synthetic polymer with a defined pore size, is used to simulate dry food. Butter and other amphiphilic foods are simulated by using a 50% ethanol (aq.).

The use of food simulants generally overestimated the actual level of migration into foods except in some special cases. An example is the migration of perfluorinated compounds into butter. Perfluorinated substances are insoluble in fat or water, but they do partition into food like butter, which has both properties. Now, 50% ethanol solution is used to simulate the migration into butter.

In addition to single and known migrants, it's also necessary to determine the entire chemical transfer efficiency from packaging into food without knowing their chemical identity. All food simulants can be used to assess the overall migration, but distilled water is the most commonly used simulants.

## 7.2 BPA

Bisphenol A (BPA) is the building block primarily used in the manufacture of polycarbonate plastics and epoxy resins. It is estimated that the EU uses more than one million tons of BPA per year. Monomers used to manufacture polycarbonate plastics (ca. 73%) and epoxy resins (ca. 26%) account for about 99% of the BPA, while the remaining 1% is used to manufacture other polymers or as an additive in manufacturing processes [10].

Due to its toughness, heat resistance, and transparency, the high-performance material polycarbonate has been used for many technical applications, such as water dispensers, molding equipment, reusable drink or food containers for food contact applications, and optical lens and syringes for medical applications. BPA-based epoxy phenolic resins are used as protective layers for food and beverage cans and as coatings for drinking water storage tanks.

When used as a monomer for the manufacture of plastics, the starting substances for the synthesis of BPA are acetone and phenol, and these may still be present as impurities in the BPA used in the manufacture of the polymer. Since BPA is regarded and listed as the monomer (starting substance), any remaining impurities of phenol and acetone as well as side reaction products of the BPA synthesis are all considered to be NIAS. Although both acetone and phenol are listed within Regulation (EU) 10/2011 [11], in this particular case, they are considered to be NIAS according to recital (18), as the starting substance used for the production of the polymer is BPA. However, independent from these considerations (NIAS or IAS), the specific migration limits (SMLs) of acetone and phenol have to be respected.

Very small amounts of residual BPA may migrate from packaging materials into food and beverages. The specific migration limit (SML) of BPA is set by food contact regulations in many countries throughout the world. In 2011, the European Commission Implementing Regulation (EU) 321/2011 took a preventive measure to prohibit the use of BPA for the manufacture of polycarbonate infant feeding bottles on the basis of the precautionary principle [12]. In 2018, based on European Food Safety Authority (EFSA) reassessment, the regulation was amended through Commission Regulation (EU) 2018/213 to expand the scope of the ban to include polycarbonate drinking cups or bottles intended for infants and young children [13]. In addition, the specific migration limit of BPA was decreased from 0.6 mg/kg of food to 0.05 mg/kg of food. Canada, the United States, and China also adopted the precautionary principle to address public concerns about BPA and restrict the use of BPA in packaging and food containers intended for children under the age of three. In France, based on a reevaluation conducted by the French National Agency for Food, Environmental and Occupational Health & Safety (ANSES), the French National Assembly and Senate voted to ban BPA from all food contact products at the end of 2012. These restrictions conflict with the harmonized EU regulations but have not been withdrawn even after the new Commission Regulation (EU) 2018/213 went into force.

Regulatory bodies and expert groups in EU, Canada, the United States, and Japan have conducted extensive risk assessment on BPA in the past 10 years. Hazard assessments were conducted using oral administration, large numbers of animals, and various doses in accordance with international guidelines and good laboratory practices (GLP). EFSA [14, 15], European Chemicals Bureau (ECB) [16, 17], FDA [18], the National Institute of Advanced Industrial Science and Technology (AIST) [19], Food Standards Australia New Zealand (FSANZ) [20], and Health Canada [21] concluded that BPA is not a concern on current exposure conditions in humans. In 1988, the US EPA set the reference dose for chronic oral exposure (RfD) at 0.05 mg/kg bw/day [22] with an uncertainty factor of 1000, based on the lowest observed adverse effect level (LOAEL) of 50 mg/kg bw/day from a rat chronic oral study. In the EU, BPA was regulated with an SML of 0.6 mg of BPA per kg of food (mg/kg) based on the evaluation by the Scientific Committee on Food in 2002 [23]. EFSA has reviewed scientific information and updated its opinion on BPA in 2006 [24], 2008 [15], 2010 [25], and 2011 [26]. In 2015, EFSA published an updated scientific opinion on the human health risk assessment of BPA and recommended a temporary tolerable daily intake (t-TDI) of 4 μg/kg bw/day [14], which was subsequently converted to an SML of 0.05 mg/kg of food in Commission Regulation 2018/213. In the EFSA 2015 assessment, the critical end point from a two-generation mice [27] study is based on systemic toxicity (kidney weight reduction). It was recalculated in a benchmark dose (lower confidence limit) $BMDL_{10}$ of 8960 μg/kg bw/day, which was extrapolated to a human equivalent dose (HED) of 609 μg/kg bw/day based on the available toxicokinetic data and physiologically based pharmacokinetic (PBPK) simulation. Moreover, for derivation of the TDI, an additional factor of six has been adopted for the uncertainty of potential health effects of BPA on the mammary gland, reproductive, metabolic, neurobehavioral, and immune systems.

Since the beginning of the 1990s, the endocrine disruption properties of BPA have been scientifically investigated. BPA was found to be an androgen receptor antagonist, which interacts with estrogen receptors. It is reported that BPA reduces the synthesis of some steroids at the molecular level [28, 29]. It was also reported to have low-dose effects on adipose, reproductive, mammary tissue, immune, and nervous systems in in vitro studies. Recent studies show that BPA may cause carcinogenesis, adipogenesis, and male reproduction [30, 31]. Epidemiological studies indicate that BPA exposure is associated with diabetes and cardiovascular disease and alters liver enzyme levels. Further, BPA can decrease semen quality, alter thyroid function, and cause sperm DNA damage, metabolic syndrome, obesity, hypertension, peripheral arterial disease, and coronary arterial stenosis [32–34]. Since the 1990s, the regulatory bodies, industry scientists, and academic researchers debated on BPA's human health risks, but no consensus was achieved. The debates focus on potential low-dose toxicity and the suspected non-monotonic dose response mechanism. In addition, GLP studies have been challenged for lacking certain end points related to the endocrine. Increased exposure sensitivity and accuracy of inference from animal models to humans are also controversial. Biomonitoring samples inadvertently contaminated by BPA in laboratories also impose challenges in quantifying biologically active BPA. The findings of academic research could not be confirmed through two multigenerational reproduction studies, which were conducted according to GLP [27, 35]. However, these studies lay the basis for regulatory risk assessment.

In order to solve the uncertainties concerning BPA toxicity, the Consortium Linking Academic and Regulatory Insights on BPA Toxicity (CLARITY-BPA) program [36] was developed by the National Institute of Environmental Health Sciences (NIEHS), National Toxicology Program (NTP), and FDA of the United States to study the full range of potential health effects from exposure to BPA. This program consists of two components: core study and grantee studies. The grantee studies used animals raised in the same conditions and exposed to the same doses of BPA as the core study and were blinded to the doses of BPA that the animals or tissues received. The core study reported by NTP was published in peer-reviewed journal in 2019. It was concluded that "in the CLARITY-BPA core study, statistical differences between BPA treatment groups, particularly below 25,000 μg/kg bw/day, and the vehicle control group detected by the low-stringency statistical tests applied to histopathology lesions, were not dose responsive, sometimes occurring in only one low or intermediate dose group, and did not demonstrate a clear pattern of consistent responses within or across organs within the stop- and continuous-dose arms and sacrifice times" [37, 38]. The published raw data from core study was also analyzed by other researchers for potential endocrine disruption and non-monotonic dose responses, which were not observed according to the criteria recognized by EFSA [39].

With new data available, the EFSA's Panel on Food Contact Materials, Enzymes and Processing Aids (CEP) has started to reassess the potential hazards of BPA in food and review again the temporary safe level set in EFSA's previous 2015 full-risk assessment [40]. Once completed, the Commission will assess the findings and decide what if any further action is necessary to protect consumers as regards BPA in food contact materials.

## 7.3 General Remarks and Conclusions

Inertness and safety of materials are two basic principles of food contact materials, which are also key principles of most regulations in the world. Substances migrating from food contact materials into food must be safe. In terms of safety, the migrants shall not result in exposure level, which could harm human health. Regulations and guidance documents are available for migration testing of plastics and other food contact materials. These regulations and guidance apply to all "intentionally added substances" (IAS) and "non-intentionally added substances" (NIAS). IAS includes authorized substances with migration limits, as well as materials not on the positive lists of authorized substances and substances with no migration limits. For the case that no migration limits or exposure is set according to the Framework Regulation, producers of the food packaging and/or food packer shall conduct a risk assessment and define a migration level, which does not harm human health. For the use of IASs, it is limited to varying degrees in different areas due to the local regulations. Thus, a reasonable threshold needs to be set for different regulations and exposure. For NIAS, it is the producer's responsibility to conduct a risk assessment.

Bisphenol A (BPA) is a chemical produced in large quantities for use primarily in the production of polycarbonate plastics used in various products and epoxy resins used as protective layers for food and beverage cans and as coatings for drinking water storage tanks. In the past 10 years, regulatory bodies and expert groups worldwide have conducted extensive risk assessment on BPA. In order to solve the uncertainties concerning BPA toxicity, the Consortium Linking Academic and Regulatory Insights on BPA Toxicity (CLARITY-BPA) program was developed by the National Institute of Environmental Health Sciences (NIEHS), National Toxicology Program (NTP), and FDA of the United States to study the full range of potential health effects from exposure to BPA. The EFSA's Panel on Food Contact Materials, Enzymes and Processing Aids (CEP) has started to reassess the potential hazards of BPA in food and review again the temporary safe level set in EFSA's previous 2015 full-risk assessment.

## References


1. Belgian Royal Decree (Koninklijk besluit/Arrêté royal). May 11th 1992.
2. Hawkes C. Food packaging: the medium is the message. Public Health Nutr. 2010;13:297–9.
3. Coles R. Introduction. In: Coles R, McDowell D, Kirwan MJ, editors. Food packaging technology. London: Blackwell Publishing, CRC Press; 2003. p. 1–31.
4. Marsh K, Bugusu B. Food packaging--roles, materials, and environmental issues. J Food Sci. 2007;72(3):R39–55. Directives 80/590/EEC and 89/109/EEC (L338/4)
5. Codex Alimentarius Commission. 2004. Report of the 20th session of the Codex committee on general principles. Joint FAO/WHO food standards programme. 2004 May 2–7. Paris, France. 44p.
6. Golan E, Krissoff B, Kuchler F, Calvin L, Nelson K, Price G. Traceability in the U.S. food supply: economic theory and industry studies. In: Agricultural economic report nr 830. Washington, DC, Economic Research Service, U.S. Department of Agriculture; 2004. p. 48.
7. EC. 2004. Regulation (EC) No 1935/2004 of the European Parliament and of the Council of 27 October 2004 on materials and articles intended to come into contact with food and repealing Directives 80/590/EEC and 89/109/EEC (L338/4).
8. ILSI. 2015. Guidance on best practices on the risk assessment of Non Intentionally Added Substances (NIAS) in Food Contact Materials and Articles.
9. E EU-RAR (European Union Risk Assessment Report). 2008. Risk assessment report: 4,4′-isopropylidenediphenol (Bisphenol-A). Human Health, Part 2, 1–168.
10. https://bisphenol-a-europe.org/wp-content/uploads/2017/07/Production-and-demand-volumes.pdf
11. Commission Regulation (EU) No 10/2011 of 14 January 2011 on plastic materials and articles intended to come into contact with food. OJ L 12, 15.1.2011, p1–89.
12. Commission Implementing Regulation (EU) No 321/2011 of 1 April 2011 amending Regulation (EU) No 10/2011 as regards the restriction of use of Bisphenol A in plastic infant feeding bottles .OJ L 87, 2.4.2011, P 1–2.
13. COMMISSION REGULATION (EU) 2018/213 of 12 February 2018 on the use of bisphenol A in varnishes and coatings intended to come into contact with food and amending Regulation (EU) No 10/2011 as regards the use of that substance in plastic food contact materials. (OJ L 41, 14.2.2018, P. 6–11).
14. EFSA CEF Panel (EFSA Panel on Food Contact Materials, Enzymes, Flavourings and Processing Aids). Scientific Opinion on the risks to public health related to the presence of bisphenol A (BPA) in foodstuffs: Executive summary. EFSA Journal. 2015;13(1):3978.
15. EFSA (European Food Safety Authority). Scientific opinion of the panel on food additives, Flavourings, processing aids and materials in contact with food (AFC) on a request from the Commission on the toxicokinetics of bisphenol a. The EFSA Journal. 2008;2008(759):1–10.
16. EU-RAR (European Union Risk Assessment Report). European Union risk assessment report: 4,4'-Isopropylenediphenol (bisphenol-A). EUR - Scientific and Technical Research Reports. 2003;37:1–302.
17. EU-RAR (European Union Risk Assessment Report). Risk assessment report: 4,4′-isopropylidenediphenol (Bisphenol-A). Human Health, Part 2. Luxembourg: Publications Office; 2008. p. 1–168.
18. US FDA (Food and Drug Administration). 2008. Draft Assessment of Bisphenol A for Use in Food Contact Applications. DRAFT version 08/14/2008. 105 pp.
19. Miyamoto K, Kotake M. Estimation of daily bisphenol a intake of Japanese individuals with emphasis on uncertainty and variability. Environ Sci. 2006;13(1):15–29.
20. Food Standard Australia New Zealand (FSANZ). http://www.foodstandards.gov.au/science/monitoring/surveillance/documents/BPA%20paper%20October%202010%20FINAL.pdf
21. Health Canada. Health Risk Assessment of Bisphenol A from Food Packaging Applications. Ottawa, ON: Health Canada; 2008. August. Available from: http://www.hc-sc.gc.ca/fn-an/securit/packag-emball/bpa/bpa_hra-ers-eng.php/

22. US EPA. 2002. Integrated Risk Information System: Bisphenol A
23. Opinion of the Scientific Committee on Food on Bisphenol A (SCF/CS/PM/3936 Final)
24. EFSA. Opinion of the scientific panel on food additives, Flavourings, processing aids and materials in contact with food on a request from the Commission related to 2,2-bis (4-hydroxyphenyl)propane (Bisphenol A). EFSA J. 2006;2006(428):1–75.
25. EFSA. Scientific opinion on bisphenol a: evaluation of a study investigating its neurodevelopmental toxicity, review of recent scientific literature on its toxicity and advice on the Danish risk assessment of bisphenol A. EFSA Journal. 2010;8(9):1829.
26. EFSA. EFSA panel on food contact materials, enzymes, Flavourings and processing aids (CEF) statement on the ANSES reports on Bisphenol A. The EFSA Journal. 2011;9(12):2475.
27. Tyl RW, Myers CB, Marr MC, et al. Two-generation reproductive toxicity study of dietary bisphenol a in CD-1 (Swiss) mice. Toxicol Sci. 2008;104:362–84.
28. Somogyi V, Horváth TL, Tóth I, et al. Bisphenol A influences oestrogen- and thyroid hormone-regulated thyroid hormone receptor expression in rat cerebellar cell culture. Acta Vet Hung. 2016;64(4):497–513.
29. Wetherill YB, Akingbemi BT, Kanno J, et al. In vitro molecular mechanisms of bisphenol A action. Reprod Toxicol. 2007;24:178–98.
30. Gies A, Soto AM. Bisphenol A: contested science, divergent safety evaluations, Lessons from health hazards. European Environment Agency; 2012. p. 247–71.
31. Allard P, Colaiacovo MP. Bisphenol A. In: Gupta RC, editor. Reproductive and developmental toxicology. München: Elsevier; 2011. p. 673–86.
32. Lang IA, Galloway TS, Scarlett A, et al. Association of urinary bisphenol A concentration with medical disorders and laboratory abnormalities in adults. JAMA. 2008;300:1303–10.
33. Shankar A, Teppala S, Sabanayagam C. Bisphenol A and peripheral arterial disease: results from the NHANES. Environ Health Perspect. 2012;120:1297–300.
34. Ye X, Zhou X, Hennings R, et al. Potential external contamination with bisphenol A and other ubiquitous organic environmental chemicals during biomonitoring analysis: an elusive laboratory challenge. Environ Health Perspect. 2013;121:283–6.
35. Tyl RW, Myers CB, Marr MC, et al. Three-generation reproductive toxicity study of dietary bisphenol A in CD Sprague-Dawley rats. Toxicol Sci. 2002;68:121–46.
36. CLARITY-BPA Program. https://ntp.niehs.nih.gov/whatwestudy/topics/bpa/index.html
37. The CLARITY-BPA Core Study. A Perinatal and Chronic Extended-Dose-Range Study of Bisphenol A in Rats. https://ntp.niehs.nih.gov/ntp/results/pubs/rr/reports/rr09_508.pdf
38. Camacho L, et al. A two-year toxicology study of bisphenol A (BPA) in Sprague-Dawley rats: CLARITY-BPA core study results. Food Chem Toxicol. 2019;132:110728. https://doi.org/10.1016/j.fct.2019.110728. https://www.ncbi.nlm.nih.gov/pubmed/31365888
39. Badding MA, et al. CLARITY-BPA Core study: analysis for non-monotonic dose-responses and biological relevance. Food Chem Toxicol. 2019;131:110554. https://doi.org/10.1016/j.fct.2019.06.001. Epub 2019 Jun 15. https://www.ncbi.nlm.nih.gov/pubmed/31207305
40. BPA plan ready for new EFSA assessment in 2018. http://www.efsa.europa.eu/en/press/news/bpa-plan-ready-new-efsa-assessment-2018

# Chapter 8
# Genetically Modified Food

**Haibin Xu**

**Abstract** Transgenic food or genetically modified (GM) food, or genetically engineered (GE) foods are foods derived from organisms whose genetic material has been modified in a way that does not occur naturally, e.g. through the introduction of a gene from a different organism. The primary goal of the development of genetically modified food was to increase crop yields and facilitate management by altering the composition of organisms. So far, transgenic plants have become the largest category of genetically modified organisms. However, as the technology develops, the concerns about its potential adverse health outcome to human or unintended effects on the environment have been rising as well. This chapter reviews the development and application of transgenic food, and focuses on the potential health problems caused by it as well as the current safety evaluation and regulatory concern of GM food.

**Keywords** Transgenic food · Genetically modified food · BT toxin · Genetically modified animals · Toxicity allergenicity

## 8.1 The History of Genetically Modified Food

Transgenic food or genetically modified (GM) food, or genetically engineered (GE) food as it is sometimes called, is the most controversial food issue of our era. Over the last 25 years GM technology has advanced to such an extent that it is now commonplace and is used to produce pharmaceuticals in cell culture systems, to study biochemical and molecular mechanisms in cells to help fight disease and to produce food more efficiently, but more of this later. They first applied in the 1970s, is one of the newest methods to introduce novel traits to microorganisms, plants and animals. Unlike other genetic improvement methods, the application of this technology is strictly regulated. Before any genetically modified organism (GMO) or

H. Xu (✉)
China National Center for Food Safety Risk Assessment, Beijing, People's Republic of China
e-mail: hbxu1231602@cfsa.net.cn



L. Zhang (ed.), *Nutritional Toxicology*,
https://doi.org/10.1007/978-981-19-0872-9_8

product can be released into the food market, it has to pass an approval system in which the safety for humans, animals and the environment is thoroughly assessed.

The GM story began in 1973 when Stanley Cohen and Herbert Boyer at Stamford University in the United States made a DNA molecule outside a cell, then incorporated it into a cell's genome (transformation) and witnessed the cell making the protein coded for by the alien (transformed) DNA. In 1973 Cohen and Boyer made it scientific fact—they invented genetic modification. The scene was set and the possibility of modifying a cell's genes became reality, but it was not until 1986 when Kary Mullis of Cetus Corp., Berkeley, USA, modified (engineered) the genes of a prokaryotic cell. The first GM crop to be approved was *Flavr Savr* tomatoes in the United States in 1994. These tomatoes were genetically modified – by inserting the polygalacturonase gene – to ripen more slowly which meant that they built up more sugars and other flavour agents and so tasted better than their conventional fast ripening counterparts [1].

From 1996 to 2015, the commercialization of genetically modified crops was just 20 years ago. During this period, the total planting area of GM crops reached 2 billion hectares, and the total benefits of farmers exceeded $150 billion. At present, the world's largest genetically modified crop planting country is still the United States, its planting area of 70 million 900 thousand hectares, accounting for 39% of the global planting area. In addition, the top ten countries are Brazil, Argentina, India, Canada, China, Paraguay, Pakistan, South Africa and Uruguay. The world's top two genetically modified plants are corn and soybeans. The international organization for agricultural biotechnology applications (ISAAA) released the newest information that global commercial development of biotechnology/genetically modified crops in Beijing in 2016, the global growth of genetically modified crops reached a peak (185 million 100 thousand ha); GM crops provided consumers with a wider range of options; the planting area of transgenic crops has increased by 110 times since 1996, reaching 2 billion 100 million hectares; 26 countries (19 developing countries and seven developed countries) have planted genetically modified crops; GM soybeans account for 50% of the world's genetically modified crops; compound characters of transgenic crops accounted for 41% of the total area of GM crops in the world, second only to herbicide resistant transgenic crops (accounting for 47%); the planting area accounts for 91% of the five largest genetically modified crops, 3 of which are developing countries (Brazil, Argentina and India), and two are developed countries (the United States and Canada); 80 million hectares of genetically modified crops have been planted in 10 Latin America countries; 18 million 600 thousand hectares of genetically modified crops are grown in eight countries in the Asia Pacific region; the four nations of the European Union continue to plant over 136 thousand hectares of genetically modified corn; South Africa and Sultan have increased the cultivation of genetically modified crops; in 2016, the world's genetically modified seed market was worth as much as $15 billion 800 million [2].

## 8.2 Application of Transgenic Technique in Agricultural Area

The primary goal of the development of genetically modified food was to increase crop yields and facilitate management by altering the composition of organisms. In order to achieve this goal, through the introduction of herbicide resistance, pest resistance, anti-virus, anti-bacteria, anti-fungi or resistant to environmental stress (drought) gene, transgenic crops with the corresponding characteristics of these genetically modified crops is known as the first generation of genetically modified crops.

The research of transgenic technology has been changing with each passing day, and the research and development speed has gradually accelerated. So far, transgenic plants have become the largest category of genetically modified organisms, involving food crops, vegetables, fruits and trees. The cultivation of these "first--generation" crops not only increased production, reduced farmland production costs, increased farmers' profits, but also reduced the health hazard and environmental pollution caused by pesticide use. Among them, the most important are herbicide-resistant and insect-resistant transgenic crops. With the development of transgenic technology and mature, agronomic characteristics and production characteristics of the researchers have not only improved crops, the focus began to shift to the research on genetically modified crops have the added value of output characteristics, such as GM is to improve the quality of crops for the purpose, such as to improve the taste of food, increase the nutrition of the food and reduce the trans fatty acids in food and oil crops to improve oil content, etc. In essence, GM crops are grown from seeds that have had their genes modified in some way that benefits crop production. For example, a gene can be inserted that codes for resistance to a particular herbicide (e.g. glyphosate) which means that when the crop is being grown the farmer can spray it with the herbicide to kill weeds without harming the crop itself. Or, perhaps, a gene that codes for the production of a particular sugar in a fruit might be inserted into the fruit plant's genome—the crop grown would then be very sweet and command a good price at market [3].

### *8.2.1 Glyphosate-Resistant Crops*

Glyphosate-resistant crops are the most widely utilized GM crops worldwide. The most important food crops are glyphosate-resistant corn (maize), soy and rape (canola). Plants synthesize amino acids in order to make proteins. If they can't synthesize amino acids they can't make proteins and they die. The herbicide glyphosate inhibits a key enzyme, 5-enolpyruvylshikimate-3-phosphate synthetase (EPSPS) involved in the synthesis of aromatic amino acids (e.g. tyrosine) in plants. Glyphosate inhibits EPSPS, so preventing the plant synthesising aromatic amino acids and thus killing it. GlyR is the gene that codes for a form of EPSPS in

A. *tumefaciens* that is unaffected by glyphosate. Therefore, when GlyR is inserted into a plant the plant is able to make its aromatic amino acids even in the presence of glyphosate. Therefore, glyphosate-resistant GM crops express A. *tumefaciens* GlyR and so can synthesize aromatic amino acids in the presence of glyphosate. This means that glyphosate can be used to kill weeds amongst GlyR (i.e. glyphosate-resistant) crops. New GM crops are being developed all the time because they represent huge commercial opportunities for the international agrochemicals companies that engineer and market them. The following are some of the glyphosate-resistant crops available at the time of writing: Soy, Rape (canola), Corn, Sugar beet, and Cotton [4].

### *8.2.2 Insect-Protected Crops–BT Toxin*

Two of the big problems that farmers face are weeds and insect pests. Glyphosate-resistant crops address the former, and the insertion of genes that code for an insect toxin addresses the insect problem. The gene used for insect protection is derived from the bacterium *Bacillus thuringiensis*. It codes for a protein toxin (BT toxin) that kills insects by interfering with their digestive process, thus starving them to death. BT toxin is insect-specific which means that it does not have major safety implications for the human consumer of BT toxin-expressing crops. The only real safety concern for humans is the possibility of allergy to BT toxin which could occur in sensitive individuals. Corn, potatoes and cotton are all successful GM BT toxin-expressing crops. It is possible to insert more than one gene into a GM crop, so giving the crop multiple desirable properties. For example, both the BT toxin and glyphosate resistance genes have been inserted into corn which is now grown commercially [5].

### *8.2.3 GM Crops with Enhanced Flavour or Improved Processing Quality*

The first GM crop was *Flavr Savr* tomatoes which have the polygalacturonase gene inserted which means that the tomatoes ripen slowly and taste better than conventional crops (at least this is what their marketers claim). Since then there have been numerous gene inserts that code for particular enzymes that produce either flavour molecules or molecules with nutritional value; for example, high oleic acid soy which has enhanced genes that code for enzymes in the oleic acid synthetic pathway, and therefore the soy oil produced is rich in oleic acid. Oleic acid is a commercially important fatty acid (it is a major component of olive oil). The possibilities are endless and it would be feasible to produce crops that might be able to solve global nutritional problems. For example, imagine rice with a gene for enhanced vitamin

B1 (thiamine) synthesis; this could prevent the development of beriberi (neurodegeneration due to lack of thiamine—its name is Sinhalese (Sri Lankan) for 'extreme weakness') in rice-eating undernourished communities. In the study of transgenic wheat, Chinese scientists have developed a new type of transgenic wheat which can improve the quality of starch. These kinds of starch metabolic enzyme genes including Granule-bound Starch Synthase (GBSSI), Starch branching enzyme IIa (SBE IIa), ADP-glucose pyrophosphorylase (AGP), and two kinds of storage protein genes including High Molecular Weight Glutenin Subunit (HMW-GS) gene and gliadin genes were transformed by RNA interference and over-expressing, and a series of new germplasms and transgenic lines were produced by the method of transgenic breeding combining with traditional breeding [6].

### 8.2.4 Improved Nutritional Properties

A promising GM rice crop has been developed that could prevent disease due to malnutrition in a large proportion of the world—the crop is Golden Rice. Golden Rice is a GM rice which expresses the *psy* and *lyc* genes from daffodils and *crt1* gene from the bacterium *Erwinia uredovora. Psy* codes phytoene synthetase, *lyc* for lycopene $\beta$-cyclase and *crt1* for phytoene desaturase—three enzymes important in carotenoid synthesis (carotenoids make daffodils yellow and carrots red). Rice expressing these three genes synthesizes carotenoids and looks yellow (or golden—hence its name), but more importantly it provides its consumer with carotenoids which are the precursors of vitamin A. Therefore, Golden Rice prevents vitamin A deficiency which affects a large proportion of the Third World—vitamin A is a key part of the biochemistry of sight and therefore its deficiency affects vision, particularly night vision [6].

### 8.2.5 Genetically Modified Animals

Genetically modified animals are still at the experimental stage. Most studies are still utilizing small animal models (e.g. mice) to develop the technology, but it will not be long before GM farm animals are a feature of farmyards in some parts of the world. Indeed, in 2002 the first cow to produce human-like milk was developed by inserting genes for the synthesis of specific proteins (e.g. lactoferrin) into the cow's genome. The idea of producing 'human' milk in cows is very interesting because it would solve some of the problems of feeding babies conveniently. Currently cow's milk is used, but it has numerous disadvantages when compared to human milk because it contains proteins (e.g. lactoglobulin) that human milk does not contain and some babies develop allergic responses to these proteins which might lead to hyperallogenicity (e.g. asthma) in later life. Transgenic farm animals might also be used to manufacture pharmaceuticals. For example, cows expressing the gene for human insulin excrete insulin in their milk; so it is conceivable that 1 day farming might become 'pharming' and 'pharmers' will milk their GM human insulin cows and sell the milk to the pharmaceuticals industry where the insulin will be extracted for the treatment of human diabetes. This is far beyond the scope of this book but is an interesting thought that is fast gaining credibility—in just a few years it is likely to be reality [7].

## 8.3 Safety Problems of Genetically Modified Foods on Human Health

Genetically modified food is the most controversial food issue in food safety era. Different from the traditional method of breeding, transgenic breeding technology has broken the reproductive isolation barriers between species, can gene fragments from different species into a receptor genome in order to change the genetic traits, the exchange of genetic material between animal, plant and microorganism. In the process of genetic manipulation, there may be some harm to human health with the gene into receptor biology, or biological genome changes caused by receptor to produce some detrimental to human health than expected changes. Therefore, since the advent of genetically modified food, its security has been paid close attention to by international organizations, governments, academia and the public. In view of a series of gene manipulation processes involved in transgenic technology, the potential health risks of genetically modified foods are mainly as follows [8].

### *8.3.1 Foreign Gene Expression Protein*

The most significant difference between the toxicity of foreign gene expression products and their parent genes is the introduction of foreign gene fragments. Safety product of exogenous gene expression is the first issue to be considered, including the expression product is safe to eat, the history of exogenous gene donor organism itself is toxic, whether the structure of the recombinant DNA containing the virus activity sequence, whether encoding pathogen or toxin. Toxicity of exogenous gene expression products can be evaluated by a series of toxicological tests [3].

### *8.3.2 Sensitization of Heterologous Gene Expression Protein*

The sensitization of heterologous gene expression protein, the expression of foreign gene and the sensitization of protein is one of the important safety problems of transgenic modified food. If the exogenous gene encoding protein is a known human allergen or with known allergens in amino acid sequence homology, or belonging to the family of proteins are members of some human allergens, then the exogenous gene expression of the protein may cause allergic reactions. Therefore, in evaluating the safety of genetically modified foods, it is necessary to evaluate the sensitivity of new proteins expressed by foreign genes [9].

### *8.3.3 Antibiotic Marker Gene*

During the process of gene manipulation, antibiotic marker genes often use antibiotic marker genes to help the screening and identification of transformants. Consumers eat genetically modified food containing antibiotic marker gene, there is a possibility of horizontal gene transfer to intestinal epithelial cells or pathogens in the genome of antibiotic resistance, the problem will produce antibiotic resistance, although currently tend to think that the small probability of this happening, but in the process of safety assessment of genetically modified food when eating, still should consider the potential the health risk of antibiotic marker gene [9].

### *8.3.4 Changes of Nutrients Content and Composition*

The transfer of foreign genes may lead to changes in the composition of genetically modified foods, including changes in nutrients and anti-nutritional factors change. At the same time, we should consider the change of toxic ingredients, and these changes are also worthy of attention [8].

### *8.3.5 Unintended Effects*

What is the unintended effect of genetically modified foods? At present, there is no scientific consensus on the answer. Some scholars believe that the 'Unintended effects' refers to when the genetically modified food crop/corresponding traditional food crops in the same growth/environment, due to other effects of genetically modified food/transgenic operation gene into receptor genomic loci caused crop in phenotype and composition in gene expressed by the expected effect; some scholars believe that the 'Unintended effects' mainly refers to the effect of exogenous gene in the transgenic operation into recipient genomic loci caused by the uncertainty that may lead to a series of unexpected changes in transgenic organisms. 'Unintended effects' is one of the most controversial academic concepts in the evaluation of the area of genetically modified food safety. The evaluation of 'Unintended effects' need developed new methods, it may be the answer to what is GM food 'Unintended effects' and how to evaluate the 'Unintended effects' [10].

## 8.4 Safety Evaluation of Genetically Modified Foods on Human Health

Risk analysis consists of three basic elements: risk assessment, risk management and risk communication. A risk assessment comprises four steps: hazard identification, hazard characterization, exposure assessment and integrative risk characterization.

Hazard identification is the first step in risk assessment and in case of genetically modified food is focused on the identification of differences between the GM plant and its appropriate comparator. The hazard characterization step is defined as the quantitative or semi-quantitative evaluation of the nature of the possible adverse health effects to humans following exposure to genetically modified food. This step is focused on a possible quantification of the toxicological/nutritional potential of identified differences between the genetically modified food and non-GM comparator. The aim of the exposure assessment is the quantitative estimation of the likely exposure of humans to genetically modified food. With regard to humans, an exposure assessment characterizes the nature and size of the populations exposed to a source and the magnitude, frequency and duration of that exposure. For exposure assessment, it is necessary that every significant source of exposure is identified. In particular, it is of interest to establish whether the intake of the genetically modified food and new constituents are expected to differ from that of the conventional product which it may replace. In this respect, specific attention will be paid to that genetically modified food which is aimed at modifying nutritional quality. The final risk characterization of genetically modified food is focused on the evaluation of all available data from hazard identification, hazard characterization and exposure/intake with respect to their safety and/or nutritional impact for humans. A comprehensive risk characterization considers all the available evidence from several approaches including molecular analysis, agronomical and compositional analysis, toxicity and allergenicity testing to potential adverse or nutritional effects of genetically modified food on humans.

Due to the potential health risks of genetically modified foods, each GM food must be systematically and comprehensively evaluated avoiding cause health damage to consumers before entering food china. GM is different from traditional food, it is difficult to find a dose-response relationship, and therefore, traditional evaluation methods of food toxicology are not fully suitable for genetically modified food. According to the safety problems of genetically modified food may exist, the Food and Agriculture Organization (FAO) and World Health Organization (WHO) FAO/WHO sets up the codex ad hoc intergovernmental task force that develops a series of technical guidelines on the safety evaluation methods of genetically modified foods. In 1993, Organization for Economic Cooperation and Development (OECD) first proposed the 'substantial equivalence (substantial equivalence)' concept. The principle of substantial equivalence means that the existing food or its source organism can be used as a basis for comparison when evaluating the safety of new food and food components produced by biotechnological means. Codex Alimentarius Commission (CAC) released 'modern biotechnology food risk analysis principles' for genetically modified food risk analysis that provides the basic framework in 2003. Further, CAC had released the principle of 'Guidelines for safety evaluation of recombinant DNA from plant food" (CAC/GL 45-2003), "Guidelines for the conduct of food safety assessment of foods produced using recombinant-DNA microorganisms' (CAC/GL 46-2003) and 'Guideline for the conduct of food safety assessment of foods derived from recombinant-DNA animals' (CAC/GL 68–2008), these documents were used to guide the genetically modified plants, microorganisms and animal food safety evaluation.

Except FAO/WHO, other international organizations and some developed countries have also carried out a great deal of research work on the safety of genetically modified foods and put forward some evaluation methods. The contents of safety evaluation focused on gene donor, receptor gene, and recombinant vector process, toxicity and allergenicity of new expression substances, unintended effects of insert genes, change the content/component of nutrient and anti-nutrient on genetically modified food, and influence cooking process on food safety of genetically modified food.

According to the principles of the safety evaluation of genetically modified foods and the various health-related factors that need to be considered, the safety evaluation of genetically modified foods should mainly focus on the following aspects.

### 8.4.1 Toxicity Evaluation

For protein expression products should be carried out with the known toxin and anti-nutritional factors in the sequence similarity of amino acids and protein conformation analysis; analysis of stability and thermal stability, digestion process. If the foreign gene expression product is a kind of new protein, it also needs to undergo oral toxicity test. For nonprotein expression products, if there is no food safety history, according to the characteristics of molecular structure, biological function and dietary exposure according to the traditional methods of toxicology of food safety evaluation, such as metabolite analysis, toxicokinetics, subchronic toxicity, chronic toxicity test, reproduction test and teratogenicity test.

Molecular and biochemical characteristics were descried using bioinformatics technology based on databases of toxic proteins, data of thermal stability and in vitro digestibility were obtained using experiment and data on equivalence to the proteins introduced into the genetically modified plants if proteins expressed in vitro are used as test materials.

Toxicological data on novel proteins: If the novel proteins expressed have no history of safe use and their safety data is inadequate, acute oral toxicity data must be provided, while the need for the 28-day feeding study depends on protein expression levels in plants and population intake levels. If necessary, evaluation of immunotoxicity must be conducted. Justifications must be provided for a lack of data on acute oral toxicity or an omission of the 28-day feeding study. Toxicological evaluation may include toxicokinetics, genetic toxicity, subchronic toxicity, chronic toxicity or carcinogenicity, reproductive and developmental toxicity, and the like. Specific requirements must be analysed case by case.

Estimation of dietary intake of genetically modified food. Exposure assessment of genetically modified food is an important basis for risk assessment, the general intake amount and maximum intake amount should be considered at dietary intake assessment, the special populations should also consider, such as infants, children, pregnant women, chronic diseases and other special groups. The impacts is must considered of Storage, processing and cooking process of genetically modified food.

### 8.4.2 Allergenicity of Protein Derived from Genetically Modified Food

Any single test cannot predict the allergenicity of a newly expressed protein accurately. Therefore, it is necessary to evaluate the allergenicity using step by step and case by case analysis. The basic strategy of genetically modified food allergen assessment has been released by CAC. The following must be first considered, whether the gene donor contains allergens, whether the inserted gene encodes allergens, and expression levels of the new proteins in the parts of the plant as human food or animal feed; bioinformatics analyses based on allergen databases; data on thermal stability and in vitro digestibility; serological test data with known allergens as antibodies if the donor contains allergens or new proteins have sequence homology with known allergens; and analyses of allergens and their levels if the recipient plant contains allergens [11].

### 8.4.3 Key Component Analysis of Genetically Modified Food

Analysis of genetically modified foods compared with no-genetically controlled foods, the key nutrients, anti-nutritional factors, toxic substances and allergenic substances in the type and quantity have changed. Once a change is found, toxicology, nutrition, immunology and other methods are needed to assess the health effects of these changes. Basic information about the GM food to be tested must be provided, as well as data on samples grown at different times and locations. Natural variations of key ingredients and literature references must also be provided. Key components include nutrients, toxins and other harmful substances, anti-nutritional factors, other components (water, ash, and other substances), and unintended components.

### 8.4.4 *Safety Evaluation of Whole Genetically Modified Food*

As part of the safety assessment of genetically modified food, mammalian toxicity studies have been conducted with whole genetically modified food. By far the most commonly reported study is the 90-day rodent dietary toxicity study based on study designs used for traditional chemical toxicity testing. These studies have generally included standard response variables such as body and organ weights, clinical chemistry, haematology and pathology. The 90-day feeding study of whole food in rats is required; where applicable, data on chronic toxicity and reproductive toxicity in rats and other whole food feeding studies must also be provided.

In most cases, whole genetically modified food toxicology studies have been conducted in the form of repeated-dose 90-day studies with rodents, with the intention of identifying potential unintended changes that could present a risk. There are review articles that show that continued routine usage of animal testing to address the question of unintended hazards that might result from modern biotechnology is not scientifically or ethically justified.

### 8.4.5 *Other Evaluation Data of Genetically Modified Food*

Nutrition evaluation must be provided for genetically modified food that develops purpose to change and enhance the nutritional value of food. The nutrition assessment is one of the important aspects of genetically modified food which nutritional composition changed greatly. Other data to be considered include content data of harmful substances, such as heavy metals, which come from cultivated soil where genetically modified food, data of processing and cooking methods affect the safety of genetically modified food, and residual data of target herbicide for tolerant herbicide transgenic plants [8].

## 8.5 The Development Prospects of Genetically Modified Food

Genetically modified foods have received a lot of bad publicity. Most concerns about genetically modified food fall into three categories: environmental hazards, human health risks, and economic concerns. Critics have claimed that altered foods may be unsafe. Ironically, although many issues surround genetically modified food, safety and nutrition should not be among them. These foods are as safe and nutritious as their conventional counterparts.

Labelling of genetically modified food is also a contentious issue. On the whole, agribusiness industries believe that labelling should be voluntary and influenced by the demands of the free market. Consumer interest groups, on the other hand, are demanding mandatory labelling. Many questions must be answered if labelling of genetically modified food.

Genetically modified food is one of the most innovative achievements of crop technology, and has been successful and unprecedented cultivation. At present, the planting rate of four genetically modified crops (maize, soybean, cotton and rapeseed) grown on a large scale still has considerable growth potential. Genetically modified food has the potential to solve many of the world's hunger and malnutrition problems and to help protect and preserve the environment by increasing yield and reducing reliance on chemical pesticides and herbicides. Yet there are many challenges ahead for governments, especially in the areas of safety testing, regulation, international policy, and food labelling. Many people feel that genetic engineering is the inevitable wave of the future and that we cannot afford to ignore a technology that has such enormous potential benefits. However, we must proceed not only with caution to avoid causing unintended harm to human health and the environment as a result of our enthusiasm for this powerful technology but also with better education and information to the public about the usefulness of genetically modified food [12]. With the development of biotechnology, genetic engineering technology has developed from a single trait gene transformation to the development of complex trait gene transformation; compared with the single trait of transgenic crops, complex traits of transgenic crops has extra advantage in a lot of aspects. The use of advanced science and technology to cultivate more types of genetically modified crops is an essential driving force in future [13].

## References

1. Garrett RH, Grisham CM. Chapter 10: Nucleotides and nucleic acids. In: Biochemistry. 4th ed. Boston: Brooks/Cole; 2010.
2. International Service for the Acquisition of Agri-biotech Applications (ISAAAA). 2016 global biotechnology and transgenic commercial development trend. China Biotechnol. 2017;37(4): 1–8.
3. Federici BA, Siegel JP. Chapter 3: food safety of proteins in agricultural biotechnology. CRC Press; 2008.
4. Kammermeyer K, Clark VL. Chapter 7 recombinant techniques. In: Genetic engineering fundamentals. New York: Marcel Dekker; 1989.
5. Parekh SR, editor. The GMO handbook. Totowa: Humana Press; 2004.
6. Ruse M, Castle D, editors. Genetically modified foods. Prometheus Books; 2002.
7. Hammond B. Food safety of proteins in agricultural biotechnology. CRC Press; 2008.
8. Principles for the Risk Analysis of Foods Derived from Modern Biotechnology, CAC/GL 44-2003.
9. Delaney B, Astwood JD, Cunny H, Conn RE, Herouet-Guicheney C, MacIntosh S, Meyer LS, Privalle L, Gao Y, Mattsson J, Levine M. ILSI International Food Biotechnology. Evaluation of protein safety in the context of agricultural biotechnology Committee Task Force on Protein Safety. Food and Chemical Toxicology. 2008;46:S71–97.

10. Ladics GS, Bartholomaeus A, Bregitzer P, Doerrer NG, Gray A, Holzhauser T, Jordan M, Keese P, Kok E, Macdonald P, Parrott W, Privalle L, Rayboould A, Rhee SY, Rice E, Romeis J, Vaughn J, Wal JM, Glenn K. Genetic basis and detection of unintended effects in genetically modified crop plants. Transgenic Res. 2015;24:587–603.
11. McClain S, Vieths S, Gary A. Chapter 8: Bannon. In: Food safety of proteins in agricultural biotechnology. CRC Press; 2008.
12. Nicolia A, Manzo A, Veronesi F, Rosellini D. An overview of the last 10 years of genetically engineered crop safety research. Crit Rev Biotechnol. 2014;34(1):77–88.
13. Pew Research Center. Americans, Politics, and Science Issues 2015.

# Chapter 9
# Tolerable Upper Limits of Nutrients

**Jiao Huo and Lishi Zhang**

**Abstract** Unlike non-nutrients, the risk for nutrient substances will increase with both inadequate and excessive intakes when homeostatic balance is broken in human bodies. Upper level of intake (UL) is the maximum level of habitual intake from all sources of a nutrient or related substance judged to be unlikely to lead to adverse health effects. Nutrient risk assessment has been adopted as fundamental roadmap in establishment of UL, in which some differences in process are presented because, unlike non-nutrients, nutrient substances are biologically essential or have a favorable impact on health at specified levels of intake. The nutrient risk assessment framework generally involves four overarching steps comparable to that of non-nutrient risk assessment: (1) nutrition hazard identification and characterization; (2) dietary intake assessment; (3) nutrient risk characterization; (4) discussion of implications and special concerns. The process is typically not so straightforward because of data gaps and variation in the type and amount of evidence for each nutrient. This chapter has reviewed the toxic profile of nutrients, and the principles and methods for establishment of tolerable upper limits of nutrients, which aims to offer help to risk manager and professionals in risk management of excessive nutrient intakes, and establishment of corresponding ULs.

**Keywords** Upper levels of intake · Nutrient risk assessment · Dietary intake · Intake-response relationships

J. Huo
Department of Nutrition and Food Safety, Chongqing Center for Disease Control and Prevention, Chongqing, China

L. Zhang (✉)
Department of Nutrition and Food Safety, West China School of Public Health, Sichuan University, Chengdu, China


L. Zhang (ed.), *Nutritional Toxicology*,
https://doi.org/10.1007/978-981-19-0872-9_9

## 9.1 Introduction

Nutritional deficiency is a universal problem among different countries and regions, which can result in a variety of health problems or diseases, such as skin problems, bone abnormalities, and even dementia. In past decades, national governments and international agencies/organizations put their focus on the strategies to defeat nutritional deficiencies by adopting various public health interventions, such as provision of universal or targeted fortification of foods and micronutrient supplementation. However, the risk of excessive intake after the increased use of fortified foods, dietary or food supplements, especially formulated foods and functional foods, has attracted the attention in the recent years.

In 1994, the Institute of Medicine (IOM) program unit (now the Health and Medicine Division) of the National Academies of Sciences, Engineering, and Medicine (NASEM) developed the Dietary Reference Intakes (DRIs) by incorporating the Upper Levels of Intake (ULs) and nutrient risk assessment into their process of establishing for the United States and Canada [1]. The DRIs comprise a set of nutrient reference values: Recommended Dietary Allowances (RDAs), Estimated Average Requirements (EARs), Adequate Intake (AI), ULs, and two additional DRIs [i.e., Acceptable Macronutrient Distribution Range (AMDR) and Estimated Energy Requirement (EER)] which have been established in 2005 [2]. Moreover, nutrient risk assessment has been adopted as fundamental roadmap in the process of decision making by the Food and Agriculture Association of the United Nations (FAO) and the World Health Organization (WHO) as well as many national authorities.

The definitions of nutrient risk assessment and ULs were derived from the processes and definitions that have been applied to the risk assessment of human exposure to xenobiotics such as food additives, natural toxicants, and environmental contaminants in the food. However, it is important to underscore that the classic non-nutrient risk assessment approach is not suitable for nutrient, because of the biological effects of nutrient on health at specified levels of intake. Furthermore, ULs are often carried out in the range of intake levels that provide the nutritional benefit. As a result, establishment of the specific standards and guidelines for ULs via nutrient risk assessment approach has been addressed by FAO and WHO, and other authoritative scientific bodies.

For over 25 years, most of the risk assessments about ULs setting drew on the considerations of vitamins and micronutrients. In recent years, there is an international collaboration on safety assessments of intakes of amino acids, fatty acids, or nonessential nutrients, which also termed "dietary bioactive components" (e.g., phytochemicals).

## 9.2 Toxicity of Nutrients

### *9.2.1 Key Terminology*

Because nutrients are different from toxic substances in that they have beneficial effects as well as a risk, the definitions of key terms have been developed and modified to take this factor into account. The following terms are used frequently in establishing ULs. Most of the terms are derived from the definitions of risk assessment on toxicants set by International Programme on Chemical Safety (IPCS) of WHO, and modified by FAO/WHO. Other international bodies and national authorities have also developed the definitions of these terms for specifying ULs or associated terms.

- Hazard (FAO/WHO, 2006) [3]: "inherent property of a nutrient or related substance to cause adverse health effects depending upon the level of intake".
- Adverse health effect (FAO/WHO, 2006) [3]: "a change in morphology, physiology, growth, development, reproduction or life span of an organism, system, or (sub) population that results in an impairment of functional capacity, an impairment of the capacity to compensate for additional stress, or an increase in susceptibility to other influences".
- Risk (IPCS, 2004) [4]: "the probability of an adverse effect in an organism, system or (sub) population caused under specified circumstances by exposure to an agent".
- Upper level of intake (FAO/WHO, 2006) [3]: "the maximum level of habitual intake from all sources of a nutrient or related substance judged to be unlikely to lead to adverse health effects in humans".

"Nutrients" in this book refer to substances as inherent constituents of food that are biologically essential or have a demonstrated favorable impact on health, which do not encompass food additives or substances such as food contaminants, pesticides, microbiological pathogens, or other food-borne hazards. In comparison to the hazard of xenobiotics, the hazard of nutrient was stressed on the two aspects' properties. The "inherent property of a nutrient" in the definition of "hazard" refers that the nutrient is responsible for the risk at the high levels of intake, meanwhile it also has a favorable impact on health at a different level of intake. It is noted that the "hazard" here is not responsible for the risk of deficiency states associated with inadequate intakes.

Adverse health effects comprise a series of markers indicating the effects of different types of impairments via different approaches. More information about adverse health effects is provided later in this chapter.

"Risk" is the probability of an adverse effect occurring. This likelihood is not an inherent feature of nutrient or non-nutrient, which is depending on comparing the outcomes of "hazard characterization" with the estimation of intake dose in exposed populations. Hence, the term "risk" in nutrient risk assessment is defined the same as that in classic risk assessment.

For illustrating the definition of ULs, one more term needs to be clarified. "Habitual intake" means the long-term average daily intake of the nutrient or substance. This ingestion pattern of nutrient is different from that of xenobiotics in food. It is impractical that a person intakes nutrient at a similar rate and in a similar manner during a lifetime. Persons are subject to a complicated homeostatic mechanism of nutrient substances, which varies with age, sex, and lifestage. The homeostatic mechanism could control or alter absorption, utilization, storage, and transport of the substance. Therefore, the long-term intake pattern characterized by considerable day-to-day and seasonal variation both among and within subpopulations was closest to the actual intake pattern.

For nutrient substances, ULs are a set of reference values for different subpopulations. This is mainly because of the differences in the physiological profile of age/sex/lifestage subpopulations. Physiological characteristics vary with age, sex, lifestage (e.g., pregnancy and lactation), disease state, and some specific occupations. This difference results in various intake–response relationships for the nutrient substance, and is sometimes reflected in the different manifestations of adverse health effects. For example, teratogenicity has been identified as the primary endpoint when setting UL of vitamin A for women of childbearing age, whereas liver abnormalities and hepatotoxicity were the main adverse health effects for all other adults. Moreover, ULs could be established for a short-term or acute effect in some occasions. It should be emphasized that ULs are a set of values to indicate the potential risk of excess nutrient intake, which means that it is not in itself an estimation of the magnitude of risk, nor a recommended intake.

### *9.2.2 Homeostasis of Nutrients*

Unlike non-nutrient substances, nutrient substances reflect a tendency to stabilize in the normal body states of the organism. The homeostasis for nutrient substances in vivo is achieved by a system of control mechanisms activated by negative feedback, which keeps the nutrient concentrations in a variable range under normal conditions. The homeostasis for nutrient is attributed to different factors, e.g., up−/down-regulation of enzyme systems, bio-accumulation/excretion of the storage organs, and increase/decrease in absorption rate in the gastrointestinal tract, so that the blood concentrations of nutrient substances do not change significantly with changes in intake. For instance, storage of vitamin A will increase in the liver when the storage capacity is exceeded.

Homeostatic adaptations can be triggered by low or high intakes of nutrients, and the responses may vary by different age/sex/lifestage. Nevertheless, the capacity of homeostatic adaptations is limited, the dynamic balance can be disturbed and corresponding risks will increase with inadequate or excessive intakes.

Figure 9.1 illustrates the dual curves of intake-response relationship for nutrient substances. The nature of nutrient substances and homeostatic mechanisms determine the dual risks. As shown in Fig. 9.1, the risk for nutrient substances is a

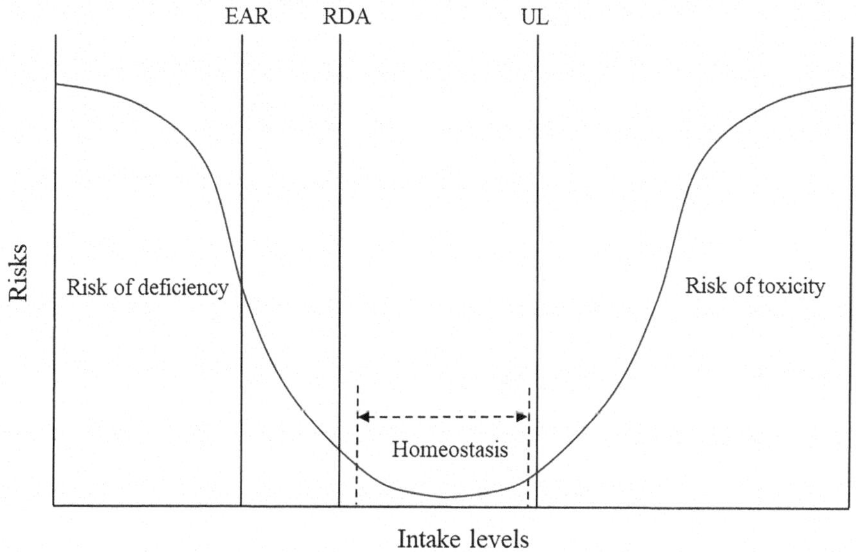


**Fig. 9.1** Intake-response curves and the relationships between DRIs of nutrients

"U-shaped" curve, in which the increases in risk are accompanied by increases in both inadequate and excessive intakes. In contrast, the classic risk characteristics for non-nutrient substances (e.g., contaminants and food additives) is an increasing function of exposure, i.e., risk increase with intake. It should be noted that this curve is not a symmetric curve as shown, as the risks for inadequate and excessive intakes are attributed to two different mechanisms and pathways. In fact, this curve is a combination of two different curves which have quite different shapes and degrees of steepness depending on the nutrient substance and the subpopulation. The range between the two curves is often referred to the "range of safe intake" due to homeostatic mechanisms.

### *9.2.3 Adverse Health Outcomes*

The adverse health effects result from the failure of homeostatic adaptations, which comprise various degrees of adverse effects of high intakes of the nutrient. As defined by most guidelines and literatures, adverse health effects include changes in morphology, physiology, growth, development, reproduction or life span of an organism, system, or (sub) population that results in an impairment of functional capacity, an impairment of the capacity to compensate for additional stress, an increase in susceptibility to other influences, and as highlighted in recent years, an adverse effect of the nutrient on the health benefits of another nutrient (i.e., an adverse nutrient-nutrient interaction).

Figure 9.2 lists the different degrees of adverse health effects. The nature of the adverse health effects is a developmental process and mainly includes the following aspects: (1) effects on normal physiological functions; (2) impairment of functional capacity; and (3) impairment of organs. An adverse health effect can range from biochemical effects without functional significance (e.g., enzyme activity) to clinical effects that irreversible impairment of organs (e.g., kidney stones). In assessment, adverse health effects should be considered separately for each adverse outcome. For example, the risks for reduction in bone mineral density, teratogenicity, hepatotoxicity, and chronic toxicity are main adverse endpoints when setting UL for Vitamin A for adults, which needs to be characterized and mapped separately for each endpoint.

### 9.2.4 Biomarkers

Biomarkers are regarded as substances, structures, or processes that can be measured in the body or its products. Because of the special nature of nutrient substances, the identification and utilization of proper biomarkers are main differences between traditional chemical risk assessment and the nutrient risk assessment. Generically, biomarkers were divided into three classes: biomarkers of exposure, biomarkers of susceptibility, and biomarkers of effect. The former two types of biomarkers are usually used in exposure assessment for non-nutrient substances, while the latter is relevant for the hazard identification and hazard characterization. For nutrient risk assessment, there are some differences in definition, scope, and application of these three biomarkers due to the nature of nutrient substances. For example, erythrocyte measures could be viewed as markers of status in nutrient risk assessment while they are usually served as markers of effects in chemical risk assessment. Examples of various types of biomarkers or indicators for nutrient risk assessment are presented in Fig. 9.2. Broadly, clinical surrogate endpoints, biomarkers, or risk factors are all measures that serve as the basis for estimating intake or indicating adverse endpoints in nutrient risk assessment [5].

Generally, "exposure" usually represents a passive process for hazardous substances while "intake" represents an initiative process for food. As a consequence, biomarker of intake commonly appears in the nutrition literature rather than biomarkers of exposure. Distinctions between "exposures" and "intake" were not made in this book because the distinction between these two terminologies is not clear for ULs. Biomarkers of intake/exposure for nutrient substances usually fall into five categories: (1) nutritional compound itself (e.g., vitamins, minerals, carotenoids); (2) integrative compound involving metabolic processing (i.e., metabolomics); (3) functional measures of nutritional status or enzyme saturation; (4) food contaminants (e.g., aflatoxin, polycyclic aromatic hydrocarbons, acrylamide). Food contaminants are less, but may be relevant or directly related to nutrients. In a broad sense, self-report measures (e.g., 24-h dietary recalls, food frequency questionnaire) are one type of biomarkers of intake, which may be subject to some types of random

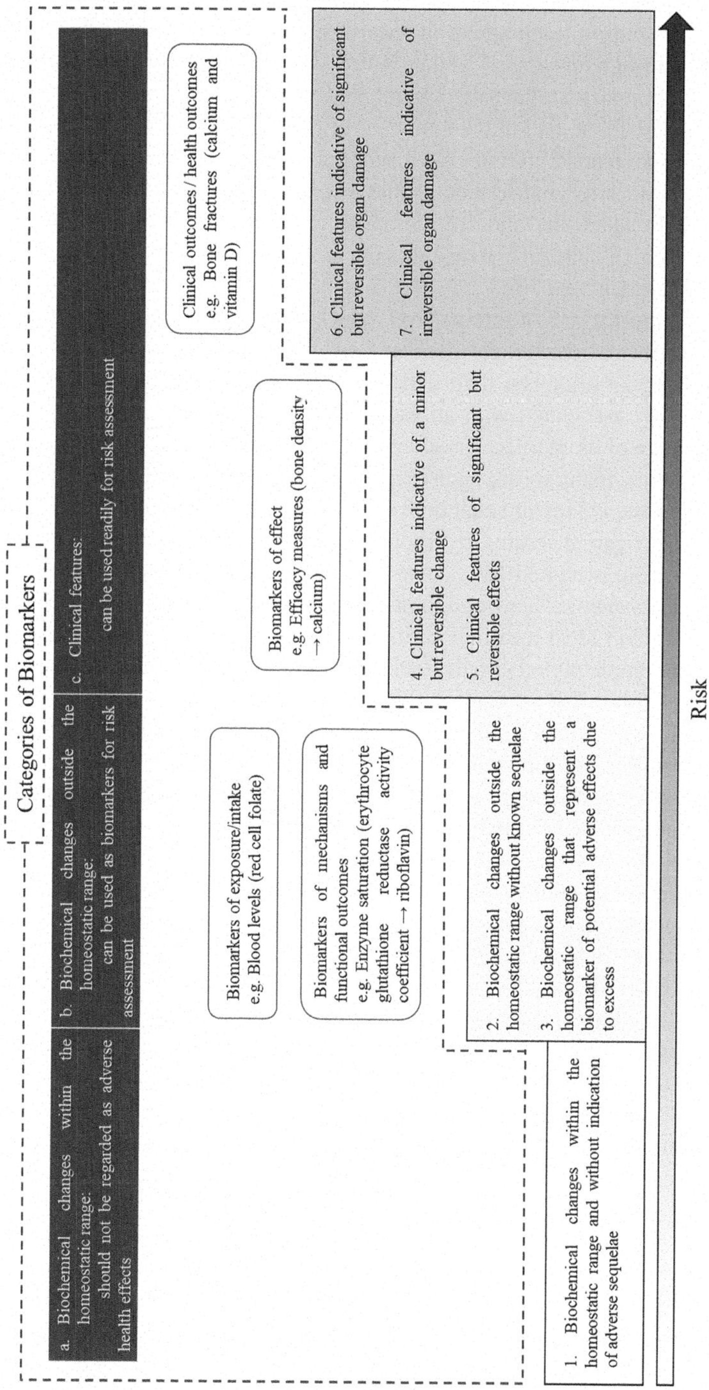


**Fig. 9.2** Sequence of adverse health effects and corresponding biomarkers in nutrient risk assessment

error or systematic bias such as recalls bias and underreporting in 24-h dietary recalls. For this reason, biomarkers of exposure may be viewed as more objective measures of nutrient substances intake. However, biomarkers of exposure are also not accurate enough to reflect dietary intake due to differential factors affecting absorption, metabolism, and utilization (e.g., ultraviolet exposure could lead to high concentration of serum 25(OH)D even without vitamin D intake). They are also subject to random error and limited by the testing methods. In the past few years, self-report calibrated with validated biomarkers of exposure represented an important methodological advance over-reliance on self-report alone to improve the accuracy of the intake data [6, 7].

In terms of biomarkers of effect, FAO/WHO defined the biomarkers of effect as "those that reflect a measurable biochemical, physiological, behavioral or other alteration within an organism that, depending upon the magnitude, can be recognized as causally associated with an established or possible health impairment or disease". A range of biomarkers for adverse effects could be used in risk assessment, which may include many forms, such as a physiological process (e.g., the glomerular filtration rate of the kidney or small tidal volume of lung), biochemical indexes (e.g., level of urea nitrogen in serum), psychological or cognitive functions (e.g., remembering nouns from a recited list), or an indicator of the presence of a disease. In classic risk assessment, "hard endpoints" (i.e., steps 4 through 7 and stage c in Fig. 9.2) are used as ideal measures when developing reference dose. However, it is impractical to completely rely on data related to the clinical manifestation of adverse health effects for nutrient substances given the state of existing data (especially for humans), and expectations for data likely to be available in the near future. This is mainly because that nutrition studies that explore associations between intakes and chronic diseases usually adopt prospective cohort studies, rather than randomized controlled trials (RCTs). Furthermore, the endpoints and the exposure of interest are subjected to various factors in nutrition studies.

As stated above, studies that employ RCTs have the greatest likelihood of establishing causation compared to observational study designs. However, using disease events as outcome measures may not always be feasible due to study expense, the rarity of the disease in question, time imperatives, or the complexity of diagnosis. In those situations, some studies may resort to surrogate markers. Biomarkers as surrogates for adverse health effects are highlighted as desirable for the purposes of nutrient risk assessment. A surrogate marker is a type of biomarker that "predicts clinical benefit (or harm or lack of benefit or harm) based on epidemiologic, therapeutic, pathophysiologic, or other scientific evidence. A surrogate disease marker is qualified for its intended purposes" [8]. Surrogate markers usually meet the following characteristics: (1) measures deemed to be on the causal pathways to illness, whose occurrence would often lead to that illness should more time and larger sample sizes be available; (2) the status of the surrogate marker, or its process history, should be able to explain a substantial portion of any relationship between the nutrients and the chronic disease risk; (3) surrogate marker may be any type of biomarker. One example of surrogate markers is in exploring the contribution of nutrition to cancer, colorectal adenomas were used as candidate surrogates because they occur earlier in the disease pathway and relatively frequently compared to the incidence of colorectal cancer. Surrogate markers could be considered with the goal of using the findings as supporting information of results based on the disease of interest. To be considered, surrogate markers should meet the qualification criteria for their purpose. Qualification of surrogate markers must be specific to each nutrient or other food substance, although some surrogates will be applicable to more than one causal pathway.

Besides, for all biomarkers, the guiding principle for selecting biomarkers for nutrient risk assessment is that they are feasible, valid, reproducible, sensitive, and specific.

## 9.3 Establishment of Tolerable Upper Limits of Nutrients

Risk assessment for scientific policy-making has been stated in many guidelines and literatures, and adopted broadly as a common paradigm for ULs by most authorities and international organizations. It results in a process that is flexible, transparent, and suitable for making decisions related to DRIs. Classic non-nutrient assessment usually consists of four general steps: (1) hazard identification; (2) hazard characterization; (3) exposure assessment; and (4) risk characterization. However, modifications to the classic non-nutrient risk assessment approach are needed for nutrient substances because, unlike non-nutrients, nutrient substances are biologically essential or have a favorable impact on health at specified levels of intake. The nutrient risk assessment framework generally involves four overarching steps comparable to that of non-nutrient risk assessment: (1) nutrition hazard identification and characterization; (2) dietary intake assessment; (3) nutrient risk characterization; and (4) discussion of implications and special concerns. Each step is described in detail in Sects 9.3.1.1–9.3.1.5. In reality, the process is typically not so straightforward because of data gaps and variation in the type and amount of evidence for each nutrient. However, because of their importance to health, establishing DRIs for nutrients has been considered necessary, regardless of the certainty in the evidence.

### *9.3.1 Nutrition Hazard Identification and Characterization*

In the process of classic risk assessment for non-nutrients, hazard identification and characterization are two separate steps. In nutrient risk assessment, however, they were merged into one single step for two reasons. For one thing, these processes in hazard identification and hazard characterization are closely inter-linked and iterative in nature. For another, they were both globally relevant steps, as data of the two steps are applicable to all populations, whereas in dietary assessment and nutrition risk characterization data are specific to subpopulation of differed dietary intakes. Thus, participants in the Joint FAO/WHO Nutrient Risk Assessment Workshop: A

Model for Establishing Upper Levels of Intake for Nutrients and Related Substances suggested to considered hazard identification and characterization as one step [3].

Nutrition hazard identification and characterization is comprised of the following four steps:

- Step1. Identifying adverse health effects associated with intake.
- Step2. Selecting the critical adverse health effect.
- Step3: Quantifying the upper level.
- Step4. Characterizing the hazard and identifying vulnerable subgroups.

This main body of this section is based on the above four steps. Before them, documentation that is clearly and comprehensively formed helps to develop a transparent and clarifies nutrient risk assessment process. This section is thus started from an iterative process of data search and evaluation.

#### 9.3.1.1 Preparation of Data: An Iterative Process

Studies specifically designed to assess the safety of nutrient substances and to characterize their hazards are insufficient. Thus, the primary evidence of nutrient risk assessment is from researches which investigate the benefits and potential mechanisms of nutrient substances. These evidences may, inevitably, lack key information that is important for nutrient risk assessment, data sets refining and combining are therefore generally involved in hazard identification and characterization process. To standardize data, scientists usually conduct an iterative process to obtain qualified evidence, in which additional or revised data searches are constantly performed.

The "evidence-based systematic review" (EBSR) is a systematic scientific review process which has been widely used in identifying research gap and developing clinical guidelines. It is usually characterized by a study question inquiring about the relationship between a specified intervention and hypothesized outcomes within a group of population. Based on the question, EBSR includes a detailed demonstration of search strategy and description and rating of the quality of data, methods, and results of searched evidence. Thus, EBSR usually generated comprehensive evidence to a study area.

Due to the comprehensive and clear nature of EBSR, it has been noted to practice in the process of nutrient risk assessment. Several components of an EBSR are very useful in developing nutrient risk assessment, the prior definition of the criteria of search strategy, the table formats of summarizing existing evidence, as well as the evaluation of the methodological quality of reviewed studies.

However, limitation is also worth noting in using of EBSR in nutrient risk assessment, which is primarily the inconsistent between the nature of outlined and practiced question addressed in EBSR and the scientific judgment that employed in nutrient risk assessment. A previous report from the U.S. Agency for Healthcare Research and Quality, Systems to Rate the Strength of Scientific Evidence [9] indicated that criteria such as "coherence" and questions of specificity and temporality are not applicable to EBSR, however, it is proper to be used in measuring risk. In addition, the key questions in nutrient risk assessment and EBSR are different. EBSR questions usually focused on the validity and effect size of a relationship in a subgroup, whereas questions in nutrient risk assessment consider what adverse health effects are associated with the intake of a nutrient and at what level do they happen. Accordingly, the nature of data important to nutrient risk assessment also differed from that to EBSR. Another vital inconsistent lies on the refinement and iterative of literature review in nutrient risk assessment, as unanticipated adverse health effects are expected to be revealed and subsequently triggers a new search. In EBSR, however, iterative search is not anticipated.

Overall, the FAO/WHO workshop [3] concludes that under three conditions the approach outlined for literature reviews should be adapted when employing EBSR in nutrient risk assessment: (1) open-ended starting question in nutrient risk assessment, (2) the need to merge a variety of data to form sufficient evidence to explore intake-response relationships and to reveal uncertainty factors, and (3) the need for refinement and iterative of review. In other cases, data preparation and dealing process in nutrient risk assessment would benefit from the clear description and documentation of search strategy and criteria and the rating of the quality of evidence in EBSR.

#### 9.3.1.2 Step 1: Identifying Adverse Health Effects Associated with Intake

To address the starting question: what are the adverse health effects associated with the intake of the nutrient substance, and what are the threshold values of intake do that happen? The first move of nutrition hazard identification and characterization should be a detailed documentation of previous evidence with objective rating of study and clear summary of results of the association between adverse health effects and nutrient intakes.

##### 9.3.1.2.1 Data Preparation

As mentioned above, studies specifically designed to explore the relationship between adverse health effects and nutrient substance intake is inadequate, thus, one major challenge is to find sufficient data to build the link between intake and response. Due to the limited human data failed to provide enough information on causal linkages, animal and in vitro data are needed to be combined to establish a causally associated link. This is because the link between nutrient substance and adverse health effects is not only a direct one, but also an indirect relationship via biomarkers, which makes it possible to complete human data with animal and in vitro data.

Notably, in using the combining data, the IPCS [10] proposed five critical aspects that needed to be taken into consideration when interpreting and combining study findings: (1) variability of the efficiency of intestinal uptake and transfer of the nutrient substance; (2) variability in intake and in metabolic states arising from age, sex, physiological conditions, and nutritional status; (3) person-to-person variability of unknown origin; (4) known genetic differences in the absorption and metabolism of the nutrient substance; and (5) short-term exposures during critical periods that greatly alter the risk and nature of adverse health effects. Information from the above aspects is desirable and is suggested to be collected from iterative data searches.

**Human Data** Ideally, human data that is strictly designed to investigate the association between nutrient and adverse health effects provide high-quality evidence for nutrient risk assessment; however, it is limited by ethical issues as the study substance might cause damage to human beings. Evidence from human data mainly includes two types of evidence: the experimental studies and observational surveys. Generally, experimental studies with study design suitable to address the key question in nutrient risk assessment provide reliable evidence, as they defined the intake levels of nutrient substance. Unfortunately, this type of evidence is in absence. Observational studies such as surveys and case-control studies can provide less reliable evidence, as the intakes are not controlled and lacking specific intake data as well as the commonly existed reporting bias.

**Animal and In Vitro Data** When properly conducted, animal and in vitro studies can be very helpful for nutrient risk assessment. For instance, experiments that were designed based on relevant guidelines and standards, such as the Organisation for Economic Co-operation and Development guidelines and China National Standards for Food Safety [11, 12] can provide important evidence. However, most exploration in animal and in vitro studies are not primarily designed to investigate the relationship between nutrient intake level and adverse health effects. Even designed for the relevant purpose, evidence provided from animal and in vitro studies also need careful screening. For example, if irrational intake levels such as very high amounts are set in experiment, exploiting to lower levels and potential interfere causing to the gastrointestinal absorption of another nutrient both could be challenges to nutrient risk assessment. Despite these difficulties, these experiments can also provide useful information to risk assessment.

#### 9.3.1.2.2 Identification and Selection of Data

Due to the iterative nature of data search in nutrient risk assessment, search criteria should be clearly defined during the whole period of search. Depending on differed conditions on the clarity of studying adverse health effects, the data search strategy can be conducted as follows: If relevant adverse health effects are clearly defined in authoritative reports, search strategy may concentrate on identifying more recent studies and on obtaining clarifying information to improve characterization of the hazard and to reduce uncertainties. If definitions of studying adverse outcomes are not clear, then the search could focus on relevant outcomes, biomarkers of effect, and measures to supplement the information.

After ensuring the search strategy, it can be moved to the step of data search. The latter started from recently published reviews with acceptable quality and scope. Later, information in the above reviews would be updated by searching for more recent data. If there is no existing review, broad-based searches are needed. Furthermore, inclusion criteria and approaches to weighting literatures are needed. Although they have not been clearly defined, attempts in developing criteria to define the rules of admissible evidence to address the assessment questions have been conducted. Guidelines proposed by EFSA [13] summarized some useful methods in this area, but guidance on study inclusion and exclusion criteria still requires more consideration.

#### 9.3.1.2.3 Initial Review of Data

**Rating Data Quality** Rating data quality is critical for nutrition hazard identification, which requires to evaluate and assign a rating to the study design of searched literature, rather than assess the relative validity of studies of differed design. It is difficult to rating data quality. Although quality scales are the commonly used method to assess adverse effects in the area such as drug safety, it owns defects such as subjective weights may cause contrary conclusions when using different scales.

When rating the quality of data, the criteria should be specific to the study design. For instance, rating the data from experimental or intervention studies considers the number of drop-outs, type of blinding, accuracy of measurement, and so on. Things are different in observational studies. They could be considered including if the number of participants meets the minimal inclusion criteria, as observational studies are also in absence. Once included, evaluating methodological quality is important.

Since the quality of evidence consists of various aspects, giving individual components to a study is more useful than scoring by a single metric. Furthermore, employing one single scale for all evidence causes potential problems. For example, randomized controlled trials are more reliable than cohort studies, but what if the randomized controlled trials are seriously flawed? In that case, results from cohort studies are less biased.

Additionally, giving a grade, e.g., A, B, or C, to each study based on some rating systems is also a useful way to rate the quality of the study. For example, studies with the least bias and valid results rated as grade A, studies that are susceptible to some bias but not sufficient to invalidate the results rated as grade B, and studies that significant bias that may invalidate the results rated as grade C. However, it can't completely substitute the way that multi-components. The choice of way of rating the quality of the study is based on specific cases. Notably, the above approaches can be used in rating the quality of animal and in vitro evidence. However, other differences between these studies and human exposure are need to concern, such as the way of intake, the studying effects to human. These issues should be reflected in the rating.

**Meta-Analysis** Meta-analysis is increasingly used in hazard identification and characterization. In hazard identification, it can be helpful to reveal candidate adverse health effects by combing data from different studies that have the same study design and similar key questions, when adverse health effects are lack of clear definition from the authorized report. In hazard characterization, due to the quantitative nature of results of meta-analysis, it can properly synthesize literatures and thus provide robust evidence on intake-response relationships. Previous studies have clearly discussed the employment of meta-analysis in nutrient risk assessment [14, 15]. Before deciding to use meta-analysis, its potential mechanism must be considered.

**Uncertainty** Uncertainty is an important concern in nutrient risk assessment. IPCS defined it as "imperfect knowledge concerning the present or future state of an organism, system or (sub)population under consideration" [4]. Notably, uncertainty is not variability. Uncertainty may be caused from data deficiencies and variability in nutrient risk assessment. Additionally, data uncertainties are more noteworthy for nutrient substances than that for non-nutrient substances. In non-nutrient uncertainties, conservative uncertainty factors are employed as standard approaches [16], which may not be appropriate for many nutrients, as ULs should not fall below the recommended intakes of nutrients. Thus, nutrient hazard identification is in vital important for uncertainty and approaches to deal with it. Detailed discussion on this point is displayed in Sect. 9.3.1.4.

The use of a strictly structured method of hazard identification, which is based on a comprehensive search of data from all sources, containing appropriate evidence ranking, and including a detailed explanation of the judgements made in identifying hazards, can reduce uncertainties with the potential to affect hazard identification. However, such a method is largely limited by the absence of literatures.

#### 9.3.1.2.4 Summarizing and Presenting Results

Identifying adverse health effects is the basis of the establishment of ULs and subsequently hazard characterization. Risk assessors are suggested to provide clear and detailed results on candidate adverse health effects in summary form. The summary form contained all relevant information on the process of identifying, which at least presents information on the following: Subjects' age, sex, health, race (species and strain in animal studies); Size of study; Nature and characteristics of studied nutrient substance; Range of intakes; Duration of intakes; Background diet and intakes from (as applicable) food, supplements, and water; Intake assessment method(s); Endpoints investigated; Relationship between intake and response (i.e., adverse health effect); Nature of critical adverse health effect (validation and quality criteria for the selected endpoint, i.e., biomarker of effect or clinically observable effect) and why selected; Effect size (relationship with intake, subgroups, other factors); Confounders (e.g., susceptibility, use of medications) and effect modifiers.

Summary table was recommended to organize and present evidence, as it can orderly present detailed information on findings from available studies and intake levels associated with adverse health outcomes, therefore helping to specify the critical adverse health effect. Besides, the rating on the quality of available evidence is also suggested to present by tables. The key information for a summary of quality assessments could include the number of subjects, intake level and duration, background diet, effect assessment method and the study quality score.

#### 9.3.1.3 Step 2: Selecting the Critical Adverse Health Effect

In order to build a UL, a critical adverse health effect should be selected. IPCS defined that critical effect is the adverse health effects judged to be most appropriate for deriving the ULs [4]. The intent is to provide public health protection by maximizing the protection of the population. In practice, it is usually associated with the effect of the lowest intake within investigated intake range. Although other factors of judging effect, such as seriousness and reversibility, are not considered in the light of selecting critical adverse health effect, they may have an impact on uncertainty and thereby influence the value of ULs.

After initial selection of the critical adverse health effects, additional data searches are needed to complete the data related to intake-response relationship and information that lacked in initial identification. If necessary, different critical adverse health effects could be selected for subpopulations of specific age or sex. In addition. If several candidate adverse health effects all provide the desired level of protection, they could be all selected as critical adverse health effects, therefore, producing a set of candidate ULs, and through further carefully selection to choose the final critical health effects.

#### 9.3.1.4 Step 3: Quantifying the Upper Level

After identifying the critical adverse health effect, it can be moved to establish the ULs. Four steps are needed in quantifying ULs. The first step is called the intake-response assessment, which means to analyze and present the association between intake level of nutrient and the occurrence of critical adverse health effect. Then followed by the specification of the outcomes of intake-response assessment: benchmark dose (BMD), no observed adverse effect level (NOAEL), or the lowest observed adverse effect level (LOAEL). Their definitions are:

- Benchmark dose (BMD): the intake of a substance that is expected to result in a prespecified level of effect. The abbreviation BMD is used for this chapter, other risk assessments sometimes refer to this value as benchmark intake.
- No observed adverse effect level (NOAEL): the greatest concentration or amount of a substance, found by experiment or observation, which causes no detectable

adverse alteration of morphology, functional capacity, growth, development, or life span of the target organism under defined conditions of exposure [14];

- Lowest observed adverse effect level (LOAEL): the lowest concentration or amount of a substance, found by experiment or observation, which causes a detectable adverse alteration of morphology, functional capacity, growth, development, or life span of the target organism under defined conditions of exposure [4].

The third step is to deal with uncertainties and establish a UL, and finally if needed, risk assessors can make adjustment of ULs for unstudied subpopulations of differed age, sex, or lifestage.

#### 9.3.1.4.1 Intake-Response Assessment

The intake-response assessment is the key process for risk assessment. Unlike intake-response assessment in dietary intake assessment, the aim of that in nutrient risk assessment is to characterize the relationship between intake of nutrient substance and the critical adverse health effect. Notably, not all nutrient substances can specify intake-response curve. Varied types of data including human, animal and in vitro data can be used in assessment, sometimes data from the most sensitive animal can provide important input to nutrient risk assessment, as human data is inadequate.

Not only the quantitative intake data of nutrient but also the biomarkers of the internal amount of that nutrient can be considered as intake level. The former value is the widely used measure of intake or "exposure", as to date, biomarkers generated for specific nutrient are not well understand. However, difference existed between the quantitative intake data and the "true intake" of nutrient substance into body, such as intestinal uptake and bio-chemical transfer. The WHO report recommended five key considerations for the assessment of intake-response [3], including the level and duration of intake, nature and size of studied population, method of estimating intake level, and use of meta-analysis. But basically, detailed documentation of information about all aspects of searched and selected data is the premise of practicing nutrient risk assessment.

#### 9.3.1.4.2 Specification of BMD, NOAEL, and LOAEL

Given that the resources and practices in developing BMDs remained to be explored, to date, practices on establishing ULs focus primarily on the specification of NOAEL and LOAEL. However, ULs that derived from BMD have greater certainty, as neither NOAEL nor LOAEL considered the shape of intake-response from intake levels that have not been explored in searched studies.

Since 1987, NOAELs and LOAELs have been put into use in risk assessment. As NOAEL is an intake below the level at the onset of adverse health effect, it is a proxy of the biological threshold. However, in practice, it is possible that the level of NOAEL is prominently lower than the biological threshold [4]. The LOAEL, on the other hand, is the lowest intake level with measured response, it may be less accurate in indicating biological threshold than NOAEL. However, challenges still existed in estimating LOAELs and LOAELs due to insufficient available data related to the relationship between nutrient intake and adverse health effect. Besides, in synthesizing existing studies, quality issues such as the size of study, sensitivity if measurement, duration and selection of intake levels in human data, as well as species, strain, sex, age, and developmental status in animal data needed careful consideration.

Further efforts on developing BMD are desired, especially when NOAEL cannot be identified such as nutrient substance like sodium. Due to the lack of data, mathematical modelling can be used to estimate a BI, which requires studies with various intake level and graded responses. In that case, it would be proper to employ regression function to estimate the intake level at the occurrence of specific effect, that is, the BI. The 95% confidence lower bound on the intake (BMDL) would then be used to calculate statistical uncertainties.

#### 9.3.1.4.3 Uncertainties and UL

Unless in studies that are fully representative of the exposed population and without uncertainties and errors, in most occasions, BMD, NOAEL or LOAEL cannot be considered as ULs. To establish the ULs, they need to be adjusted for uncertainties.

**Quantitative Adjustment** The first move is quantitative adjustment for data, which is using objective data to build factors that is the premise of adjustment of NOAEL or LOAEL. Before quantitative adjustment, risk assessors need to consider the sufficiency of available data. In fact, available data is usually adequate to allow few quantitative adjustments for developing ULs. Once the data is sufficient, it can be used to determine uncertainties in difference in body size between humans and animals, bioavailability for multi-forms of a nutrient substance, such as nicotinic acid and nicotinamide.

**Uncertainty Factors** Furthermore, uncertainty factors are needed. They can be used to address the interspecies adjustments, inter-individual variability in humans, inadequacy of the data, and the nature of the adverse health effects to ensure that levels under ULs are harmless.

Uncertainty factors are commonly used in the risk assessment of non-nutrient substances, it usually defined a conservative value, such as ten-fold corrections for species differences and ten-fold corrections for human variability, producing a 100-fold uncertainty factor. However, although the use of this widely recognized and conservative value could provide enough assurance on protection, it may cause a UL value lower than the recommended intake of some nutrient substances such as iron, zinc, copper, etc. Thus, in determining the uncertainty factors for nutrient substance, the established intake requirements need to be considered.

The following listed several conditions that uncertainty factors cannot be quantitatively adjusted [3]:

- Human variability: traditionally a factor of 10 has been used for a NOAEL or LOAEL that is based on a small study in healthy volunteers; however, a factor of 10 may be too high in the case of specific nutrients in more sensitive subpopulation (e.g., vitamin A in fertile women, vitamin D in adults and young children).
- Interspecies differences: typically associated with absorption, metabolism, and/or excretion nature of uncertainty factor. Scaling according to metabolic body weight ($BW^{0.75}$) is more suitable for nutrient substances than is the ten-fold default uncertainty factor associated with non-nutrients.
- Use of a LOAEL rather than a NOAEL: An uncertainty factor may be necessary to account for the differences introduced using a LOAEL rather than a NOAEL. Usually, a value between 1 and 10 would be used. If there is sufficient data on the intake-response data, the estimation of the BMD or BMDL would be proper.
- Short duration of study: study duration might be too short to show the adverse health effect, using an uncertainty factor for this purpose may be particularly important for nutrient substances with storage mechanisms.
- Inferior quality of the study: in this case the magnitude of the factor should reflect the extent to which the study data may have over-estimated the true threshold.
- Specific population groups: “at risk” groups—for example, subgroups with phenylketonuria or those with different nutritional status than the studied subpopulation—may require separate consideration and separate values.

**Composite Uncertainty Factor** Taking uncertainties for all aspects into consideration, a set of uncertainty factors are developed, thus, a composite uncertainty factor should be built. The likelihood of obtaining a UL lower than nutrient requirement is lower when using a composite uncertainty factor, comparing with an individual uncertainty factor derived from NOAEL or LOAEL, as it considered both the toxicity and essentiality of nutrient. The following are the steps recommended to derive a composite uncertainty factor:

- Identify uncertainties that are not addressed by quantitative adjustment.
- Rate above uncertainties by expected impact level on the total uncertainty.
- Build a list or table of rankings including the corresponding rationale.
- Incorporate the above uncertainties into a composite uncertainty factor based on the ranking.

A detailed explanation is needed to be recorded in each step.

###### 9.3.1.4.4 Adjustment of ULs for Unstudied Subpopulations of Different Age, Sex or Lifestage

Data on the relationship between nutrient substance and adverse health effects are scarce, not to mention data stratifying by age, sex, or lifestage of subpopulation. To complement ULs in subpopulations, the practical way is to adjust ULs calculated from relatively sufficient data on “normal” adults. There are three ways to adjust adult ULs to estimate ULs relevant to children:

- Adjusted by reference body weight for age group, the formula is:

$$(UL_{child}) = (UL_{adult})\ (weight_{child}/weight_{adult})$$

This method does not consider intermediary metabolic rates, energy intake, and basal metabolic rate, thus ULs speculated by this way may be lower than that extrapolated based on body surface area and energy requirement. However, these issues inevitably exist in nutrient risk assessment, thus the following two methods may more appropriate. Notably, with increasing age, the difference in ratios among the adjusting methods becomes smaller.

- Adjusted by body surface area, the formula is:

$$(UL_{child}) = (UL_{adult})\ (weight_{child}/weight_{adult})^{0.66}$$

- Adjusted by energy requirement, and sometimes referred as metabolic body weight, the formula is:

$$(UL_{child}) = (UL_{adult})\ (weight_{child}/weight_{adult})^{0.75}$$

Adjustment of the ULs by basal metabolic rate appears to be a more appropriate way compared to that adjusted by body weight, as the turnover of energy may mostly parallel to nutrient substance. However, it also has several limitations. This approach neither considered the efficiency of nutrient absorption and elimination nor the variance in metabolism and body composition during growth. Besides, it may not be appropriate for pregnant women and sometimes the elderly population. As they are in special lifestage, the basal metabolic rate may not parallel to body weight.

#### 9.3.1.5 Step 4: Characterizing the Hazard and Identifying Vulnerable Subgroups

The final step of nutrient hazard identification and characterization is to characterize the hazard of studying nutrient substance through a summary of conclusions and related arguments, as well as identifying vulnerable groups and other pertinent information. It can be formed by table, narrative, or comprehensively display conclusions by combing both forms. In hazard characterization, an overview of available data and uncertainties, including severity and reversibility, genetic diversity, vulnerabilities and relevance of ULs to vulnerable subgroups, are displayed.

## *9.3.2 Dietary Intake Assessment*

Dietary intake assessment is the process of exposure assessment in risk assessment for non-nutrients. This section starts from giving an overview of definitions, principles and harmonization of the method of dietary intake assessment, then moves to a six-step approach to harmonize the approach of dietary intake assessment.

### 9.3.2.1 Overview

#### 9.3.2.1.1 Definitions

The definitions of terms in this section varied in regions, thus, the WHO report [3] unified the definition of major terminology of dietary intake assessment as follows.

- Dietary intake is the quantitative amount of the nutrient substance ingested from sources that generally include foods (and beverages), fortified foods, especially formulated foods (sometimes called functional foods), dietary/food supplements, water, and other non-drug products such as botanicals and plant extracts. Dietary intake refers to the ingested amounts of the nutrient substance obtained through dietary consumption.
- Dietary consumption refers to the amounts of products (e.g., food, supplements, water) consumed that provide nutrient substances.
- Habitual intake refers to the long-term average daily intake of the nutrient substance, consistent with the term “usual intake”.

#### 9.3.2.1.2 Objectives and Key Principles

Dietary risk assessment provides information on the quantitative estimate of intake of nutrient, which makes it possible to evaluate the proportion of population whose intake level of specific nutrient is higher than the UL. Combining with nutrient hazard identification and characterization, they build the basis of nutrient risk characterization.

Concisely, dietary intake assessment is to calculate intake of specific nutrient intake from composition data and consumption data, further to compare calculated intake to the UL. To improve the quality and credibility of dietary intake assessment, many principles are proposed. The WHO recommended six principles for dietary intake assessment, they are:

- Be conducted and presented in a manner that answers the questions posed by the risk manager.
- Be based on a distribution of habitual intakes when possible.
- Be developed to reflect the age/sex/lifestage groupings for which the upper-level intake has been established;

- Take into account the uncertainties associated with estimating intakes of nutrient substances.
- Use reasonable strategies to produce needed estimates when data are not available.
- Contain full documentation for all aspects of the intake assessment.

Inevitably, dietary intakes assessment add uncertainty to nutrient risk assessment, resulting from the validity of method of measurement, statistical process of estimating, quality of composition data, as well as the difference between food intake and nutrient that pass through the biochemical process in the body.

##### 9.3.2.1.3 Harmonization of Methods for Dietary Intake Assessment

The method of dietary intake assessment varied worldwide. For instance, European Food Safety Authority–Scientific Committee on Food (EFSA SCF) collect consumption data by 2-day, 7-day, and 8-day dietary records; 24-hr dietary intake recalls; and household surveys. They report the mean/median intake and the 97.5th percentile of intake by males and females, respectively. Expert Group on Vitamins and Minerals (EVM) collect data by weighing 4-day or 7-day records from a 1986–1987 national nutrition survey, using the manufacturers' information on the nutrient content of supplements available in the United Kingdom to estimate nutrient intake, they additionally identified subpopulations with potential high intakes. Institute of Medicine (IOM), on the other hand, collect dietary information based on 24-h recalls and quantitative assessments of water and supplement intakes from nationally-representative U.S. surveys and provincial survey from Canada, and estimated the distributions of habitual nutrient intakes for specific age, sex, and lifestage subpopulations and compare it to their ULs.

To improve the comparability of dietary intake assessment among differed regions, the FAO/WHO workshop [3] developed a six-step approach to harmonize that, which is:

- Specifying the type of dietary intake assessment.
- Using composition data.
- Using consumption data.
- Estimating intake.
- Dealing with uncertainties.
- Reporting the dietary intake assessment.

The following sections described the above approach.

#### 9.3.2.2 Specifying the Type of Dietary Intake Assessment

It is important to specify the type of dietary intake (total or targeted) and the time frame of assessment of dietary intake. Generally, the total dietary intake from all sources during a long period would be appropriate, but in practice it is difficult to obtain such data.

##### 9.3.2.2.1 Specifying the Type of Dietary Intake

Upon establish a UL for a specific nutrient substance, to assess the nutrient risk for a (sub)population, risk assessors need to consider the total intake of the nutrient substance in that (sub)population. Nutrient from all sources, such as foods and beverages, supplements, functional foods, etc., should be estimated; however, due to the nature of dietary intake assessment, it is usually hard to obtain enough information on dietary intakes to assess total dietary intake.

In some cases, nutrient substance may be primarily obtained from a particular source, such as supplements, that is, targeted dietary intake. Thus, before nutrient risk assessment, adequate background information of studying nutrient substance and (sub)population are required to comprehensively assess dietary intakes.

##### 9.3.2.2.2 Specifying the Time Frame of Dietary Intake

Mechanisms of nutrient substances varied with age, sex, and lifestage; meanwhile, adverse health effects also alter with differed characteristics of subpopulation. Thus, it is necessary to predetermine the time frame of dietary intake.

Habitual intake is defined as the long-term average daily intake of the nutrient substance. Theoretically, it is calculated from the mean intake over a large number of days at different times of a year for every individual in the studying population; however, in practice, it is usually calculated by data from a representative sample of individuals during a shorter period of time.

At times, the studying adverse health effect may occur during an acute intake of a nutrient substance. In these conditions, time frame of dietary intake can be confined to the time period associated with the development of adverse health effect, and considered case by case.

#### 9.3.2.3 Composition Data

##### 9.3.2.3.1 Source of Data

The composition data describes the amount of nutrient substance in specific amounts of dietary items. It is an increasing challenge to obtain composition data accurate enough for dietary intake assessment, due to the variance in regions, food supply, new source of food developed by advanced food science, government involvement, and so on.

There are mainly two ways to formed composition data: the most accurate way, chemical analysis, and the indirect way, calculation from known composition data.

Chemical analysis acquired the most accurate composition data only if it is built on sufficient analysis from large amounts of samples, however, it has limitations like time-consuming and expensive. The calculations, on the other hand, can supply composition data on multi-ingredient food, by synthesizing the composition data from ingredients based on respective proportion. Thus, the recipe of this food is in a vital role. More defects exist in this way of generating composition data, such as the frequently updating recipe, the mismatch between real food and recipe, variance of receipt among individuals. Pre-packed food may to some extent reduce these defects; however, it also can introduce additional uncertainties. For example, the food label is under control of local regulations, in the US the manufacturers tend to label the low end of the distribution of content whereas in the UK manufacturers list the average content, and in some cases, considering the decay of nutrient over time, manufactures labelled higher level of nutrient comparing with content.

Furthermore, to obtain accurate and sufficient dietary data, the quality of composition also deserves consideration. A desirable composition data should be complete, comprehensive, and detailed; accurate, current, and in accordance with standard international nomenclature; well documented and available in electronic form. Existed composition data cannot reach these goals, which required that in dietary intake assessment, consideration of uncertainties is noted.

#### 9.3.2.3.2 Modifying and Adjusting Composition Data

To get composition data of enough information and accuracy, modifying and adjusting existed composition data is desired. The major process for modifying and adjusting composition data is adding missing data, updating information, and adjusting for bioavailability. Details are described below.

**Adding Missing Data** In the region of interest, if existing composition data is insufficient, data set that which best matches to local dietary intakes of residents should be chosen to be the basis of worked composition data. Additional items added to the composition data should be documented clearly and completely. Ideally, additions are best to be obtained from chemical analysis; however, in most cases, information acquired from other ways, such as acquiring data from other sources, contacting manufacturers for detailed labels, replacing data with similar foods, and calculating from recipes, is used.

**Updating Information** If there is any updating information regarding dietary item that already existed in composition data, commonly seen in fortification food, update is needed to promote the accuracy of dietary intake assessment. For single food, it should be updated item-by-item; however, for mixed food with ingredients fortified, updating should be focused on ingredient basis, then re-calculated the overall nutrient content for the mixed food. The IOM report Dietary Reference Intakes: Applications in Dietary Planning [17] practiced this approach.

**Adjusting for Bioavailability** Nutrient substance that is absorbed and ingested by the human body is a growing field. With increasing evidence, it may be possible to adjust for the bioavailability of absorption of the specific nutrient substance. In doing this, information on the correction factor of studying nutrient substance or specific form of it, the proportion of the total nutrient substance (form) per dietary item, and if necessary, the amount of potentially competing substances is needed.

#### 9.3.2.4 Consumption Data

Apart from composition data, consumption data is another basis of dietary intake assessment. There are two types of resources for consumption data, the individual-reported data and the aggregated data. Generally, individual data from the representative sample is appropriate to assess intakes, when these data is unavailable, other types of data are used after adjustment, even if the latter introduce uncertainties into assessment.

Ideal consumption data should include characteristics such as unbiased estimates of representative consumption data of population if interest, sufficient information on dietary items, data from subgroups of differed age, sex, or lifestage, detailed documentation, and providing electronic format. However, in practice, it is not widely available, other types of data are used by making adjustments and considering corresponding uncertainties. Despite all, detailed documents are highly required in all types of data to supply enough background information to the scientific judgement for risk assessors.

##### 9.3.2.4.1 Data on Individuals

Data on individuals are usually provided by routine conducted nationwide surveys, and can provide relatively accurate and sufficient information of a large sample. Methods that collect individual data include 24-h dietary recalls, weighted dietary intakes, food diaries, as well as food frequency questionnaires. Details on these methods can be found in numerous publications [5, 18], they have respective strengths and limitations, thus, the proper way taken to collect dietary intake data is carefully considered based on specific situations. Multiple days of 24-h dietary recalls and dietary records can provide detailed data to estimate the distribution of intakes of the population of interest, but they tend to underestimate energy and protein intake. Besides, time frame of collecting dietary data by dietary recalls and records are limited to short-period due to the time-consuming nature of these methods, they thus may neglect intermittent consumption information, and can be less representative of the habitual intake of a population. In that case, food frequency questionnaire may be more appropriate, even if it lack accuracy on collecting dietary information.

#### 9.3.2.4.2 Aggregated Consumption Data

Aggregated consumption data includes household data, nationwide data and regional-wide data, regional diets, and marketing and sales data.

The aggregated data on the household level may involve information on purchasing, food inventories, food disappearance on the basis of the household. This type of data cannot reflect dietary intakes of family members, residual of foods, and food that intake outside the home. It can estimate the nutrient intake of individual by adjusting household data with information on the household demographics.

The aggregated data at the national or regional levels are data that develop to reflect the availability of food of a nation or a region. For instance, the FAO Food Balance Sheets [19]. The food balance sheet displays the source and use of both primary and processed commodities during a specific time in a nation or region, it addresses the production and consumption of foods. Combining data on the population consuming the food, food composition factors can be used to estimate the energy and nutrient provided by these foods.

Food availability and dietary patterns of a region can be used to establish the representative diets of that region. The representative diet can be useful when more detailed data is not available. The Global Environmental Monitoring System proposed five regional diets: African, European, Far Eastern, Latin American and Middle Eastern diet. They are used to assess the long-term dietary intake of the specific nutrient substance of a region at the international level.

Marketing and sales data, at times, may be useful in filling data gaps and assessing the availability of nutrient in some degree. However, such data is highly aggregated, and is likely to overestimate intake, uncertainties regarding them are worth noting.

#### 9.3.2.4.3 Combining Consumption Data to Estimate Intake from all Sources

When using consumption data of different sources, it is necessary to integrate them. To combine consumption data from all sources, in particular, it requires to add the information on dietary supplement intake to a major database that reflects the consumption of foods and fortified foods, due to the increasing use of supplements. At times, when information of supplements intakes is missing, either to individual consumptions or to selected percentiles of the estimated habitual nutrient intake distribution, can be used to substitute the consumption of supplements in the population of interest.

*Analyses are needed to examine selected methods of combining data on food and supplement consumption* An analysis of the consumption data from the third U.-S. National Health and Nutrition Examination Survey (NHANES III) [20] explored the effect of various methods of estimated supplements consumption on the distribution of total nutrient intake. Basically, two methods are used to combine consumption data on food and supplements: method based on combining individual data on nutrient intake from food and from supplements, and method based on combining individual data on nutrient intakes and aggregated data regarding supplement intake. Generally, the former method is suggested, however in practice, the latter method is, at times, employed.

If there is no available consumption data on supplements, the following process may estimate the effect: estimating of the distribution of available doses on available foods, then estimating habitual nutrient intake distribution by various amounts corresponding to selected percentiles of the dose distribution, and then to compare the results of two former methods to assess the effect of lacking consumption data on supplements. In most cases, since not every participant in a population consumes supplements, using the habitual nutrient intake distribution by fixed amount as a substitute may overestimate the consumption. Furthermore, the supplement consumers also tend to intake higher amount of nutrients, and thus to result in a left tail distribution of total nutrient intake but a similar median intake to those of the habitual nutrient intake distribution that is based only on food consumption.

**Acceptable Approaches to Combining Consumption Data** Problems like overestimation may exist when adding the median supplement dose to individual intake from food sources. It would be useful to collect information on the proportion of consuming supplements in the population of interest. The approaches recommended to combine consumption data to estimate intake are described below.

If the information on consumption data from all sources is available. It would be appropriate to assess the total nutrient intake by adjusting the intake from food sources and adding the self-reported habitual intake from supplements in the food frequency questionnaire. If both consumptions from both sources are collected by food frequency questionnaires, then a habitual daily intake is derived from each questionnaire, and then the two estimated habitual daily intakes are added to evaluate the total intake.

However, if consumption data of nutrient is obtained from different subpopulations, other ways are proposed. A conservative approach is to add to individual intake from food sources with overestimation of individual supplement intake such as the 90th percentile of the distribution in the supplement marketing data. A less conservative approach is to add the median or a representative dose with the greatest sales. Besides, if information on the supplement consumption is available, aggregating different amounts of nutrient intake from the supplement source to corresponding subgroups of population may reduce overestimates of intakes among non-users.

#### 9.3.2.4.4 Strategies to Obtain Additional Consumption Data

In addition to the approach of using marking and sales data to complement consumption data, another practical way is to conduct a small-scale survey on the population of interest. Food frequency questionnaires are most useful in such surveys. The following way may increase the accuracy of estimation: conduct a short time-frame survey to reduce recall bias, conduct repeated surveys to reduce the effect of changing season, ask about the minimum items that focus on lacking sources. Although the accuracy of estimating food intakes by food frequency questionnaire is undoubted, it is still more reliable than related information derived from national or regional food availability data.

#### 9.3.2.5 Intake Estimate

##### 9.3.2.5.1 Based on Individual Data

High-quality composition data as well as consumption data from individual source, especially data collected by 24-h recalls or food records provide the most reliable basis for estimating dietary intakes. The purpose of this process is to evaluate the proportion of individuals whose intakes are higher (or lower) than the UL, and to assess standard errors for those estimates taking generate from the survey design.

Statistically, adjustment is an indispensable component for intake estimates, as potential confounding may exist even with the best study design. The day-to-day variability is an important potential confounding to be adjusted, which is highly relevant to the estimated upper percentiles of intake distribution as well as assess intakes for nutrient varied greatly among days. Using data from 24-h recalls or multiple-day records are two ways that may address the day-to-day variance. They rely on a simple measurement error model that evaluated daily intakes of an individual deviating from habitual intakes.

If potential confounders are not well adjusted, distributions of total daily intake may have long tails, resulting in bias on the estimation of the proportion of population with higher/lower intake level than UL. Besides, bias that generated from nutrient takes during conservative days is also worth noting.

##### 9.3.2.5.2 Based on Other Types of Data

**Household Data** At times, when data on individual level of intake is unavailable, household data may be useful, such as food accounts, list-recalls inventories (or disappearance), and household food records data. In this case, the aim is to get per-capita data on the food consumption of household members. To disaggregate household-level data into individual-level data, the following process is called:

During survey, household member records the eating occasions and the number of meal participants (including visitors) on each occasion. Then, based on the above information, risk assessor calculates four kinds of data to obtain the distribution of per-capita intake of a nutrient substance as the distribution of mean per capita intakes over households in the survey: the consumption per person and per meal of each item during the survey period, the total amount of the nutrient consumed during the survey period divided by the number of person-meals, the daily intake for each person, and when information on age, sex, and weight of each houschold member is available, the per-capita intake according to the average caloric requirements of a household member.

This approach has limitations such as the requirement for some degree of education level of at least one household member, and impossible to account for the differential allocation of foods within the household. If possible, as mentioned above, prorates the calculated overall per-capita intake of nutrient based on age, sex, weight, or even role in family of household of interest may provide more accurate information.

**National or Regional Availability Data** When the above types of consumption data are unavailable, the national balance sheets or regional diets may be the best source of consumption data, and to estimate the habitual intake distribution from mean per-capita intake of that nation or region based on strict assumptions, which could cause large inaccuracy and uncertainty to nutrition intake assessment, and further to the results of nutrient risk assessment.

The assumption is usually hard to practice, such as an assumption of normally distributed per-capita intakes in studying population and a coefficient of variation (CV) borrowing from other known population. In that case the standard deviation (SD) of intake could be computed. However, for most nutrient substances, rather than normally distributed, their intake distributions are skewed, then a more complicated approach based on other distributions such as log-normal can be used to estimate the intake level. Overall, neither approach given is likely to result in relatively accurate estimates of the upper intake levels of a nutrient substance in a population, they can only be employed unless there is no other choice.

**Marketing and Sales Data** Marketing and sales data usually provide information on specific products, thus, they can only be used as complemented data for other formed estimations of intake. If these products are the major or only source of intake of nutrient of interest and they are widely taken by individuals of studying population, they can be used directly as the source of dietary intake assessment, even though, uncertainties still remain.

If intake data from the above source are all unavailable, there still exist other ways to roughly estimate the intake of nutrient. For instance, using data on dietary guidelines for a country or region, as these guidelines usually to some extent represent the availability and consumption of food and related products. Another way when no any other approaches are feasible, is to create a diet based on observations for subpopulation and then to estimate the intake of the nutrient. These cases provide the least accuracy to dietary intake assessment.

#### 9.3.2.6 Uncertainties

Specification of uncertainties is a crucial component for any type of risk assessment, so as to nutrient risk assessment.

Due to the nature of the nutrient substance that with a preferable intake level to contribute to health, source and generation of uncertainties in classic risk assessment and nutrient risk assessment are different. In assessing the risk of non-nutrients, in many cases, conservative uncertainty factors are encouraged. However, in nutrient risk assessment, they cannot be too conservative to meet the requirement of maintaining the health of individuals. Thus, they are usually worked by the incorporation of both types of uncertainties.

Uncertainties and bias for dietary intake assessment can be generated from the composition data, consumption data, and the statistical analysis method regrading them, and they may raise some impact on the next step of nutrient risk assessment, the nutrition risk characterization.

#### 9.3.2.6.1 Composition Data

In this process, uncertainties may result from inaccurate average content of a dietary item as even the best approach to generate composition data, chemical analysis, can have limitations such as sample may not large enough to ensure enough representativeness of dietary item. Besides, the mismatch between dietary items in composition data and actual intake in population of interest, outdated composition data, bioavailability of the nutrient substance, reliance on labelled content information may all cause uncertainties.

#### 9.3.2.6.2 Consumption Data

In this process, uncertainties are caused by the inaccurate measurement of actual consumption data. For instance, data collected from limited sample size may not represent the overall population, decreased accuracy on estimating the upper tails of distribution, recall bias on self-reported dietary intake information, and lacking representative measurement also may introduce uncertainties to dietary intake assessment.

#### 9.3.2.6.3 Analytical Methods and Corrections

The choice of appropriate statistical analysis method to estimate the upper-tail intakes is important. Even with the most accurate consumption data with representative individual data, it is important to employ analysis method to reduce measurement bias, such as day-to-day variance. When using the less accurate data, such as data at household level or nation level data, they have to be combined with complemented data. The analytic methods in these cases are, inevitably, introducing uncertainties.

#### 9.3.2.7 Reporting

As always, the reporting should be documented detailed enough. A reporting on dietary intake assessment is recommended to include the following components:

Provide intake distributions as percentiles for subpopulations and for appropriate nutrient substance form, specify adjustment factors or other calculations, reasons that distribution cannot be estimated, identify the database and analytical method as well as the reasons, specify uncertainties, and provide guidance on how results should be interpreted relative to ULs.

### *9.3.3 Nutrition Risk Characterization*

This section describes the functions and key elements of nutrition risk assessment. Notably, sufficient information should be provided to ensure the communication between risk assessors and managers to ensure the reliability of decisions made.

Nutrition risk characterization is a process to integrate information and results from nutrition hazard identification and characterization, dietary intake assessment to estimate the risk of specific nutrient in the population of interest. It is the final step of nutrient risk assessment. Information, including quantitative information such as ULs and estimates of risk at different levels of intake, and qualitative information such as specification of (sub)populations believed to be at highest risk, reasons for the risk, descriptions of the nature and severity of the risk, and importantly, uncertainties, are finally provided to risk managers to scientifically support their decisions on risk management.

Every time when a set of steps of nutrient risk assessment is done, the risk manager put up with a decision, which declared the end of the assessment. However, if there appears a new study that reveals an adverse health effect that has not been identified before, a new round of nutrient risk assessment is started again. Thus, in principle, the nutrient risk assessment, may, never end.

#### 9.3.3.1 Component of Nutrient Risk Characterization

##### 9.3.3.1.1 Basic Components

A standard approach or identified process could ensure the completion of risk characterization. These are the key components of nutrient risk characterization:

The nature of adverse health effects especially critical adverse health effects, the severity and reversibility of adverse effects, mechanisms and conditions of effect, intake-response relationship and the threshold level, uncertainties and quantitative factors, derivation of UL, approach of dietary intake assessment, proportion of population that exceeding UL, population of high risk, specification of the magnitude of the risk above the UL, and exceptions for which exceeding UL may be warranted.

Notably, to facilitate decision-making process for risk managers, above information should be displayed in a way that can be easily understood and utilized for risk managers, such as categorizing key components by decisions.

The FAO/WHO workshop provided a model [10] of elements of risk characterization to inform key decisions made by risk managers, which consists of five decisions, including reducing the level in food supply, product labelling, and education, as well as corresponding. It can be used as a general model for nutrient risk characterization to serve a wide international use; however, in some cases, not all components in that table are equally important, the final model should be depended case by case. When possible, risk managers could ask for more detailed information on some issues that are closely related to a particular policy option or a risk management decision to be made; such activity may introduce the iterative process from literature review to nutrition risk characterization again. For example, risk managers may inquire what would happen to the distribution of intake of a nutrient substance by different subpopulations if the fortification level of a food were increased by specified amounts to address problems of inadequacy?

#### 9.3.3.1.2 To Improve Interface Between Risk Assessor and Manager

Meaningful communications are encouraged between risk assessors and managers. When risk characterization is based on the above-mentioned general model, information that has the potential to impact the management decision may be overlooked without sufficient communication between risk assessors and managers. For example, information on the limitation of data in intake-response assessment, and on the indication on data gaps, and on the further discussion about behaviors that may change risk may be all of importance. Besides, problem formulation, the communication between stakeholders and risk managers may also be useful on the utility and appropriateness of the risk characterization for the risk manager. However, only very few studies take it into practice, more attention should be paid in this area.

### *9.3.4 Model and Applicability of Nutrient Risk Assessment*

This section (1) describes the general model that synthesizes the process of nutrition hazard identification/characterization, dietary intake assessment, and nutrition risk characterization, (2) presents key questions and activities related to the model, and (3) discusses the use of a data-driven model.

#### 9.3.4.1 Model of Nutrient Risk Assessment

The FAO/WHO workshop build a general model for nutrient risk assessment, linking three major processes of nutrient risk assessment and related key activities [3]. As described in the model, the process of nutrient risk assessment involves many questions addressed by decision-making steps, generating from the processes of problem formulation, nutrition hazard identification and characterization, dietary intake assessment, and nutrition risk characterization. The corresponding key questions and activities are discussed below.

**Key Questions and Activities for Problem Formulation** These questions and activities are important in preceding the assessment and building the basis. Questions in this process include the goals and reasons for the nutrient risk assessment and the nature of available prior knowledge, and the activities are conducting dialogues among stakeholders to ensure understanding of the problem and the purpose of the assessment, as well as collecting and evaluating key elements of prior knowledge to determine whether a risk assessment is necessary.

**Key Questions and Activities for Nutrition Hazard Identification and Characterization** Nutrition hazard identification and characterization are carried out but risk assessors, this process highlights the importance of combination and iteration nature of establishing UL and hazard characterization. Key questions in this part involve the identification of adverse health effects and the intake levels, conduction and summarization of searching, background information, validity and reliability of biomarkers, establishment of UL, data selecting criteria, identification of critical adverse health effect, quantitative of results and uncertainties, and subpopulation of interest. Activities are defining data search strategy a priori, obtaining and summarizing relevant information, ranking the nature and validity of biomarkers, specifying selection and excluding criteria, and so on.

**Key Questions and Activities for Dietary Intake Assessment** Dietary intake assessment provides estimation on the intake level of the population of interest. Key questions in this process involve the type of dietary intake estimation and approaches of estimating. Related activities are determining the type of intake estimation and improving the utility of composition data and the validity of consumption data.

**Key Questions and Activities for Nutrient Risk Characterization** In this process, key questions are the nature and magnitude of risk, and to determine the most useful information for risk manager. Activities are characterizing the nature, identifying vulnerable subgroups, determining the proportion of population whose intake exceeds UL, and putting forward the most useful information to risk managers.

The essence of nutrient risk assessment is to make use of available data, thus it is usually called a data-driven model. The following implications are worthy to consider.

Detailed documentation of scientific judgement is required. Though it is preferable to use high quality and complete data when conducting nutrient risk assessment, however, in practice, available data are limited. Thus, the use and interpretation of data usually involve scientific judgement. With detailed documentation of the process and reason for scientific judgement, the transparency and utility of the outcomes of nutrient risk assessment can be ensured. However, this is usually overlooked in existed reports.

ULs should be adequately adjusted for uncertainty. As there is generally lacking of sufficient and high-quality data regarding nutrient risk assessment, uncertainties should be addressed carefully to ensure the confidence of protection on the population of interest. Practically, UL value is lower than the lowest value of intake level on the distribution that is related to adverse health effect, the degree of uncertainties determines the gap between UL and the lower end of observed intake that generated adverse effect. Lower ULs should be established with greater uncertainties to confirm population with intake level of nutrient lower than the UL is not likely to develop adverse health effect. It is to some extent similar to the establishment of recommended intake level for nutrient. The latter is based on the high end of intake distribution of the population of interest.

ULs should be built even if data are limited. UL is a major reference for risk managers to make decisions, even with limited data, the risk of excessive intake of specific nutrient still remains in the population. With limited data, the UL should be built with detailed documentation of limitation and uncertainties to assist risk managers when making decisions.

The absence of evidence does not mean evidence of absence. As mentioned before, many studies referred in nutrient risk assessment are not primarily designed for investigating the relationship between nutrient intake and adverse health effect, they are usually designed to explore the "good" health effect of a nutrient substance. In this situation, using priori protocols may contribute to identifying adverse health effect. Thus, the confined evidence may turn to be useful evidence instead of lacking evidence.

The model of nutrient risk assessment discussed in this section outlines key questions and activities so as to identify and characterize nutrient hazard, assess dietary intakes, and characterize nutrient risks. It supports the process of decision-making by risk managers with acceptable and detailed quantitative and qualitative information on the nutrient risk assessment, as well as fully understanding of uncertainties. Notably, uncertainties about ULs have been corrected when establishing UL, users of UL do not have to correct UL again. Finally, due to the lacking of adequate studies exploring the nutrient intake and adverse health effect, the model should be used with thorough consideration of uncertainties of all sides.

#### 9.3.4.2 Applicability of Model to Nutrient Substances

This section discusses the applicability of the model of nutrient risk assessment to a wide range of nutrient substances. In addition to general applicability of the model, three special conditions that have limited applicability are also discussed.

The definition of nutrient substances is not consistent around the world, in some cases that only essential nutrients are acknowledged while in other cases substances that contribute to the health but are not indispensable to body, such as certain carbohydrates or fatty acids are broadly considered to be nutrient. The above model can be used in nutrient substance of both definitions. Essential substances are generally well identified and with established recommended intakes, whereas the latter defined nutrient substance is an emerging studying area. The recommended intakes for substances that may have a favorable impact on health are usually considered as higher or lower intake instead of a specific value. In any case, the model for nutrient risk assessment should adequately consider uncertainties regarding the NOAEL, LOAEL or BMD to make sure a UL value that higher than recommended intakes. For some nutrient substances, their homeostatic mechanisms are also needed to be considered. Notably, a wide range of adverse health effects can be identified in the model, such as diseases, adverse organ functions, as well as cancers, which again confirm the general applicability of the model.

In three conditions the applicability of the above model is limited: in nutrient substances without any identified adverse health effects, in nutrient substances without any identified safety intake level that may not result in adverse health effect, and in nutrient substances that their favorable level maintaining health overlaps with intake levels which may cause risk. However, they are probably to be assessed when possible adjustments are made.

In some nutrients, such as Vitamin $B_{12}$, even the highest intake level of used or observed cannot result in significant adverse health effects. This could be because of the lacking of studies or a very high threshold for generating adverse health effects compared to known studies. If necessary, the highest observed intake (HOI) can be used. As specified [3], the highest observed intake (HOI) is derived only when no adverse health effects have been identified. It is the highest level of intake observed or administered as reported within (a) study(ies) of acceptable quality. In doing this, uncertainties and scaling of using HOI in model of nutrient risk assessment need careful consideration. The risk of nutrient may generally be overestimated in that case, as the actual UL may be higher than HOI.

Though evidence indicate that increasing intake of some other nutrients, such as saturated fatty acid, is associated with higher risk of adverse health effects, such as coronary heart diseases, the threshold at which adverse health effect started is unclear. Apparently, a zero value cannot be set as the UL, as a wide range of foods contains those nutrients. One possible method is to establish a specific dietary pattern that provides sufficient essential nutrients but contain the least level of the studying nutrient. More practices are needed in this area.

Furthermore, the applicability of the model is also limited in nutrients with overlap between intake level may cause adverse health effects and recommended intakes, such as vitamin A. Scientific judgement is an important process for nutrient risk assessment of these nutrients. In the need of supporting risk managers for decisions, detailed documentation of reason and approach in setting a UL and hazard characterization as well as risk characterization should be added to ensure risk managers fully understand the limitation and uncertainties of this approach.

For non-nutrients, especially for those without similar non-threshold effects with nutrient substances, such as germ cell mutagens and genotoxic carcinogens, approaches on the adjustment of the model of nutrient risk assessment remain unclear. In this situation, even a dose-response relationship is not helpful for setting the ULs.

The model built for nutrient risk assessment is appropriate in assessing the risk of a wide range of nutrient substances; however, in some certain circumstances, such as in nutrients without identifying adverse health effects, or lacking of threshold level, or have overlap intake level between levels that generate "harmful" and "good" health effect that the applicability of the model is limited.

#### 9.3.4.3 Applicability of Model to Inadequate Nourished Populations

The applicability of nutrient risk assessment in adequate nourished population is well acknowledged. Meanwhile, the application of this model to inadequate nourished population whom may have different homeostatic situations to adequate nourished companions wait to be considered. As inadequate nutrition intake is usually associated with higher risk of infectious disease. Thus, the impact of infectious diseases on nutrient risk assessment is also worth noting. Overall, the general process of establishing UL is applicable based on considerations of the metabolic status of inadequate nourished population.

**Homeostatic Considerations** Homeostatic mechanisms in inadequately nourished population, such as decreased nutrient transport and reduced liver function, may largely affect the results of nutrient risk assessment. However, strong evidence is generally lacking on the exploration of high intake of specific nutrient in inadequately nourished population. Here present several examples of existing evidence. In population of protein deficiency, decreased amount of transport and binding protein may prevent Vitamin A from normal absorption and mechanism, thus the UL of Vitamin A in a well-nourished population may be too high for undernutrition population. Another example lies in the diet of this population, it usually contains a large amount of phytates that may interfere the absorption of specific minerals, the establishment of UL of these minerals, are thus noted.

**Establishing UL for Inadequate Nourished Population** Comparing with ULs established for adequately nourished population, ULs that build for inadequately nourished population are different, which could be lower, such as UL developed for patients with acquired immune deficiency syndrome, and also might be higher, such as UL for iron in iron-replete children. Therefore, although the overall process of establishing a UL for adequately nourished or inadequately nourished population is consistent, the former value cannot be transferred to the latter value, as the effect of physiology characteristics remains unclear in inadequately nourished population. Instead, a different set of ULs is needed to be develop specifically for them, in the light of uncertainties on nutritional status.

**Impact of Infectious Disease** The high intake of a specific nutrient substance may increase the risk of infectious diseases. For instance, to deal with the widespread nutritional issues of iron deficiency, many countries initiate various programs regarding improving the iron status of pregnant women and their children. However, evidence reveals that supplement of micronutrients might increase the risk of infectious diseases [21]. Some programs of zinc supplementation also reveal the same trend. Thus, it is worth noting that a high intake of specific nutrient substance related to nutrition supplemental initiatives might be associated with the adverse health outcomes such as infectious diseases. Notably, such relationships may not exist in fortified foods or foods that naturally risk in a specific nutrient, only to supplements that increase the risk. More data are needed in this area.

The ULs established for adequately nourished population may not be appropriate for inadequately nourished populations; a new round of nutrition hazard identification and characterization are called for the development of ULs in such populations. However, potential mechanisms still remain unclear, more studies are needed to reveal the appropriate ULs for inadequately nourished populations as well as potential adverse health impact that nutrient supplement may have.

## References


1. Institute of Medicine. How should the recommended dietary allowances be revised? [B]. Washington, D.C.: The National Academies Press; 1994.
2. Institute of Medicine. Dietary reference intakes for energy, carbohydrate, fiber, fat, fatty acids, cholesterol, protein, and amino acids [B]. Washington, D.C.: The National Academies Press; 2005.
3. The Food and Agriculture Organization of United Nations (FAO) and the World Health Organization (WHO). A model for establishing upper levels of intake for nutrients and related substances: report of a joint FAO/WHO technical workshop on nutrient risk assessment, 2–6 May 2005. [R]. 2006. Available from: http://www.who.int/ipcs/methods/nra/en/
4. International Programme on Chemical Safety (IPCS). IPCS risk assessment terminology [R]. IPCS Risk Assess. Terminol. 2004. Available from: http://www.who.int/ipcs/methods/harmonization/areas/ipcsterminologyparts1and2.pdf?ua=1
5. Kumanyika S, Oria PM. Guiding principles for developing dietary reference intakes based on chronic disease [B]. Washington, DC: The National Academies Press; 2017.
6. Tinker LF, Sarto GE, Howard BV, Huang Y, Neuhouser ML, Mossavar-Rahmani Y, Beasley JM, Margolis KL, Eaton CB, Phillips LS, Prentice RL. Biomarker-calibrated dietary energy and protein intake associations with diabetes risk among postmenopausal women from the Women's health initiative [J]. Am J Clin Nutr. 2011;94(6):1600–6.

7. Zheng C, Beresford SA, Van Horn L, Tinker LF, Thomson CA, Neuhouser ML, Di C, Manson JE, Mossavar-Rahmani Y, Seguin R, Manini T, LaCroix AZ, Prentice RL. Simultaneous association of total energy consumption and activity-related energy expenditure with risks of cardiovascular disease, cancer, and diabetes among postmenopausal women [J]. Am J Epidemiol. 2014;180(5):526–35.
8. Yetley EA, MacFarlane AJ, Greene-Finestone LS, Garza C, Ard JD, Atkinson SA, Bier DM, Carriquiry AL, Harlan WR, Hattis D, King JC, Krewski D, O'Connor DL, Prentice RL, Rodricks JV, Wells GA. Options for basing dietary reference intakes (DRIs) on chronic disease endpoints: report from a joint US-/ Canadian-sponsored working group [J]. Am J Clin Nutr. 2017;105(1):249S–85S.
9. West S, King V, Carey TS, Lohr KN, McKoy N, Sutton SF, Lux L, West S, et al. Systems to rate the strength of scientific evidence [J]. Evid Rep Technol Assess (Summ). 2002;2002(47): 1–11.
10. International Programme on Chemical Safety (IPCS). 2002. Principles and methods for the assessment of risk from essential trace elements. Environmental Health Criteria 228 Geneva, World Health Organization [R] Available from: http://www.inchem.org/documents/ehc/ehc/ehc228.htm.
11. OECD Guidlines for the Testing of Chemicals No. 407 Repeated Dose 28-Day Oral Toxicity Study in Rodents, OECD [S]. Available from. https://www.oecd-ilibrary.org/environment/oecd-guidelines-for-the-testing-of-chemicals-section-4-health-effects_20745788. 2008.
12. GB 15193.1-2014 National Standards for Food Safety: Toxicity Assessment Protocol for Food Safety, National Health Commission of the People's Republic of China [S]. 2014. https://sppt.cfsa.net.cn:8086/db.
13. EFSA. Application of systematic review methodology to food and feed safety assessments to support decision making [J]. EFSA J. 2010;
14. Zhao JG, Zeng XT, Wang J, Liu L. Association between calcium or vitamin D supplementation and fracture incidence in community-dwelling older adults: a systematic review and meta-analysis [J]. JAMA. 2017;318(24):2466–82.
15. Palacios C, De-Regil ML, Lombardo KL, Peña-Rosas PJuan. Vitamin D supplementation during pregnancy: updated meta-analysis on maternal outcomes [J]. J Steroid Biochem Mol Biol. 2016;164:148–55.
16. EFSA. Guidance on uncertainty analysis in scientific assessments [J]. EFSA J. 2018;
17. IOM. Dietary reference intakes: applications in dietary planning [B]. Washington, DC: The National Academies Press; 2003.
18. Mejía-Rodríguez F, Sotres-Alvarez D, Neufeld LM, García-Guerra A, Hotz C. Use of nutritional supplements among Mexican women and the estimated impact on dietary intakes below the EAR and above the UL [J]. J Am Coll Nutr. 2007;26(1):16–23.
19. FAO. FAO Statistical Databases: Food balance sheets [D]. 2005.
20. National Center for Health Statistics (NCHS). 2005. National Health and Nutrition Examination Survey [R]. Available from: http://www.cdc.gov/nchs/nhanes/participant.htm.
21. Gombart AF, Pierre A, Maggini S. A review of micronutrients and the immune system-working in harmony to reduce the risk of infection [J]. Nutrients. 2020;12(1):236.

# Chapter 10
# Risk Assessment and Risk-Benefit Assessment

**Jinyao Chen and Lishi Zhang**

**Abstract** The framework of risk analysis has become the principal procedure for dealing with food safety issues. Risk analysis consists of three components: risk management, risk assessment, and risk communication, while risk assessment is defined as the scientific evaluation of possibility and consequences of adverse health outcomes resulting from food-borne hazards exposure in the case of food safety issues. Risk assessment is a scientifically based process consisting of the following steps: hazard identification, hazard characterization, exposure assessment, and risk characterization. The procedures of risk assessment of chemical hazards and microbiological hazards are a little bit different. This chapter would focus on the chemical hazards. On the other hand, positive and adverse effects may be induced concurrently by a single food item, e.g., fish, whole grain products, or even a single food component, e.g., folic acid, phytosterols, in which scenarios the risk-benefit assessment should be adopted. The principles and main steps of risk-benefit assessment are the same with risk assessment. Risk-benefit assessment comprises three parts, i.e., risk assessment, benefit assessment, and risk-benefit comparison, among which risk-benefit comparison is the trickiest one, usually a common metric of the health outcome is needed. Risk-benefit assessment is a valuable approach to systematically integrating the current evidence to provide the best science-based answers to address complicated questions in the areas of food and nutrition, especially in evaluating nutrient fortification policy, developing a tolerable upper intake of nutrient, and recommending a particular dietary pattern.

**Keywords** Risk assessment · Risk-benefit assessment · Risk analysis · Food safety

J. Chen (✉) · L. Zhang
Department of Nutrition and Food Safety, West China School of Public Health, Sichuan University, Chengdu, China


L. Zhang (ed.), *Nutritional Toxicology*,
https://doi.org/10.1007/978-981-19-0872-9_10

## 10.1 Overview of Risk Analysis and Risk Assessment

### *10.1.1 Overview of Risk Analysis*

Due to extensive globalization, more frequent international food trade, extensive worldwide distribution of food products, and continuing evolution in consumer dietary patterns, known or newly discovered food safety hazards have aroused attention. Therefore, national and international food safety authorities are under great pressure for the guarantee of consumer's health and global economic stability.

In 1991, a joint Food and Agriculture Organization of the United Nations (FAO)/ World Health Organization (WHO) conference strongly recommended measures that should be taken to increase awareness of food safety principles [1]. Subsequently, FAO and WHO convened a series of expert consultations to address the three components of risk analysis of food safety, i.e., risk assessment, risk management, and risk communication, which basically established the framework of risk analysis. In 1995, the first joint FAO/WHO expert consultation depicted basic terminology and principles of risk assessment. Then in 1997 and 1998, a risk management framework for food safety and guidelines for risk communication was proposed, respectively [1, 2]. Gradually, risk analysis has become the universally recognized framework for coping with food safety issues and developing food safety standards.

Risk analysis is defined by the Codex Alimentarius Commission as "a process consisting of three components: risk management, risk assessment, and risk communication" [2]. The currently universally recognized framework of risk analysis is the organically integrated three components, which interacted with each other even though it is recommended that each part should be conducted by relatively independent experts. One of the most outstanding advantages of "risk analysis" is that it guarantees a structured, objective, and transparent decision-making process in food safety [3].

Principles and guidelines of each component of risk analysis, i.e., risk assessment, risk management, and risk communication, have been elaborated with further details. The risk assessment process consists of four steps, i.e., hazard identification, hazard characterization, exposure assessment, and risk characterization. The frameworks of risk management are presented in various forms from different nations/ organizations. However, most of them emphasize the following steps: commonly risk profiling, if necessary, risk assessment, options weighing, decision implementation, and monitoring & review through the whole process. Risk communication is emphasized throughout the entire process, not just the ongoing inter-dialog between risk managers and risk assessors, but also the multichannel communication with food industries, consumers, and all the other stakeholders [2, 3]. The risk assessment process is commissioned by the risk managers, and the risk managers also define the purpose and scope of the risk assessment. It should be strengthened that to promote effective and efficient communication between the risk manager and risk assessor and to guarantee the objectiveness of risk assessment, the task and staff of risk management and risk assessment should be separated as far as necessary [1, 3].

Based on the principles of protection for human health as well as fair food trade, risk analysis is quite important in the regular running of the World Trade Organization (WTO). In 1995, the WTO Agreement on the Application of Sanitary and Phytosanitary Measures (WTO SPS) agreement entered into force, calling for international standards based on risk assessment. To establish a set of principles regarding health and safety regulations for food, the SPS agreement requires "members to ensure, as appropriate to the circumstances, that their SPS measures are based on an assessment of the risks to human, animal or plant life or health, taking into account risk assessment techniques developed by the relevant international organizations". That is to say, WTO members are required to ensure and prove that risk management decisions are transparent and science-based, and member countries are obligated to follow the standards and guidelines of Codex Alimentarius in food trade unless they could justify their divergence based on principles of risk analysis [2, 4].

To sum up, food safety risk analysis is important for the protection of human health and fair food trade. Risk analysis has become the foundation for standards establishment at both international and national levels, while Codex Alimentarius has become mandatory. Codex has developed a series of guidance documents to promote the application of risk analysis for national authorities.

### *10.1.2 Overview of Risk Management*

Risk management is defined in the Codex Alimentarius framework as "the process, distinct from risk assessment, of weighing policy alternatives, in consultation with all interested parties, considering risk assessment and other factors relevant for the health protection of consumers and for the promotion of fair trade practices, and, if needed, selecting appropriate prevention and control options." [2]. Food safety authorities and organizations worldwide have adopted several risk management protocols; however, the frameworks and steps are more or less the same with the procedures of risk assessment initiation, options weighing, implementation of risk management decisions, and monitor & review throughout the process [5].

Acted on the food safety incidents, the risk manager should gather resources to prepare a preliminary risk profile on the foodborne hazards based on available information as soon as possible, and then decide whether to initiate a risk assessment or not; if a risk assessment is needed, the purpose and scope of the risk assessment should be defined by risk managers and delivered to risk assessors. After the designed risk assessment is completed, the risk managers would weigh possible policy alternatives based on the results of the risk assessment and other factors, such as social, economic, and religious factors. Then, if necessary, the risk managers should select an appropriate risk management measure and make sure it gets implemented. The risk manager should also establish and implement correspondent monitoring activities to evaluate the effects of the selected measures and, if necessary, to make certain adjustments [5, 6]. In a food safety emergency, risk management decisions should be achieved quickly and the implementation of them should gather all the efforts from multiple agencies, industries, and other stakeholders [7, 8].

Typical risk management measures include setting/revising regulatory standards, codes and guides, measurable limits, compulsory use of specific control measures, increasing food inspection, certification, and approval procedures, educational and training programs, prohibiting foods with a documented history of contamination or toxicity, food recall, etc. Several key factors should be considered when weighing risk management options, including these listed below [5, 6].

- Capacity: The capacity to implement the risk management options of the national food safety authorities should be evaluated. For instance, if it is not feasible to expand inspection or to perform larger sampling and testing, the national food safety authority may choose to turn to international organizations for assistance, or other more cost-efficient options.
- Uncertainty: Uncertainties may have great impacts on the selection and implementation of risk management decisions, and risk assessors should fully depict the uncertainties and communicate with risk managers.
- Risk perceptions of the public: Risk perceptions of the public have considerable influence on their reaction to the food safety issues and satisfaction with the national authorities. Public demands tend to be much higher during emergency incidents; hence more stringent actions might be needed under these circumstances. Risk communication is particularly important in the cases where the selected risk management option may not as stringent as expected by the public.
- Legal considerations: The extent to which the laws and regulations support or constrict risk management implementation should be considered.
- Industries involvement: Industries cooperation and engagement in some cases could be a key factor in determining which options are feasible or to what extent the expectation could be met.
- Consideration for international organizations or other countries: Risk management measures undertaken by other countries or international organizations should be referenced; and related import and export issues should also be considered.
- Other factors: For example, foodborne hazards related to vulnerable subpopulations should be paid more attention, and the measures might be more stringent; the responses to previous similar events should be considered to improve the consistency of policies.

Several principles of risk management framework have been proposed by different organizations [5, 6, 8], commonly including the following points: protection of human health was usually the primary consideration; a structured approach should be followed; the scientific integrity of the risk assessment process should be ensured; the decision-making process should be transparent; interactive communication with consumers, industries and other stakeholders should be involoved; and finally, risk management should take into account all updated data as a continuous process during which monitoring and reviewing matters.

Risk managers should also take on the following tasks in a broader perspective: systematically identify problems and options at the earliest stages; expand the range of evaluated effects beyond endpoints of only human health to include broader outcomes of social impact and environment protection; increase the understanding of the strengths and limitations of risk assessment by decision-makers; make the risk communication among all the stakeholders efficient and transparent.

### 10.1.3 Overview of Risk Communication

Risk communication is defined by the Codex Alimentarius as "the interactive exchange of information and opinions throughout the risk analysis process concerning hazards and risks, risk-related factors and risk perceptions, among risk assessors, risk managers, consumers, industry, the academic community and other interested parties, including the explanation of risk assessment findings and the basis of risk management decisions" [1, 9].

Risk communication is an indispensable part of the risk analysis framework. The goals of food safety risk communication include the following aspects: to enhance the understanding among all interested parties, to enable the public to make informed choices regarding foodborne hazards, and to improve the overall effectiveness of the risk analysis. While the main goal is to increase understanding among various stakeholders regarding the reasons behind the efforts made to assess and manage food safety risk. To elaborate, risk communication is essential in helping risk managers to understand the process and results of the risk assessment and the impacts of their different risk management measures, and thereby to better assess weighing the management options. To reduce the risk of foodborne hazards, food safety risk communication sometimes as a risk management measure itself is also an ongoing process (e.g., promotion of hygiene practices among general public). Furthermore, food safety risk communication may also involve communication of benefits, which would enable the public to make more informed food choices [10].

"Stakeholders" is a commonly used expression in the documents of risk analysis. Stakeholders refer to people and organizations that may be affected by the relevant risk management decisions. Usually, stakeholders mainly include the general consumers affected by the foodborne hazards and food-related industries that are responsible for the hazards and/or the implementation of the risk management measures. Food safety risk communication among all stakeholders should be based on the following principles, which include openness, transparency, responsiveness, and timeliness, all of which contribute to the maintenance and development of trust of consumers and effective communication [6, 9]. Openness means all stakeholders could contribute if they are willing to, including "risk takers" as well as "risk makers". Communicating in a timely manner is essential for building the trust with the consumers and establishing the first-hand information source, and can prevent the spread of rumors and misinformation. If the waiting time is too long, the questions of the public about food safety incidents become focused on “Why didn’t you tell us sooner?”, rather than on the risk itself. Even when there is indeed little information to deliver, the communication should still be initiated sooner than later, as to communicate about how the authorities are investigating the event and to tell people more information will be available. Transparency refers to the stakeholders and the public having access to the process of how the risk assessment conducts and how the risk management decisions are made, as well as how their opinions are valued. This means that all the stakeholders and the public should have certain ways to access information regarding documentation about the decision-making process. Responsiveness is the extent to which food safety authorities address the communication expectations of target audiences in the early stages. Unresponsiveness to public concerns may lead them to turn to other sources for information, and also jeopardize the trust of the public to authorities. From another perspective, risk communicators should also be aware of the changes of the external environment and internal process, and take into consideration of the unforeseen events as well as uncertainties, to revise or adjust communication strategies accordingly.

The WHO guidelines provide the template of risk communication messages [11], which include: description of the risk; the advice to consumers to mitigate or avoid the risk; quote from a reputable and trusted source; explanation of what is being done to reduce the risk; additional relevant information. Different food safety issues require different approaches to risk communication. An important distinction to consider is whether it is an urgent issue or a more enduring food safety problem. Emergency food safety events (e.g., outbreaks of food-borne illness) require a rapid response. In cases of food safety emergencies, the initial public communication at the onset of a food safety emergency is critical. Risk communication key points and strategies may need to change rapidly and accordingly as more information about the risk is obtained or as risk management measures are implemented. It is important to consider perception of the risk and consumer behavior in the specific food emergency and design the communication messages accordingly. It is essential that risk communication points are as accurate and simple as possible during emergencies, and demonstrate confidence. It would be more preferred that the messages are well targeted to the audience and address their concerns. It is also important to identify potential different target audiences, and to find efficient ways of communicating with different groups. Responsiveness matters a lot during emergency, instead of waiting for solid information, it is important to start public communication activities early. During an emergency food safety incident, messages are often direct, and are delivered frequently and urgently. While enduring food safety issues (e.g., persistent heavy metal contamination in food, genetically modified food) require continuing communication. The following items should be indicated as clearly as possible during communication in emergencies: what food products are involved; what is known about the concerned food safety issue; what the health outcomes are and what levels of exposure could induce the effects; what the public should do if they have consumed affected products and what actions they could take to protect themselves; what measures the government is taking; and how to access additional information.

It may also be necessary to develop messages to counter inaccurate statements or misleading messages from the media or other groups.

Choosing appropriate communication channels is important, which generally include print media, television, websites, broadcast, new social media, opening days, etc. Communicators should choose wisely which communication channels are most appropriate for communication to each group of the target audiences under different circumstances. Efficient dialogue between risk assessors and risk managers can be achieved through frequent/regular meetings by using all available channels such as phone calls, e-mail, and teleconference [10, 11].

Risk perception determines the reaction to the risk. It means what people think about the characteristics of the risk, and the likelihood and severity of the outcomes. How people perceive risks influences their attitudes and behaviors [10]. It is important to understand risk perceptions of the public in order to develop effective risk communication strategies. Understanding the food safety issue and the target audiences' risk perception would certainly benefit the development of communication messages.

### *10.1.4 Overview of Risk Assessment for Food Safety*

Risk assessment is a scientific process to evaluate the nature and likelihood of adverse health outcomes resulting from human exposure to harmful agents, the purpose of which is to provide a scientific basis for risk management measures. Risk assessment utilizes all relevant scientific data, as well as identifies the uncertainties inherent to the risk assessment. From a food safety perspective, the focus is on the nature and probable health outcomes related to the intake of foodborne hazards. The following steps are recognized as the standard process of risk assessment: hazard identification, hazard characterization, exposure assessment, and risk characterization. The four steps are specified as follows [1, 2].

Hazard identification is "the identification of biological, chemical, and physical agents capable of causing adverse health effects, which may be present in a particular food or group of foods."

Hazard characterization is also called dose-response relationship derivation, which is "the qualitative and/or quantitative evaluation of the nature of the adverse health effects associated with biological, chemical and physical agents which may be present in food." And to further specify, a dose-response relationship should be derived for chemical agents, while for biological or physical agents, a dose-response assessment is preferable if the data are available.

Exposure assessment is "the qualitative and/or quantitative evaluation of the likely intake of biological, chemical, and physical agents via food as well as exposures from other sources if relevant."

Risk characterization is "the qualitative and/or quantitative estimation, including attendant uncertainties, of the probability of occurrence and severity of known or potential adverse health effects in a given population based on hazard identification, hazard characterization and exposure assessment."

Risk assessment strongly relies on the availability, selection, and processing of data. Hazard identification and characterization would need high-quality toxicity data, and exposure assessment depends on data from food contamination survey and food intake investigation. Data gaps can be partially filled with non-traditional approaches such as structure-activity research, read-across, and default values. The validity of the methods, the reliability of the results, and the relevance and adequacy of the data should be carefully evaluated.

A core value of risk assessment is to adopt the best scientific methods to make use of the highest-quality evidence. A shared recognized ideal scenario in science-based risk assessment is that all uncertainties through the evidence should be fully described. The responsibility of consideration of uncertainty is ultimately taken by risk managers. Thus, the nature and magnitude of uncertainties must be clearly depicted.

To ensure that risk assessments are maximally used in risk management decision-making, the questions that risk assessments need to address must be raised before risk assessment is conducted. Planning/scoping and problem formulation before the initiation of risk assessment has been emphasized, and it should be formalized and implemented, ensuring their extensive application in risk assessment [1, 3].

## 10.2 Risk Assessment for Food Safety

### *10.2.1 Hazard Identification and Characterization*

The likelihood of bad consequences due to exposure distinguishes risk from hazard, which means no exposure, no risk, while hazard is the nature of the concerned agent. Hazard identification involves systematically searching for relevant literatures and evaluating data on the types of health effects that may be induced by the specific hazard. In the data collection process, hazard identification and characterization should not be divided into two separate parts. Hazard identification involves characterization of the behavior of a chemical within the organism and its interactions with organs, cells, or genetic material; during hazard characterization, the dose-response relationship of the critical adverse health outcomes would be derived and a health-based guidance value should be derived [3, 12].

Currently, the data needed for hazard identification and characterization can be divided into:

- Human epidemiological and intervention data.
- In vivo studies: Performed in laboratory animals.
- In vitro studies: Performed in laboratory cultured cells, tissues, or organs of animals or humans.
- In silico studies: Performed to denote computer simulations.

In data evaluation, usually, human data weighs the most since it directly provides evidence on human exposure and adverse health effect; however, it couldn't give more information on dose-response relationship, which could be derived from designed animal studies. While in vitro studies could give hints about the target organs of the test substances or provide information on their mode of action, in silico studies are more efficient and have the potential to be used as screening tools for new chemicals [12, 13].

#### 10.2.1.1 Toxicokinetics Data

It is universally recognized that all toxicological effects of xenobiotics are induced through interactions with the substance at target sites. Sufficient description of these processes is essential for hazard characterization. A response could occur if enough of a chemical (or its active metabolites) reaches a target organ. Thus, toxicokinetics data on the absorption, distribution, metabolization, and excretion (ADME) is quite important. The ADME process determines the fate of the chemical in the organism, resulting in the internal dose at the target organ [14].

"Toxicokinetics" is an integrative interdiscipline, which depicts not only the processes of ADME, but also exposure and outcome in the organism. Basic principles of pharmacokinetics are adopted to determine the relationship between exposure of a chemical and its fate in the body. Toxicokinetics data could also be used to estimate the range of internal target organ doses resulting from realistic human external exposure and to depict the difference of ADME among species. Furthermore, toxicokinetic studies are also important in another way that they could provide information in the selection of species to be tested further and the design of doses for toxicity studies.

The design of a toxicokinetic study may depend on the scientific questions and involve several different strategies. Controlled acute and repeated-dose toxicokinetic animal studies can be adopted to identify a chemical's target organ, biological persistence, blood and whole-body half-life, as well as its potential for bioaccumulation. Of course, toxicokinetic profiles of one chemical are closely related to exposure duration and dose. Therefore, mathematical techniques have been developed in order to extract the maximum amount of information from data of a toxicokinetic experiment. These mathematical techniques could simulate the biological behavior of a chemical in the biological systems. Currently, two types of models can be distinguished, i.e., compartment models and physiologically based kinetic models (PBK models) [14]. "Compartment" is an important unit in modeling the concentration-time profiles of the targeted substances. The one-compartment model assumes the mammalian organism to be a well-mixed compartment. Two-compartment or more may be more appropriate to describe the kinetics of most chemicals. Actually, it is should be kept in mind that the number of compartments reliably distinguished strongly depends on the extent of the data. These mathematical compartment models cannot be extrapolated to interspecies, usually a new model has to be developed for each species.

PBK modeling has been greatly enhanced as computer technology advances. PBK is based on three groups of parameters: physiological parameters, metabolic parameters, and partitioning parameters. The concepts of compartments are also used in this type of modeling, but in contrast to compartment modeling, the number of compartments is physiologically based. PBK modeling can be used in various occasions, e.g., dose extrapolations, route-to-route extrapolations, and interspecies/intraspecies extrapolations [14, 15].

Physiologically based toxicokinetic (PBTK) is currently regarded as the most adequate approach to simulating human toxicokinetics and has the potential to be used to extrapolate between in vitro and in vivo experiments [15]. However, the need for high-quality in vitro and in silico data to predict human dose-response curves is urgent for risk assessment.

#### 10.2.1.2 In Vivo Toxicological Tests

In vivo toxicological studies with laboratory animals are quite important in hazard identification and characterization. Acute toxicity, repeated-dose toxicity, chronic toxicity, genotoxicity, reproductive toxicity, and carcinogenicity, are all listed in the toxicity evaluation profile of food contaminants or food-related hazards. However, not all toxicological tests must be compulsorily conducted. The acute, repeated-dose, chronic toxicity tests are more often to be performed to assess systematic toxicity, and to identify target organs. The systematic toxicity evaluation may further indicate the need for more specific toxicity studies, e.g., neurotoxicity and immunotoxicity [16].

International harmonized test guidelines, e.g., International Conference on Harmonization of Technical Requirements for Registration of Pharmaceuticals for Human Use (ICH) Guidance documents, OECD guidelines, are of prime importance to guarantee high-quality, universally accepted toxicity data. Key points in designing toxicological tests include chemical properties, animal selection, route of administration, dose selection, diet and caging, environment such as temperature, humidity, parameters, duration, data analysis, and data reports. Other issues for consideration include good laboratory practice (GLP), personnel requirements, animal welfare, etc.

Animal toxicological experiments should be conducted in a manner that best simulates the actual human exposure scenarios. The dose design should take into account the estimated human intake, as well as the frequency and duration of exposure. For food-borne contaminants, the administration route is usually through diet, by gavage, or via drinking water. Several important considerations are depicted as follows [3, 16].

1. Test substance

   During the initiation of the toxicity assessment, information such as chemical and physical properties, purity, and characteristics of impurities should be known. The solution of administration should be stored carefully, and the administered dose or concentration in every batch of feed, gavage solution, or drinking water should be measured to make sure the consistency of the given test substance during the experiment.

2. Dose design

   The design of dose regimen depends on many factors. In most cases of safety evaluation of chemicals, acute toxicity studies are the first to carry out, and there are several recommended acute toxicity testing methods. In order to deduce a LD50 value, it is preferred that the dose regimen be designed in a way to simulate a dose-response curve. However, in a limit test, one dose (the potential maximum tolerable limit level) is administered which should not cause any mortality. The alternative fixed-dose procedure aims at finding a dose inducing systematic toxicity.

   Because the objectives of repeated-dose toxicity studies, i.e., sub-acute and sub-chronic studies, include the determination of dose-response relationship and the derivation of a point of departure (POD) in some cases, it requires reasonable selection and spacing of the doses in order to generate different toxicity manifestation among groups. The number of doses is usually at least three (low, middle, high) doses plus control, sometimes plus positive control when necessary. Increments between doses are in the range of 2 to 10. Ideally, the high dose should induce evident toxicity, but little mortality (below 10%); the middle dose should produce less toxic action while the low dose should induce no toxicity, which means that the low dose corresponds to the No Observed Adverse Effect Level (NOAEL), and the middle dose corresponds to the Lowest Observed Adverse Effect Level (LOAEL) [17, 18].

3. Animal Species

   The selection of an appropriate animal species and strain for toxicity studies should be carefully considered. However, it is complicated. There is particular documentation for animal choosing. Usually for regular toxicity tests, small laboratory rodents are used, while for specific toxicity testing, rabbits, dogs, guinea pigs, and non-human primates may be required. Most toxicity assessments require both sexes to maximize the opportunity to capture any changes. Sometimes, studies are conducted in two species, one rodent and one non-rodent, or in two rodent species.

4. Diet

   Composition of the diet influences the response to a chemical substance, especially in the food toxicological evaluation where the test substances are usually administered by feed, drink, or gavage. Macro and micronutrients content might influence the biotransformation of substances and/or enzyme activity. It should also be kept in mind that over-nutrition and food restriction may be associated with certain incidence of adverse health outcomes, e.g., tumor incidence. Also, relatively high doses of test substances may also influence the palatability of diets, especially when the substance is administered by feeding, which could influence the growth of the animal, and in turn growth retardation may also either reduce or increase the overall toxicity. It should be guaranteed that feeding and drinking water should be tested for the presence of naturally occurring toxins and contaminants, or other confounding factors.

5. Endpoint

   Endpoints should be observable or measurable to the very least. The biological significance of the endpoints should be universally recognized. The selection of endpoints should be decided at the initiation of the study. The endpoints should be validated to be capable to reflect the toxic effects of the chemicals.

#### 10.2.1.3 In Vitro Tests

Over the last decades, an increasing number of in vitro test systems have been developed, based on lower levels of biological systems rather than the intact organism, e.g., organs, tissues, cell cultures, and even subcellular systems. These in vitro systems have been adopted widely in exploring the underlying mechanisms of a chemical's toxicological mode of action. The European Centre for the Validation of Alternative Methods (ECVAM) has issued a series of reports summarizing the state-of-the-art for the in vitro testing systems, which can be found on the ECVAM official website (http://ecvam.jrc.cec.eu.int/index.htm). The methods described in those reports mainly include: use of hepatocyte cultures in biotransformation and hepatotoxicity studies; application of cytotoxicity parameters in phototoxicity studies; skin culture for irritancy and corrosivity studies [19]. In addition to that, bacterial mutagenicity tests to determine genotoxicity has been used in regulatory procedures for decades, and some of these methods have now been adopted by the OECD guidelines.

Based on the fact that in vitro tests can directly reflect cytotoxicity, they have great potential to replace some in vivo toxicity tests, especially for the endpoints of lethal toxicity. Furthermore, in vitro systems can be used to study early molecular events and cellular responses that may predict detectable traditional systematic responses of the target organs in the in vivo exposure [19, 20].

Even with the wide application of in vitro test systems, it should still be noted that a hazard assessment cannot be conducted without further information of the chemical's behavior in the intact organism. Therefore, data obtained from in vitro studies mostly could not be directly applied to the in vivo situation. The doses in in vitro tests may not be in the same orders of magnitude with the actual concentration at the target tissue under in vivo exposure. In addition, metabolic activation and/or saturation of specific metabolic pathways should be taken into consideration to guarantee proper interpretation of in vitro data. Toxicokinetics describes the ADME process of xenobiotics within an organism as described above, the integration which could help to extrapolate in vitro doses to in vivo exposure levels, if done well, contributing greatly to risk assessment [21].

#### 10.2.1.4 Human Data

Information obtained from clinical trials and epidemiological studies in humans is of great importance to hazard identification and characterization, including epidemiological studies, controlled experiments with volunteers, and case reports [13]. In the cases of animal data and human data being both available, well-reported relevant human data for any given endpoint is of no doubt more preferred for the hazard assessment. However, health-based guidance values derived from human data are rare.

The evaluation of the relevance and weight of human studies and animal laboratory studies should be carried out on a case-by-case basis [22]. Regarding hazard identification, human data usually lack sensitivity. Negative data from human studies will not usually be used to override the toxicity manifestation observed from laboratory animals' studies, unless the toxic action is based on a mechanism which would certainly not be expected to occur in humans.

In the food safety risk assessment, the use of epidemiologic data has been limited. There are no clear criteria for epidemiologic data integration in risk assessment for regulatory purposes. Compared to high doses administration in animal studies, epidemiological exposure is usually at physiological intake levels. Epidemiologic study is usually designed to identify and quantify associations between human exposures and health outcomes, and the exposure and health outcomes are most often hypothesis-driven. As humans are free-living and subject to many influencing factors, both exposure and outcomes are difficult to measure accurately, and systemic bias and measurement error should be carefully considered. Direct association of causal relationships should not be derived from the results of observational studies, including cohort studies and case-control studies. Combining data of different categories of studies would certainly increase the power and derivge dose-response relationships. Evaluation of validity and quality of epidemiological studies has been well documented in other monographs [22, 23].

#### 10.2.1.5 Dose-Response Relationship

The dose-response evaluation should be used to establish a dose that does not induce biologically significant toxic effects, e.g., one in a million for cancer incidence. This dose is called a "Reference Point (RP)" or a "Point of Departure (PoD)" more often currently. In the dose design, as indicated above, if more than three dose levels were set, the ideal doses regimen would be that toxic effects are produced evidently at the highest dose level tested, with lower dose level inducing mild but detectable responses, and no adverse effects were observed at the lowest dose level. The Joint FAO/WHO Meeting on Pesticide Residues (JMPR) establishes the acute reference dose (RfD), which is an estimation of a substance which can be ingested in a period equal to 24 h or shorter without observable health outcomes for the consumer, and the acute RfD refers to the risk of short exposure [3, 24].

The "No Observed Adverse Effect Level (NOAEL)" approach has been used as the standard approach for representing thresholds in past decades. NOAEL stands for the highest dose at which no (adverse) effects were observed, and the NOAEL for the most sensitive endpoints would be found for the chemical assessed. The toxicity study and endpoint associated with the NOAEL is called "critical study" and "critical endpoint" in the hazard characterization, respectively.

Currently, a NOAEL can only be derived from one critical endpoint from one critical study in which toxic effects on the endpoint were measured at higher doses. In the process, professional judgment about toxicological significance is required. For instance, the toxic manifestation may not be relevant for humans, e.g., effects in the forestomach. Also, some specific endpoints may be considered adaptive and/or reversible without actual adverse health outcomes, which would need the expertise judgment.

Apart from the basic shortcoming of the NOAEL approach is that it is often misunderstood as a "no effect level", other disadvantages of the NOAEL approach have been proposed, which are briefly summarized as follows. The value of the NOAEL depends on the number of animals per group as well as the detectable effect size. Additionally, the NOAEL can only be one of the designed doses. The derivation of NOAEL strongly depends on the study design (doses spacing and number of animals per group). Not unexpectedly, replication of a particular toxicity study using another doses design (different dose levels and/or various number of animals) would most likely result in another value for the NOAEL with the same endpoints in the same animal species. This uncertainty cannot be quantified, which leads to another disadvantage of the NOAEL approach, as quantification of the uncertainties is crucial for deriving a health-based guidance value in risk assessment. Lastly, the NOAEL approach only cares about yes or no, rather than how much of the effect is, in other words, it does not make full use of the dose-response information.

The NOAEL approach has been considered not applicable for genotoxic carcinogens. Since for this category of chemicals, it is recognized that carcinogenic risk could occur at any dose. Only a dose with an acceptable risk, e.g., a tumor incidence, $10^{-6}$ (one in a million over a lifetime) could be determined. The lowest acceptable risk level has been denoted as a *de minimis* risk. Another unignorable problem is that most carcinogenic dose-response data are originated from chronic animal toxicity studies, which usually use dose groups of 50 or 100 animals per dose, with an observable risk (carcinoma incidence) of the order magnitude of $10^{-1}$. Therefore, a *de minimis* risk of $10^{-6}$ is far below the range of observation in carcinogenic toxicity studies. The extrapolation of the dose ranges is of great concern [25].

The ALARA (As Low As Reasonable Achievable) principle is firstly adopted in the case of genotoxic carcinogens. Before Benchmark Dose (BMD) was adopted, the default approach, particularly the linearized multi-stage model, had been used frequently to fit a dose-response model, and then the fitted curve was further extrapolated to estimate a dose associated with a specified acceptable risk level, usually $10^{-6}$. The estimated dose is called the "virtually safe dose (VSD)". However, the fitted model has been questioned regarding its uncertainty with the extrapolation. Currently, BMD method is developed quickly and more widely used. The BMD is defined as a dose level that is associated with a pre-specified (slight) change in response, denoted as Benchmark Response (BMR) compared with the controls, and it can be used with endpoints showing a dose-related response. The values of the BMD model are estimated by fitting dose-response models to the observed data of all doses groups. The lower confidence bound of the BMD (BMDL) estimate is normally used as the RP or PoD. BMD can equally be well applied for non-threshold effects, as genotoxic carcinogenic effects [24].

In other words, in the BMD approach, a dose-response model is carefully selected to be fitted to the tumor incidence data according to the data characteristic, and the fitted model is then used to estimate a dose associated with a risk level that is within the observation range, usually a 10% risk. The lower confidence bound of the estimated dose at 10% risk or 5% risk is called the BMDL10 or BMDL5, which is considered as a RP or PoD for further evaluation [24].

Except for the development of mathematical modeling approach, other considerations have also been put forward. Even though animal dose-response studies will continue to contribute the most important data in establishing PoDs in chemicals hazard assessment, data on human heterogeneity (e.g., genetically, physically variance), background exposures, and data from mechanistic studies will also play a crucial role in selecting the approach to the dose-response analysis and the uncertainty evaluation. The variability assessment should consider different susceptibility due to sex, age, genetic background, health status, and other factors. Uncertainty in human variability estimates would be better described quantitatively. The assessment of background variety would provide valuable information on the shape of the dose-response relationship, including appropriate dose-response modeling methods, the possibility of low-dose linearity, and potential high-risk subpopulations. Furthermore, the uncertainty distributions for the toxicokinetic and toxicodynamic components should also be explored for the interspecies and intraspecies extrapolation.

The Mode of Action (MOA) evaluation explores molecular initiating event and other key events after chemical exposures, including metabolic activation and detoxification, initial interactions with critical cellular targets, effects on cellular and subcellular processes (e.g., cells death or apoptosis, gene expression, and signal transduction), and other types of biological action which lead to the toxic manifestation [26]. MOA exploration is quite crucial in understanding dose-response relationships, and there are many factors that might contribute to the shape of the dose-response curve, including depletion of cellular defenses, induction of enzymes by repeating dosing, potential of repair, interaction with background disease processes, and chemical interactions [27]. Meanwhile, computational research that applies systems-biology techniques to analyze omics endpoints might be more informative to the MOA. For example, data from high-throughput approaches with genomics endpoints could help locate the upstream indicators of disease vulnerability. The biological processes that gradually lead to pathologic conditions should be described as pathways composed by continuous parameters that may have the potential to be used as disease biomarkers in the human studies.

#### 10.2.1.6 Combining Toxicological and Epidemiological Data to Explore Causality for Health Outcome

The epidemiologic and toxicological data are equally important for establishing a causative link of a human disease historically. From the perspective of public health, the most notable case is tobacco smoke and lung cancer. By 1964, epidemiological evidence had suggested a causal association between smoking and lung cancer. However, at that time, similar lung tumors could not be replicated in animal models, which implied the difficulty of consistency [23]. The relationship between epidemiology and toxicology has shifted with the development of each discipline. Both disciplines contribute data regarding exposure and human health outcomes and constantly rely on each other to derive a causality. Toxicology took on a predictive/explorative rather than a confirmatory role by providing clues for potential toxicity in humans. However, in epidemiological studies, the exposure scenario is confounded by a wide variety of influences from real life. Animal experiments may be controlled to a certain extent, but the results require extrapolation from observed effects at high doses to possible health outcomes in humans at much lower doses [22].

For example, animal experiments indicated that aflatoxin was an extremely potent liver carcinogen, at that time, the only relevant data in humans were correlations of liver cancer mortality rates with differences in estimated aflatoxin intake. However, the population with the highest incidence of liver cancers also have high prevalence of Hepatitis B infection. Not until later, stronger epidemiologic evidence became available (ecologic studies with more power and nested case-control studies in which pre-diagnosis urinary markers of aflatoxin intake was adopted), the causal effect between aflatoxin exposure and liver cancer was established.

To increase the power of animal studies, a systematic approach has been first proposed by the International Programme on Chemical Safety (IPCS) and later expanded substantially with the development of frameworks for evaluating the human relevance of mode of action (MOA) in experimental animals for carcinogens. Three questions need to be addressed to establish MOA of a chemical: (1) Is there sufficient evidence in animal studies to establish a MOA? (2) If so, is that MOA feasible in humans? and (3) If so—considering toxicokinetic and dynamic characteristics—would the MoA be feasible in humans? If the answers are yes to three questions, then the outcomes and its MOA seen in animals could be plausibly extrapolated in humans. Otherwise, the biological and toxicological plausibility of the effect is still in suspension [28, 29].

#### 10.2.1.7 In Silico Studies

With the development of cutting-edge techniques, theoretical models of "structure-activity relationships (SARs)" and "quantitative structure-activity relationships (QSARs)" analysis are widely used to predict the biological and toxicological properties of chemicals from the characteristics of molecular structure. The development of (Q)SARs requires expertise in molecular modeling and statistics, as well as basic knowledge in chemistry, biology, and toxicology [30, 31].

In chemical risk assessment, the information provided by (Q)SARs and other similar relevant property estimation methods, sometimes referred to as "*in silico* tests", can be used in combination with other tests by applying stepwise and/or weight-of-evidence approaches [31, 32]. A (Q)SAR model usually consists of three main components: the characteristics of the chemical on which the model is based (structural elements); the effects of the predictions (i.e., endpoints, responses); and the algorithm converting the descriptors into the endpoints (responses) of interest.

It should be noted that in vivo tests in most cases could provide data for developing both in vitro and in silico methods. Mode of action-based in vitro models and in vivo studies could also accumulate meaningful data for the development of (Q)SAR models.

#### 10.2.1.8 Weight-of-Evidence Approach to Evidence-Based Toxicology

Weight-of-evidence approach is the application of expert judgment to evaluate the strengths and weaknesses of all relevant studies, to compare the design and findings of them, and probably ranking the evidence to finally map the toxicological profile. WHO (2009) has defined weight-of-evidence assessment as "a process in which all of the evidence considered relevant for a risk assessment is evaluated and weighed" [32]. Weight-of-evidence is involved and valued in cases that more than one piece or one side of evidence is used to address one question. On the other hand, weight-of-evidence assessment is not needed where no evaluation and integration of data is required.

Many efforts have been focused on a systematic process that has been trying to reform and improve the risk assessment process, making it more rigorous and, importantly, more transparent and even reproducible, with explanations and justifications for decisions. Although such processes are important advances, it is not simply about a matter of finding the right data; the evaluation and integration of evidence is more important. For the purpose of transparency and science-based property, each risk assessment is required to clearly document: (1) the range of evidence considered and how it was integrated into lines of evidence; (2) the principles of the selection and weighing evidence while considering reliability, relevance, and consistency of the data; (3) the range and probability of the possible answers qualitatively or quantitatively [32].

The first step of weight-of-evidence is assembling the evidence, involving searching for and gathering evidence that is relevant for addressing the question. Then the evidence should be grouped into lines of evidence [13]. The categories of weight-of-evidence can be divided by in-depth degrees, i.e., (1) Listing evidence: presentation of individual lines of evidence without integration; (2) Expert judgment: qualitative integration of multiple lines of evidence with strengths and weaknesses evaluation of each study; (3) Causal criteria: a criteria-based approach for evaluation of the evidence in cause-effect relationships; (4) Logic models: evaluation of individual lines of evidence based on qualitative logic models; (5) Scoring: quantitative assessment of each lines of evidence using a set of weighting or ranking standards; (6) Indexing: integration of multiple lines of evidence into a single measurement based on empirical models; 7) Quantification: integrated evidence integration using statistical decision analysis and simulating methods [32].

Reliability, relevance, and consistency have been recognized as three basic considerations in weight-of-evidence assessment [32]. Reliability is how closely the integrated evidence represents the characteristic or event that it is supposed to refer to, including both accuracy and precision. Relevance means the contribution a line of evidence would make to address the concern of interest, which includes biological relevance and other relevance aspects, e.g., temporal, spatial, etc. Consistency is the extent to which the different pieces or lines of evidence are in accordance with the concerned interest. Apart from the three aspects, other factors that should be also considered in the weight-of-evidence assessment include, e.g., quality, coherence, risk of bias, specificity, and biological plausibility.

Uncertainty associated with the weight-of-evidence approach itself should also be considered, e.g., the selection and screening of evidence as well as the assessment of reliability, relevance, and consistency, and uncertainty which is already addressed in the weight-of-evidence process does not need to be re-analyzed, but may be added to the framework and specified in the final document [1, 3].

To guarantee an evidence-based principle, the weight-of-evidence approach can be applied in the assessment of the parameters required by each step of the risk assessment in food safety when feasible, which would certainly ensure the transparency and repeatability of the assessment. However, because the weight-of-evidence approach is time-consuming and resource-intensive, not all parameters in the risk assessment process should be estimated with the approach. In the risk assessment process, for each of the four steps, different lines of evidence may be available and can be integrated depending on the purpose of the assessment, and availability of data, time, and resources.

Here listed the literature search range in "Hazard identification and hazard characterization": toxicokinetics: in vivo and/or in vitro studies on ADME, in silico models; acute toxicity: in vivo and in vitro studies, case reports in human; genotoxicity: in vitro and in vivo studies, in silico models; subchronic toxicity/chronic toxicity/carcinogenicity: in vivo studies mainly, epidemiological studies and clinical trials; reproductive/developmental toxicity: in vivo and in vitro studies, epidemiological investigation and clinical studies; MOA information: in vivo and in vitro "omic" studies, in silico models, read-across extrapolations, threshold of toxicological concern (TTC), etc. Exposure assessment: concentration of the chemical in food (total diet study, food contaminants monitoring program, food composition tables, etc.), default values (e.g., maximum residue levels (MRLs) for pesticides residues, maximum adding limits for food additives, etc.); food consumption data (national food consumption surveys, food consumption and composition databases, etc.) [3, 26, 32].

The most formal weight-of-evidence activity concerns the classification of carcinogens. However, the process of weight-of-evidence from epidemiology and experimental studies pertaining to specific chemicals of interest also frequently leads to controversy and even extends debate among different scientific agencies, e.g., the Environmental Protection Agency (EPA), the International Agency for Research on Cancer (IARC), European Food Safety Authority (EFSA), the Institute of Medicine (IOM), and the National Toxicology Program (NTP).

There are several advantages of using a weighing-of-evidence systematic review methodology in the risk assessment in general, as follows [22, 32]. It introduces a structured approach in the process of identifying and selecting studies, allowing this process to be reproducible and transparent. It presents as a useful and uniformed method for data searching, evaluation, and synthesis in cases where there is a large amount of (heterogeneous) data and conflicting results. It has the potential to be better applied in risk assessment in a regulatory perspective. Combining and synthesizing available studies regarding the same or related endpoints may improve the overall confidence of the whole risk assessment.

### *10.2.2 Exposure Assessment*

An exposure assessment is the process of determination of the duration, frequency, and magnitude of exposure to a particular hazard, as well as the characteristics of the (sub)population exposed. In any dietary exposure assessment, the ideal scenario would be using the concentration of the hazards, either microorganism or chemicals, in the food at the time of consumption. Currently, under most circumstances, the contamination at the time of consumption is often substituted with national food consumption data to estimate dietary exposure. However, factors that need to be considered include sampling representability, time between sampling and likely consumption, samples storage temperature, microbial profile and its growth, chemical degradation rates, and chemical inactivation or degradation through cooking or other processes. Ideally, each influencing factor should be described by a distribution as a basis for a probabilistic statistical analysis [3].

Sometimes, it is not realistic to calculate the exposure levels of all individuals of one (sub)population. At the very least, exposure from average consumption pattern and worst-case scenario due to high consumption of contaminated food (the 95th or 99th percentile) should be investigated. Apart from the high-exposure scenario, the exposure assessment should also consider vulnerable populations, e.g., pregnant, or breastfeeding women, infants, toddlers, and other physically challenged groups.

The scenario of exposure comprises the source, mode, route, and extent of contact with the hazard of concern. There are several key points of exposure assessment, as listed below [1, 3, 12].

Exposure Duration and Frequency—While frequency describes how often the exposure happens, the duration is the length of time that the exposure lasts. For some hazards, exposure to a low dose for a long duration may be of greater concern even if the dose would not exert any health risks with short-term exposures.

Exposure Routes—The information of characteristics of hazards as well as the exposed population would help develop the appropriate exposure pathways to consider in the exposure assessment. In addition, it can be helpful to think of specific exposure activities that may put someone at higher risk.

Sensitive Individuals/(sub)Populations and Life Stages – Some individuals/(sub)-populations may be more susceptible to the hazards or more likely to develop severe manifestations of concerned health outcomes. For example, while healthy adults may not show symptoms with exposure to melamine, it can be deadly to infants with targeting kidneys. Common sensitive subpopulations and/or life stages may include but are not limited to: young children; the elderly; persons with compromised immune systems; pregnant women; chronic smokers; occupationally exposed individuals; other groups based on behavioral patterns (e.g., individuals who live on fish or corn). In most cases, dietary patterns due to personal preference or health status would certainly be associated with a higher risk of exposure; therefore, subjects with special dietary habits or specific dietary requirements should also be given serious attention. Depending on the scope of the risk assessment, information in ethnic groups, different socio-economic status, and different religions should also be collected to assess their exposure since ethnicity, socio-economic status, and religion might be correlated with particular consumption patterns.

In food safety risk assessment, data of chemical levels can be extracted from the following sources. (1) Legislated Maximum Permitted Levels (MPLs) or Maximum Residue Levels (MRLs), usually regulating the levels of chemical substances intentionally added to food or plants. MPLs or MRLs can be used as maximum contamination levels for a preliminary or screening exposure assessment in order to identify those substances that may exceed health guidance values in a worst-case scenario. (2) Monitoring and surveillance programs (e.g., food contaminants), in this case, food sampling procedures preceding analysis can critically influence how representative the data is to the true value in the overall distribution. Another major limitation of using monitoring data is that often not all commodities entering the food chain are monitored, which might be insufficient to estimate exposure from all possible dietary sources. In addition, sampling design, analysis, and reporting procedures are critical for obtaining consistent and comparable data on chemical concentrations in food. (3) Experimental data provided by applicants (e.g., novel pesticides), mostly submitted by the applicants in support of the registration. (4) Estimation from mathematical modeling, in particular, mathematical modeling is used to estimate the migration of substances from food contact materials. (5) A Total Diet Study (TDS), also called a market basket study, is designed to cover the average diet pattern in a country or by a specific population group. In principle, a TDS should provide the most accurate estimation of the average amount of a chemical actually ingested through meals, since the chemical levels are measured in meals as they have been prepared for normal consumption. However, the design and implementation of a TDS needs qualified professionals and is costly. It also should be noted that the TDS method might not be suitable for the assessment of acute dietary exposures.

Data on levels of chemicals in food as consumed are more accurate than levels in primary products. However, in many cases—especially for pesticide residues, natural toxins, and environmental contaminants—occurrence data are often only available for marketed food products. Therefore, efforts should be made toward the transformation of raw commodity data to ready-to-eat food data [3, 12].

Food consumption data includes data on solid foods, beverages, drinking water, and dietary supplements that individuals or subgroups consume. Food consumption data can be retrieved through national or targeted food consumption surveys at an individual or household level or estimated through food production statistics roughly from national data. The latter two can be used to derive a gross estimate of average food consumption per capita without indicating the distribution of consumption in the population. Individual dietary surveys that provide information on the distribution of food consumption in groups of individuals are preferred for dietary exposure assessment, among the various dietary survey methods, as indicated above, and duplicating portion study is recognized to reflect actual consumption more closely.

An exposure scenario is the combination of conditions about exposure sources, pathways, levels of the hazard, and the characteristics of the exposed (sub)-populations. While an exposure assessment may be comprised of many different scenarios. Scenario analysis would lay the basis for exposure model development.

The purpose of risk assessments is not only to assess the probability of adverse health outcomes, but usually to evaluate how risk management decisions could change the risk of the adverse health outcomes. Hence from the perspective of an exposure assessment, it usually includes scenario assumption and model estimation with the proposed change. For instance, risk managers may want the risk assessors to evaluate the effect of certain risk mitigation measures on the exposure estimation. In that case, a baseline exposure evaluation would be compared to exposure estimation based on the implementation of the mitigation measure, e.g., down-regulation of a limit level for a contaminant.

Several different methods used for exposure assessment range from quick worst-case scenario estimations to more accurate methods aimed at assessing actual exposure. As the accuracy of dietary exposure assessments increases, the cost of conducting the assessments also increases. The method mainly suits different purposes and depends on a few factors, including the availability of information, the cost-effective consideration. A stepwise approach is recommended in most cases [12].

There is an increasing interest in the potential use of biomarkers for assessing internal exposure to food-borne hazards. The approach takes account of exposure from all routes, including non-dietary sources, e.g., respiratory tract, skin absorption. Thus, it is important to acknowledge the extent to which exposures from other sources are likely to contribute, especially when trying to make a comparison of the results of such internal measurements with dietary exposure estimates. A number of biomarkers for specific individual food chemicals are currently being developed. However, very few methods have been fully developed and validated.

There is a need to further harmonize the exposure assessment methods and the modeling of high-exposed consumers for calculating dietary exposure using individual data. Emerging challenges for exposure assessment are simply summed as follows: the probabilistic statistical analysis should be conducted on a more routine basis; cumulative exposure assessment should be explored for functionally or structurally related chemical substances; statistical methodologies for the estimation of long-term intake from short-term data should further be looked into; infants and toddlers should always be considered as high-risk groups, since they often consume more with per kilogram.

### *10.2.3 Risk Characterization*

Risk characterization is the step in the risk assessment process where the results of the exposure assessment and the hazard characterization are compared, and whether it could result in an estimation of the risk. Risk characterization can be quantitative when exposure and dose-response values are available in the risk equation, or semi-quantitative when only some values are available. It is the final integrative step of the iterative risk assessment process. In this step, all the information from previous analysis should be integrated into a coherent, understandable, and informative conclusion that is useful to decision-makers and stakeholders. Risk characterization should communicate the key findings as well as the strengths and weaknesses of the assessment through a conscious and deliberate transparent report [3, 12].

Commonly, risk has three main components: the nature, magnitude, and probability of the hazard. In quantitative terms, risks are often described in terms of probability estimates ranging from zero (adverse health outcome will not occur) to one (absolute certainty that adverse health outcome will occur). There is a distinction in risk characterization between chemicals with and without threshold levels of toxic effects. In the case of chemicals without threshold levels, e.g., genotoxic carcinogens, a linear relationship is modeled between exposure (dose) and effect (incidence of cancer). For chemicals with threshold levels, the exposure estimation would be compared with the threshold level; if lower, the risk would be considered negligible; if higher, it would be assumed that risk should be taken into consideration, and exposure levels should be mitigated. In such cases, the probability of additional cancer cases occurring might be presented as a risk. For substances that do not exhibit a safe threshold of exposure (e.g., genotoxic carcinogens), the concept of margin of exposure (MOE) is adopted, which is the ratio of POD (e.g., BMDL) with the theoretical or estimated exposure level, reflecting the interval between the human exposure and the dose with a "known" risk level. The MOE can be used to compare the exposure risk of various genotoxic carcinogens, particularly, to help risk managers in prioritizing chemicals. As yet, there is no consensus on the "safe" value of MOE, for instance, EFSA suggested that an MOE higher than 10,000 might be regarded as a low level of concern [10, 12].

U.S. EPA has used MOA information in dose-response assessment. When a DNA reactive and direct mutagenic chemical is assessed, or the expected exposure is relatively high, or the body burdens near doses associated with key adverse events, a non-threshold approach is recommended, which assumes that risk below the POD would decrease linearly with dose. For carcinogens with sufficient MOA data to imply nonlinearity at low doses, e.g., the chemicals acting through a cytotoxic MOA, the approach of threshold is applied [25, 29].

ALARA is an acronym for As Low As Reasonably Achievable, which is initially developed for chemicals without threshold [29, 33]. It is a policy used to minimize known risks by keeping exposures as low as reasonably possible. Currently, it is also used for chemicals with threshold, taking into consideration cost, technology, benefits to public health and safety, and other societal and economic concerns. ALARA today is mainly used in the context where risk management limits are not based on a toxicological threshold, but rather based on "acceptable risk". In these cases, it is reasonable to take protective measures to minimize the risk that can be presumed to exist even at levels below regulated limits.

Besides the main comparison with health-based guidance values, risk characterizations should also include uncertainties, variability, potential impact of alternative assumptions, strengths/weaknesses of the data, and discussions of the scenarios, models, and default parameters that may deserve further consideration. Risk characterization should include the following information.

Several key points should be taken into consideration: (1) the major risk estimates and the extrapolations; (2) the selection of default parameter values; (3) potential policy choices and risk management decisions; (4) the characteristic of the key data used; and (5) uncertainty and variability [3, 12].

- Context: (1) handling of the food safety issues or incidences and the risk management questions; and (2) discussion of current regulatory measures and regulatory requirements.
- Susceptible Subpopulations: the scope of people that may be affected, including innately susceptible populations (e.g., gender, ethnic groups, other genetic predisposition) and those with risky behaviors (e.g., socioeconomic and/or nutritional status).
- Life Stages: (1) the age groups evaluated; (2) life stages that may have particular vulnerability due to behaviors or some other situations that affect exposure patterns and/or inherent susceptibilities.
- Assumptions: (1) where key data gaps exist; (2) what are the key assumptions used during the assessment procedures; and (3) how the assumptions impact the assessment outcomes. The data sources for these key assumptions need to be cited and any adjustments in the data should be discussed.
- Key Conclusions: (1) the key finding that has to be communicated of the risk assessment; (2) the subset of findings that support the key points that is significant in the assessment outcome.

- Variability: (1) how variability arises from true heterogeneity; (2) whether the values of some variables used in the assessment could change with time and space or across the population whose exposure is being calculated.
- Uncertainty: (1) uncertainties in the assessment, e.g., data gaps, measurement uncertainty, and model uncertainty; (2) how uncertainty is addressed or presented, such as uncertainty analysis and sensitivity analysis; (3) is there any measure that could be taken to reduce scientific uncertainties in the assessment.
- Strengths and Weaknesses: the strongest and weakest evidence for the conclusions, the quality of the data used, and how the data quality pertains to variability and uncertainty.
- Alternatives Considered: (1) if there are plausible alternatives to the risk estimation and how to deal with the alternatives (e.g., alternative models that could be used, different exposure pathways); (2) the comparisons among the alternatives.
- Research Needs: (1) the key data needs, e.g., toxicological data; and/or (2) the methodology gaps, e.g., detection methods or statistical modeling methods.

### *10.2.4 Uncertainty Analysis*

Uncertainty is a lack of knowledge; therefore, uncertainty can often be partly reduced by obtaining or generating more evidence. On the other hand, variability is a natural phenomenon and cannot be reduced, but can be described. Variability refers to true differences due to heterogeneity and natural diversity. The aim of uncertainty analysis is to identify major sources of uncertainty in either hazard or exposure assessment [3, 12]. Any risk assessment carries uncertainty with it. Characterizing uncertainty and variability is significant in risk assessment process, which must make full use of the best available science in the presence of uncertainties and variability to help risk managers to make better-informed decisions.

One of the traditional uncertainty factors considered is that the derivation of a NOAEL or BMD of a particular substance is only the starting point in the process of deriving a human health-based guidance value, e.g., Reference Dose (RfD), Acceptable Daily Intake (ADI), or Provisional Tolerable Weekly Intake (PTWI). To achieve the goal, it is inevitable to deal with the differences between the animal experimental design and the human exposure situation, while taking into account variability and uncertainty, which is usually done by applying "uncertainty factors" (UFs). The most frequently used extrapolation factors include: interspecies differences, intraspecies differences, differences in exposure duration, extrapolation from LOAEL to NOAEL, issues related to dose-response, and quality of the database. Uncertainty in risk assessment can be divided into four types [12, 14]:

1. Lack of information. When essential data are lacking, the use of expert judgment, assumption methodologies or default values become necessary.

2. Observation uncertainties. Observation uncertainties include spatiotemporal variability, differences between natural occurrence and laboratory observation, and differences between tested/observed species and species of interest for risk assessment.
3. Measurement uncertainties. Measurement uncertainties include difficulties in indicator measurements, inaccuracy of measurement methods, low statistical power, and human error (incorrect measurements, misidentifications and computational errors, etc.).
4. Inappropriateness of models. Inadequacies of models include a lack of knowledge concerning underlying mechanisms of the action of concern, extrapolation beyond the range of observations, and instability of parameter estimates.

In risk assessment of chemicals, both deterministic and probabilistic approaches dealing with uncertainty are used. The advantage of the simple deterministic approach is quick and easy to conduct without having to specify and describe uncertainties that are difficult to estimate. It also could avoid the problem of communicating risks in terms of probability and statistics that is hard to comprehend for most people. The deterministic approach has proven to be very efficient in providing advice for regulatory decisions. However, the disadvantage of the deterministic approach is that when several reasonable worst-case assumptions could be combined to result in unrealistic assessment results, it might also give a false sense of accuracy and ignore variability.

To ensure that the judgments are consistent, explicit, and not unduly influenced by risk-management considerations, default values of UFs were recommended to be developed independently of any particular risk assessment [3, 12]. Defaults of UF can be applied in the following extrapolation: extrapolation of results between species; extrapolation across human populations; extrapolation of metabolic pathways across species, age groups, and genders; extrapolation of toxicokinetic characteristics across species, age groups, and genders; data gap in some cases.

In a quantitative uncertainty analysis, both uncertainty and variability in different components of the assessment (dose-response relationship, exposure, and pharmacokinetics) are combined by using an uncertainty-propagation method, such as Monte Carlo simulation, with two-stage Monte Carlo analysis utilized to separate uncertainty and variability to the extent possible. Uncertainty calculation depends on the quality, quantity, and relevance of data, and the selection of models and assumptions is also quite crucial.

Variations among individuals in a population with respect to susceptibility and exposure should be valued. Many approaches described above regarding uncertainty analysis are applicable to variability analysis [1, 3]. For example, probabilistic approaches, especially Monte Carlo methods, can be used to describe variability throughout all steps of a risk assessment, and in this process, expert elicitation can be used to characterize various percentiles in a distribution. However, even with similar modeling approaches, it should still be kept in mind that variability can only be better characterized, not reduced, so it often must be addressed with different strategies from those used to address uncertainty.

Susceptibility factors include any factor that could increase (or decrease) the response or toxic manifestation of an individual to a dose compared to the response of an average individual in the evaluated population [1, 3]. Difference in susceptibility and distribution of exposures would contribute to the final distribution of adverse health outcomes in a population. Taken together, variations in disease susceptibility and exposure potential give rise to potentially important variations in vulnerability to the effects of chemicals.

Here list some reasons why uncertainty should be analyzed. (1) Characterizing uncertainty in risk informs the affected public about the range of possible risk estimates from an exposure. (2) Characterizing the uncertainty in risk is also associated with risk management measures by communicating about the range of potential outcomes resulting from the risk management decisions, which in turn assists in evaluating decision alternatives. (3) Mathematically, it is often not possible to understand what may occur on average without understanding what the possibilities are and how probable they are. (4) It can guide directions of research strategies by providing the basis on how much the research is expected to reduce the overall uncertainty in the risk estimate and how the reduction in uncertainty leads to different decision options. (5) Acknowledging uncertainty adds to the credibility and transparency of the decision-making process [3, 12, 34]. Also, variability should be assessed for the following reasons. (1) Assessing variability in risk illustrates the risk among the population better, while the risk managers could focus on the people at greatest risk. (2) Understanding how the population may vary in risk can facilitate understanding of the shape of the dose-response curve. (3) By understanding how variants contribute to the risk, people might make their own changes to mitigate their personal risk, e.g., cooking thoroughly or eating fewer deep-sea fish [3, 12, 34].

U.S. EPA has made recommendations on the principles of uncertainty and variability analysis. (1) A quantitative, or at least qualitative, description of uncertainty and variability with available data should be provided. (2) In addition to describing the full population at risk, careful consideration should be directed to vulnerable individuals and subpopulations that may be particularly susceptible or more highly exposed. (3) The detailed approaches of the uncertainty and variability analysis should be commensurate to inform the risk manager about the importance and nature of the decision. (4) The risk assessment should characterize the type, source, extent, and magnitude of variability and uncertainty associated with the risk assessment. (5) To maximize public comprehension and participation, the basis and results of the uncertainty analysis should be illustrated in a clear way for the public and risk managers. (6) Uncertainty and variability should be kept separate and marked in the risk characterization [35].

When presenting the results, a balanced and impartial treatment of the information should be the goal, with the key assumptions and uncertainty highlighted. Characterization of the uncertainty will generally include a qualitative discussion in specific exposure scenarios. An independent judgment about the validity of the conclusions reached by the assessor should be informed by describing the uncertainty associated with any extrapolations and simulations used and the weight-of-evidence that led to the particular conclusions.

### *10.2.5 Statistical Approach in Risk Assessment*

From the statistical perspective, two types of uncertainties can be distinguished: quantifiable uncertainties and undefined uncertainties that cannot be described or quantified. There are some statistical approaches or softwares designed for the implementation of risk assessment, especially uncertainty assessment. Computational model is constantly referred to in the statistical analysis of risk assessment as a model that is expressed in formal mathematics with equations, statistical relationships, or a combination of the two. However, values and expert judgments are inevitably embedded in the statistical analysis, scenario assumptions, and selection of default parameters.

#### 10.2.5.1 Monte Carlo Analysis

Monte Carlo analysis is a commonly used quantitative technique for exposure assessments [36], e.g., the contamination levels of the chemicals in food, food consumption, etc. It involves the random sampling of each of the probability distributions in a model to estimate the likelihood of the model's possible results. Each recalculation of the model is an iteration; and a set of iterations constitutes a simulation.

In the Monte Carlo simulation, each variable is inputted as a probability distribution. Then numerous values are generated by repeated random sampling from the probability distributions for each variable, and the values are used for calculating the parameter of interest, in the case of risk assessment, mean exposure and high-end exposure levels could be estimated. The variation in the outcomes represents the overall uncertainty for the parameter of interest. A recognized rule of this analytical approach is that every iteration should be possible in nature.

#### 10.2.5.2 Sensitivity Analysis

Sensitivity in risk assessment means the degree to which the outputs of a quantitative assessment are affected by changes in input parameters [3, 12]. Therefore, sensitivity analysis examines how a model's each input impacts on its output. A well-designed exposure assessment model should comprise inputs that influence exposures, among which contamination levels and food intake are the necessary two kinds of inputs. There is no universal recommendation for conducting sensitivity analysis. Sensitivity analysis may be conducted as part of an exposure assessment or risk characterization.

#### 10.2.5.3 Quantitative Uncertainty Analysis

The goal of uncertainty analysis is to identify those model inputs in which uncertainty substantially contributes to the total uncertainty about exposures or risks. Although the objective of uncertainty analysis differs from sensitivity analysis as stated above, results of uncertainty analysis are usually not independent of the implications of sensitivity analysis. In other words, if uncertainty analysis showed that an input contributes to substantial uncertainty about the model's results, then it can also be suggested that the input is also quite possible to be identified as highly influential in sensitivity analysis.

### *10.2.6 Cumulative Risk Assessment*

The actual common scenario would be simultaneous exposure to mixtures of chemicals, multiple pathways of exposure, various routes of exposure, and different timeframes of exposure. Cumulative risk is defined by U.S. EPA as the combined risk from aggregate exposure to multiple agents or stressors. In fact, it is quite paramount to understand the effects of co-exposures to agents that have similar structures and/or MOAs [37].

Cumulative risk assessment, then, is an analysis, characterization, and possible quantification of the combined risks to health or the environment from multiple agents or stressors [37, 38]. Evaluation of any particular mixture of agents requires an empirical determination of how they combine to produce any concerning effect. Experiments should be conducted to decide if the mixtures may be considered to exhibit dose addition or independent action for that particular effect. Dose addition mode may apply at concerning levels of effect. It would require examination of all effect levels or at least over the range of effect levels of practical concern.

Some groups of chemicals may have similar chemical structures and act in similar ways in the body, e.g., be absorbed in the same way, be detoxified in the same organ by the same enzyme systems, etc. In such circumstances, it is possible and may be plausible to assume the dose addition action, in which each agent could be compared directly with one specific reference agent in the chemical group by using a relative potency that specifies the effective magnitude, which means the group may have parallel dose-response curves for the specific effects. In other words, a dose of one agent is equivalent to a multiple of the dose of the others in this chemical group. However, it should be noted that relative potency may differ for different endpoints.

A toxic equivalence (TEQ) approach has been adopted to estimate co-exposed risk, for example, co-exposed risk associated with 2,3,7,8-TCDD and other dioxin-like congeners. The 2,3,7,8-chlorinated dibenzo-p-dioxins and -furans and for "dioxin-like" PCB congeners, have been recognized relative to the prototype of the 2,3,7,8-tetrachlorodibenzo-p-dioxin group, and in this case, to apply to "dioxin-like" effects regulated via the aryl hydrocarbon receptor. Toxicity equivalence factor

(TEF) is applied, assuming a model of dose addition with AhR-mediated effects. Each dioxin-like congener has been assigned a TEF to represent the fractional toxicity of the congener compared to that of 2,3,7,8-TCDD. TEFs of each congener are used to transform concentration of each dioxin-like congener into an equivalent concentration of 2,3,7,8-TCDD, and combing with toxicity data of 2,3,7,8-TCDD, the risk exposed to dioxin-like congener mixtures is estimated [39].

Cumulative risk assessments of chemicals with a common mode of action involve dose-response modeling of the same effect for each chemical, which could provide the relative potencies used in the dose-additivity-based cumulative models for these chemicals. A typical example of cumulative risk assessment is quoted from EPA documents, as follows.

Benzo[a]pyrene (B[a]P) and six other polycyclic aromatic hydrocarbons (PAHs) are all categorized as B2 carcinogens by EPA. Results with cancer bioassays were consistent among B[a]P and these PAHs. However, data were not sufficient to derive cancer slope factors for each PAHs. And although with the same carcinogenic mechanism, these PAHs appeared to be less potent than B[a]P. Therefore, a relative-potency approach was applied to estimate cumulative cancer risk associated with these PAHs by comparing cancer potencies with results of skin tumorigenicity bioassays. The "order of magnitude" relative potency factors (RPFs) for the six carcinogenic PAHs were calculated based on comparison with the index chemical, i.e. B[a]P. This RPF approach can be used in the cumulative risk assessment of PAH mixtures as they often co-occur in the environment [40].

Biomonitoring data could have great potential in cumulative risk assessment if used well [12, 41]. For instance, in the case of organic phosphorus pesticides, if multiple stressors are thought to influence acetylcholinesterase inhibition, simultaneous collection of compound-specific biomarkers, and biomarkers of effect can provide extra insight into the combined effects of the exposures. Biomonitoring could provide valuable biologic samples which allow measurement of simultaneous exposure to multiple agents, even though exposures to each of the compounds individually would be difficult to model.

### *10.2.7 Thresholds of Toxicological Concern*

Thresholds of toxicological concern (TTCs) have been used in the risk assessment of chemicals to which humans are exposed at very low levels. TTCs are exposure threshold values for chemicals below which no significant adverse health outcome is expected to occur [41, 42]. The approaches of TTCs have only been adopted under circumstances where there is only limited or no information on toxicity of the chemical and where human exposure is so low that conducting further toxicity studies is considered not cost-effective. The potential validity of the TTC concept for risk assessment is largely determined and warranted by the adequacy of the exposure information. In other words, the establishment of TTCs is based on the principle of establishing a human exposure threshold value for chemicals, and assuming that the probability of health risk below the value is very low.

The establishment of TTC values is based on the systematic analysis of toxicological and structure-activity data of a broad range of chemicals. The application of TTCs relies on expert judgment on the basis that the exposure of the chemical is extremely low and toxicity tests are unnecessary. TTC approach would help concentrate limited resources on the toxicological evaluation of substances with greater potential to actually induce risks to human health. The first step of TTCs is the identification of possible genotoxic and/or high potency carcinogens, then non-genotoxic substances are evaluated, which are related to the concerns that would be associated with increasing exposure. The U.S. Food and Drug Administration has adopted the TTCs approach for substances used in food contact materials, as follows: If a substance has not been shown to be a carcinogen in humans or animals as well as in QSAR analysis, a threshold of limit is defined as a dietary concentration of 0.5 μg/kg food/drinks or 1.5 μg/person/day when assuming a consumption of 3 kg diet per day. The TTCs values might be different in other organizations for risk assessment, but the basic principles are the same [41, 42].

### *10.2.8 Comparative Risk Assessment*

Risk comparison actually has been done in daily life back in old days. Should we drink dirty water or go thirsty? Should we eat an unfamiliar mushroom or go hungry? It is only since the late 1980s that the comparative risk paradigm has entered into application in the decision-making process. It is currently more commonly used to help understand the relative impacts of different threats or hazards in the same scenario, for example, skin cancer from sunlight or vitamin D deficiency, which is worse? Like risk assessment, comparative risk assessment is a structured approach to analyzing and comparing the outcomes of different scenarios to help make the most appropriate risk-related decisions.

Comparative risk analysis supports trade-off decisions, which could be the foundation of policy weighing [41]. Therefore, the implementation of comparative risk involves a series of assumptions and value judgments. The traditional approach for comparison requires considering multiple parameters into a common unit, and mathematical optimization procedures are quite important.

From the narrow perspective of health risk assessment of chemicals, for instance, a margin of exposure (MOE) approach can be conducted to compare the risk of alcohol, tobacco, cannabis, and other illicit drugs. The MOE is calculated as the ratio between toxicological threshold and estimated human intake. Substances with MOE $< 10$ were included in "high risk" category, those with MOE $>100$ were "risk", and those with MOE $>10{,}000$ were considered to be "low risk". The toxicological MOE approach could be an example of health-based comparative risk assessment [43].

## 10.3 Risk-Benefit Assessment

### *10.3.1 Overview of Risk-Benefit Assessment*

Nutrients and foods, as well as diets, may present both potential risks and potential benefits to consumers. In other words, the beneficial and adverse effects may occur concurrently through ingestion of a single food item, e.g., fish, whole grain products, vegetables, or a single nutrient or food component, e.g., folic acid, phytosterols. The balance between risks and benefits is of interest to consumers considering dietary changes, to food/nutrition authorities developing food policy and advice, and to food industries developing novel food products. However, currently, most information on risks and benefits is presented separately. A scientific approach should be applied to aseess the risks/benefits of high or inadequate intakes in an integrated way that takes into account the optimal nutritional enhancement for relevant population groups. Therefore, risk-benefit assessment aims to scientifically evaluate the risks and benefits of the same target. In the risk-benefit assessment, the probability of an adverse health outcome as consequence of exposure is weighed against the probability of benefit from the same exposure [44, 45].

From the food safety and nutrition perspective, the implementation of risk-benefit assessment would be needed in the following situations, but not limited to these [44, 45]:

- Where a single component or food constituent has both positive and negative health outcomes, whether the effects occur in the same subpopulation, or target different subpopulations.
- Where similar levels of dietary exposures are associated with both health risks and benefits.
- Where positive and negative health effects induced from different components co-exist in the same food, e.g., DHA/EPA and heavy metals from deep-sea fatty fish, the main potential beneficial effects of which are related to prevention of cardiovascular diseases or enhancement of neurodevelopment of fetuses/infants by n-3 fatty acids, need to be weighed against the potential adverse health outcomes of environmental persistent pollutants such as dioxins, PCBs, or heavy metals in it.
- Before the initiation of community intervention, such as folic acid fortification, or iron fortification.
- Where a significant change of diet patterns has occurred or may occur, e.g., substitution of sugar by low-calorie sweeteners.
- Where the beneficial effects of novel food process methods or food additives, such as prolonged retention duration of nutritional value resulting from novel food additives, may need to be assessed against the possible negative effects, e.g., a greater survival of foodborne pathogens.
- Where new evidence emerges with major implications for either the risks or the benefits in a previous assessment, for instance, the possible association between folic acid intake and colon cancer.

Actually, risk-benefit assessments have been performed in different disciplines for a long time, even the perspectives may be different and the approaches may be various. Most similar areas are pharmaceutical risk-benefit assessment, e.g., assessment of the benefits and risks of a new drug application. Currently, most risk-benefit assessments would include socioeconomic considerations or aspects other than human health.

Similar to risk assessment, risk-benefit assessment requires substantial data or assumptions, and is affected by many uncertainties. Furthermore, risk-benefit assessment requires a high level of expertise in the disciplines of toxicology, nutrition, epidemiology, social medicine and health; and for quantitative risk-benefit assessment, excellent understanding of statistics is also required.

It is universally recognized that the risk-benefit assessment should comprise three parts, i.e., risk assessment, benefit assessment, and risk-benefit comparison [46, 47]. Similar to the risk assessment paradigm which has been well developed for food safety, the benefit assessment should also include the four main steps: positive health agent identification, positive health agent-effect characterization (dose-response assessment), exposure assessment, and benefit characterization. Identification of positive health effects is more difficult than hazard identification. A number of endpoints have been proposed for the assessment of positive health effects, e.g., DALYs/QALYs, neurologic and cognitive function, metabolic function, wellbeing, etc. The methodology for quantitatively assessing such beneficial endpoints is less well developed than that for assessing adverse health outcomes. The exposure assessment is positioned as a central part of the risk-benefit assessment and should take into account all relevant dietary and non-dietary sources. In some cases, risks and benefits may occur in the same subpopulation. However, the benefit (s) may be greater in one subpopulation, while the risk(s) may be greater in a different subpopulation, which presents as a big challenge. The risk-benefit comparison will weigh the risks against the benefits. It is expected that both hazardous and beneficial effects can be taken into account and potential risks and benefits should be evaluated by an integrated measure of health, which expresses both risks and benefits on the same scale. As the framework of risk assessment has been widely used in the area of food safety and sometimes nutrition, it is increasingly recognized that a similar paradigm is needed and should be constructed for both benefit assessment and risk-benefit assessment. The comparison is preferable to contain a means, to be quantitative, if possible, to compare/weigh the potential health risks against the potential health benefits. For this, an integrated measurement (also called "composite metric") for the risk and the benefit would facilitate the comparison and weighing.

### *10.3.2 Consideration for Stepwise Approach*

Due to the data-demanding and time-consuming of a quantitative risk-benefit assessment, the stepwise approach for risk-benefit assessment has been recognized to be scientifically efficient regarding time and resources. By following a stepwise framework, an estimation/recommendation could be reached after a qualitative or semi-quantitative assessment, and a full quantitative assessment may not be necessary.

European Union has organized experts to develop a tiered approach for risk-benefit assessment (BRAFO-approach), which comprises four Tiers, which differ mainly in the way how the risks and benefits are compared [46, 47]. At Tier1, risks and benefits are assessed separately, while in Tiers 2 ~ 4 they are integrated using increasingly sophisticated and statistically accurate approaches. With tier increasing, the net health effect is assessed with increasing accuracy. In Tier 1, each potential risk and benefit is assessed independently. If the estimated dietary exposure indicates no appreciable risks, or no appreciable benefits, then there would be no need to compare risks and benefits. In Tier 2, risks and benefits are compared in a qualitative way separately in cases, but judgments could easily be reached without the need to further quantify, seeking a common metric. In Tier 3, risks and benefits are compared quantitatively with a common metric, using deterministic statistical methods, and at this stage variability and uncertainty are not analyzed quantitatively. In Tier 4, risks and benefits are again integrated quantitatively using the same common metrics as at Tier 3, with the addition of probabilistic methods to quantify variability and uncertainty.

The tiered approach developed by the National Institute for Public Health and the Environment (RIVM) of the Netherlands also comprised stepwise instructions to allow for a timely evaluation to decide whether the gathered data is sufficient to answer the initial risk-benefit question [44, 46]. As compared with the steps of risk-benefit framework proposed by EFSA, exposure assessment has been moved upward while the dose-response deriving is moved afterward. The advantage of giving priority to the exposure assessment in risk-benefit assessment is that in case of no or very limited exposure in risk groups, the risk-benefit assessment can be terminated at an early stage, before any more source and effort is spent.

### *10.3.3 Uncertainties in the Risk-Benefit Assessment*

Uncertainty has been described in the above "risk assessment" part. Most of uncertainty results are from limitations in scientific knowledge, which can often be reduced by further investigation. Uncertainty should be characterized at each step of the assessment, and it provides essential information for decision-making and identification of data needs. The uncertainty analysis in toxicity is relatively general, and it can be applied to both adverse and positive health effects, and their net health impacts after conversion into a composite metric. The uncertainty characterization in the adverse health outcomes and the positive health effects differs depending on whether the data are from animal studies, human populations, or subgroups. The "Monte Carlo simulation" is still a generally applicable approach of evaluating uncertainties in risk-benefit assessment.

### *10.3.4 Comparison of the Common Metrics for Risk and Benefit*

Several risk-benefit assessment approaches have been developed over the last decades. However, a common metric is usually needed nearly in all approaches between risks and benefits, e.g., the disability-adjusted life years (DALY) and quality-adjusted life years (QALY). The choice of composite metrics should be made on a case-by-case basis, and the influencing factors include risk-benefit question, identified hazards, and potential positive health effects. Health effects can be evaluated in a number of different dimensions, e.g., incidence of effect, severity and reversibility of effect, morbidity and mortality rate, and DALY/QALY [48]. It is unrealistic to evaluate all dimensions of health for a risk-benefit assessment. A composite metric is a measurement expressing risks and benefits in the same unit. The outcome of the risk-benefit comparison can be expressed as a single net health impact value. An ideal composite metric should reflect every dimension of health, such as morbidity, severity and mortality of the diseases. It is meaningless to compare the incidence of a minor disease with that of a major disability. Comparison of the incidence of the same disease may also be problematic due, to some extent, to differences in severity or age groups affected. Even though mortality metrics are more directly comparable, they still have limitation that mortality does not take into account the severity of the cases [44, 45]. The DALY/QALY combines incidence with life years to obtain an estimate of years saved or lost respectively; however, it doesn't capture all the dimensions, especially considering the effects are varying in children and adults.

#### 10.3.4.1 DALY/QALY Approach

As stated above, the most used composite health metrics in food-related risk-benefit analysis are the DALY (disability-adjusted life years) and the QALY (quality-adjusted life years). The EU project has developed a detailed methodology and software for DALY/QALY assessment of dietary choices, i.e. QALIBRA (www.qalibra.eu), which can integrate the risks and benefits of dietary change into a single measure of net health impact. QALYBRA offers a mathematical and computing framework. It can model and address a wide range of dietary risk-benefit problems with the necessary inputs. QALYBRA also aims to raise awareness of the necessity to include uncertainty and variability analysis, and it provides a flexible approach, allowing gradual progression from simple assessments using point estimates to refined assessments using distributions to quantify uncertainty [49, 50].

The procedures of DALY/QALY approach of risk-benefit assessment should comprise the following steps: (1) Problem formulation, including specification of the population to be considered and food or dietary scenario of concern. (2) Identification of the adverse and positive health effects to be assessed. (3) Calculation of exposure estimates under which the adverse and/or positive health effects occur.

(4) Modelling the dose-response relationship for each effect evaluated. (5) Input of the probability of every consequence, i.e., recovery rate and mortality for affected individuals. (6) Selection of a common metric (DALY or QALY), and specification of the severity and duration of the effect. (7) Calculation of the net health effect, and quantification of uncertainty. (8) Qualitative assessment of unquantified variability and uncertainty [49, 50].

In the QALY concept, 1 represents full health and 0 represents death. Traditionally, QALYs are frequently used to measure health gains with social medicine policies, e.g., to compare two community health intervention strategies. QALYs of all individuals are summed, and the scenario with the highest number of QALY represents the preferred health promotion measures. To generate QALY values, health states reflecting physical, social, and emotional well-being are valued [48]. Individually, QALYs are calculated by multiplying the duration of the disease by a quality weight (quality reduction due to the disease) and adding this to the age of onset of disease. The life years reached at age of death or an earlier reference point are adjusted for their lack of life quality and result in an age number which is lower than the true age of death or age at an earlier reference point. In the DALY context, 0 represents no disability and 1 represents death. A set of standard weights covering specific diseases needs to be followed for calculation. The Global Burden of Disease 2010 study has revised and issued many disease weights, which are available on the WHO website. For example, the disability weight for cancer in the metastatic phase is 0.75 and for first-ever stroke is 0.92. However, disease weights might be different for specific regions and countries. DALYs are calculated initially to communicate a population-aggregate measure of loss of health—the Burden of Disease used by WHO. DALYs of all individuals of the targeted population are summed, then the scenario with the lowest number of DALY represents the highest health enhancement or lowest health loss. One DALY represents the loss of the equivalent of 1 year of full health. The years of life lost (YLL) and the years lived with a disease (YLD) are added and thus result in a number which is higher than the true years lost. Often, the choice for DALY or QALY is a pragmatic one, based on data availability or experience, or expert judgment.

It is noted that benefits and risks incorporated in one integrated measure may occur in different subpopulations. Hence it is important to always present results of the DALY or QALY with the distribution of separate risks and benefits in subpopulations. The data requirements for calculation of a common currency are high, and it is a big challenge to come up with suitable inclusion data. Furthermore, several other drawbacks of the approaches have been proposed, including the difficulty of quantifying specific beneficial and/or adverse effects, the applicability to specific target populations (e.g., children, pregnant women, elderly), and the inability to fully utilize toxicity data of animal studies.

#### 10.3.4.2 The ED50/CV Approach

Considering the disadvantages of the approach of DALY/QALY, it may not be applicable for the risk-benefit assessment for micronutrients. The approach developed by an International Life Sciences Institute (ILSI) Europe expert group in 2004 is more recommended for a risk-benefit assessment of micronutrients from dietary supplements [51, 52]. Since the purpose of taking dietary supplements is for beneficial effects, the approach basically is based on two intake-incidence relationships, one for the absence of benefit, and the other for toxicity. Each dose-response curve is derived from a 50% effect dose (ED50), which is the dose that has an effect on half of the population. A coefficient of variation (CV) represents variation among the investigated population, and is used as the slope of the effect curve. Based on the ED50 values both for benefits and risks and the CV, intake levels can be related to probability that a specific effect may occur. Taken together, the two curves could depict a shape for the dose range between benefits and risks for a given population. Furthermore, the uncertainty regarding the data used is included in this methodology. In the cases where effects are different for different subpopulations, the curve can easily be remodeled. Another advantage is that this approach can be applied to different chemical forms of the same micronutrient, and with the knowledge of bioavailability and other differences, the curve can be adjusted accordingly [51, 52].

By considering both risks and benefits, the approach explores an optimum range of dietary supplement intake, while bioavailability for sub-population of interest and the severity of effects are included in decision making. The above risk-benefit approach will greatly facilitate proportionate risk management decision-making about micronutrient fortification.

### *10.3.5 Case Studies of Risk-Benefit Assessment*

A couple of case studies have been performed to explore the various approaches of risk-benefit assessment [53–56]. These case studies deal with specific foods that may deliver both risks and benefits, e.g., vegetables, fish, whole grain cereals, soy, and food components that may confer both risks and benefits, e.g., folic acid, vitamin A, phytosterols, and dietary recommendation or food component substitution scenarios, e.g., adding sugar/artificial sweeteners. The most investigated topic is fish consumption, especially deep-sea oily fish.

Consumption of oily fish is often recommended based on its nutritional benefits, but there is a concern about a number of contaminants that can be present in different types of oily fish. The beneficial components of oily fish include long-chain n3-polyunsaturated fatty acids (n3-PUFAs), e.g., eicosapentenoic acid (EPA), docose hexenoie acid (DHA), various kinds vitamins and essential elements, and proteins that can be better digested and are less associated with saturated fat. It would not be feasible and realistic for a risk-benefit assessment to consider all potential beneficial components as well as contaminants one by one. There have been several risk-benefit assessments about fish, most of which have generally focused on the n3-PUFAs, for which fish is the major dietary source [53–55]. However, there are considerable kinds of chemical contaminants present in oily fish, e.g., persistent organic pollutants and heavy metals. Methylmercury is frequently found in large predatory marine fish, while other contaminants may result from specific pollution incidents. The risks and benefits usually target n3-PUFAs against methylmercury, and the positive and negative health outcomes with targeting subpopulation could be described as follows: for pregnant women, the adverse health concern is impaired neurodevelopment of the fetus while the benefit is enhanced neurodevelopment; for general adults, especially middle-aged and elderly, the benefits and risks are cardiovascular effects; and for children, the effect of concern is still neurodevelopment.

In several earlier studies, the estimated intake of contaminants, mostly methylmercury in this case, was simply compared to its health-based guidance value as well as recommendation of the food itself, the Provisional Tolerable Weekly Intake (PTWI) for methylmercury of 1.6 μg/kg.bw, and consumption of at least one portion of oily fish per week (the recommendation of some agencies for beneficial health effects), respectively. In other studies, dose-response relationships for MeHg and omega-3 FA effects on coronary heart disease (CHD) and neurodevelopment were identified, then an equation was developed and each endpoint was calculated, using dose-response relationship and exposure information. The increased risk (e.g., coronary heart disease from exposure to methylmercury) was then subtracted from the decreased benefit (on heart disease, through n3-PUFAs), and then net health effects were calculated in the coronary heart disease perspective. A DALY/QALY approach considering CHD was also conducted to bring out the best recommendations [53–56].

Other than only taking account of methylmercury, a composite approach describes benefits and risks by dose curve to identify a zone of benefit, which would be above the benefit threshold and below the harm threshold. Separate dose-response curves for positive and adverse effects are combined into net benefit-risk composites. Benefit Cancer Risk Ratio (BCRR) and Benefit Non-cancer Risk Ratio (BNRR) have been calculated to represent the rate of consumption of n-3 fatty acids (g/d) while controlling for the cumulative level of acceptable carcinogenic or non-carcinogenic risk of contaminants in fish [55, 56].

### 10.3.6 Challenges in the Risk-Benefit Assessment

A general problem in the disciplines underlying risk-benefit assessment is that good dose-response data are scarce, i.e., effects at relevant intake levels and suitable for the target population are almost impossible to get. Lack of data is a common reason why assessors initially aiming to perform quantitative risk-benefit analysis turn to qualitative analysis. The problem of data gap can sometimes be relieved by adjusting the initial problem formulation or by using best-case/worst-case scenarios [57, 58].

The scenarios to be compared usually focus on the food or food component of interest; however, effects of the possible substitution are usually not taken into account, e.g., the benefits of eating fish will be different if not eating fish in a diet would result in or imply eating more processed red meat, or probably, soybean and eggs.

In theory and in practice, risk-benefit assessment entails more than mirroring a benefit assessment to a risk assessment and requires methodology development in both benefit and risk assessment to provide a well-balanced estimate. Even with the progress currently achieved, a truly quantitative benefit-risk assessment will often not be possible because of a lack of data. Even with possible quantification, it is recommended that a conclusion should always be presented at least with a narrative description of the major uncertainties and assumptions in the assessment.

Quantitative uncertainty analysis for risk-benefit assessment is usually more informative but more difficult to conduct than just risk or benefit uncertainty analysis [51, 57]. Firstly, in comparison of risks or risk with benefit, the key is to determine the probability that one risk outweighs another or a moderate amount of intake should be preferred. Second, if the uncertainties in each of the items being compared are related or overlapped, the uncertainty within the whole comparison assessment can be less than that in an individual risk assessment. However, the uncertainties are usually relatively separated, which leads to the uncertainties in a comparison exceeding the uncertainties separately evaluated.

To sum up, risk-benefit assessment is a valuable approach to systematically integrating the current knowledge to provide the best possible science-based answers to complicated questions in food and nutrition. However, the methodology is still in ongoing development, and its proper utilization is quite promising and needs to be further explored.

## References

1. World Health Organization. About Risk Analysis in Food. 2010. Available at: http://www.who.int/foodsafety/micro/riskanalysis/en/
2. Codex Alimentarius Commission. 2010. Working principles for risk analysis for application in the framework of the Codex Alimentarius. Codex Alimentarius Commission Procedural Manual, 19th edition, Rome 2010. Available at: ftp://ftp.fao.org/codex/Publications/ProcManuals/Manual_19e.pdf
3. FAO/WHO. Food Safety Risk Analysis: A Guide for National Food Safety Authorities. FAO Food and Nutrition Paper No. 87. 2006. Available at: ftp://ftp.fao.org/docrep/fao/009/a0822e/a0822e.pdf
4. FAO/WHO. 1995. Application of Risk Analysis to Food Standards Issues. Report of the Joint FAO/WHO Expert Consultation. Geneva, 13-17 March 1995. Available at: ftp://ftp.fao.org/es/esn/food/Risk_Analysis.pdf
5. FAO/WHO. 1997. Risk management and food safety. FAO Food Nutr Pap No. 65. Available at: ftp://ftp.fao.org/docrep/fao/w4982e/w4982e00.pdf

6. FAO/WHO. 1998. The application of risk communication to food standards and safety matters. FAO Food and Nutrition Paper No 70. Available at: http://www.fao.org/docrep/005/x1271e/x1271e00.htm
7. FAO/WHO. 2005. Working principles for risk analysis for application in the framework of the Codex Alimentarius. In Codex Alimentarius Commission. Procedural Manual. 15th Edition. Available at: ftp://ftp.fao.org/codex/Publications/ProcManuals/Manual_15e.pdf
8. FAO. 2003. Food Safety: Science and Ethics. Report of an FAO Expert Consultation. Rome, 3–5 September 2002. FAO Readings in Ethics 1. Available at ftp://ftp.fao.org/docrep/fao/006/j0776e/j0776e00.pdf
9. European Food Safety Authority. Transparency in risk assessment carried out by EFSA: guidance document on procedural aspects. EFSA J. 2006;2006(353):1–16. Available at: http://www.efsa.europa.eu/en/science/sc_commitee/sc_documents/1494.html
10. Joint Institute for Food Safety and Applied Nutrition. Website of the Food Safety Risk Analysis Clearinghouse. A joint project between the University of Maryland and the United States Food and Drug Administration. Collection of resources related to food safety risk communication. Available at: http://www.foodrisk.org/risk_communication.cfm
11. FAO/WHO. 2016. Risk communication applied to food safety handbook. Food safety and quality series, 2. Rome. Available at: https://www.who.int/foodsafety/Risk-Communication/en/
12. EFSA scientific committee; guidance on human health risk-benefit assessment of food. EFSA J. 2010;8(7):1673. https://doi.org/10.2093/j.efsa.2010.1673. Available online: www.efsa.europa.eu
13. Weed DL. Weight of evidence: a review of concept and methods. Risk Anal. 2005;25:1545–57.
14. Dixit R, Riviere J, Krishnan K, Andersen ME. Toxicokinetics and physiologically based toxicokinetics in toxicology and risk assessment. J Toxicol Environ Health B Crit Rev. 2003;6(1):1–40.
15. Coecke S, Pelkonen O, Leite SB, Bernauer U, Bessems JG, Bois FY, Gundert-Remy U, Loizou G, Testai E, Zaldívar JM. Toxicokinetics as a key to the integrated toxicity risk assessment based primarily on non-animal approaches. Toxicol In Vitro. 2013;27(5):1570–7.
16. ECETOC. Framework for the Integration of Human and Animal Data in Chemical Risk Assessment. Technical Report No. 104 ISSN-0773-8072-104. Brussels: European Centre for Ecotoxicology and Toxicology of Chemicals; 2009.
17. James RC, Britt JK, Halmes NC, Guzelian PS. Evidence-based causation in toxicology: a 10-year retrospective. Hum Exp Toxicol. 2015;34(12):1245–52.
18. Rodricks JV, Levy JI. Science and decisions: advancing toxicology to advance risk assessment. Toxicol Sci. 2013;131(1):1–8.
19. Jennings P, Corvi R, Culot M. A snapshot on the progress of in vitro toxicology for safety assessment. Toxicol In Vitro. 2017;45(Pt 3):269–71.
20. Sauer UG, Deferme L, Gribaldo L, Hackermüller J, Tralau T, van Ravenzwaay B, Yauk C, Poole A, Tong W, Gant TW. The challenge of the application of 'omics technologies in chemicals risk assessment: background and outlook. Regul Toxicol Pharmacol. 2017;91 (Suppl 1):S14–26.
21. McMullen PD, Andersen ME, Cholewa B, Clewell HJ 3rd, Dunnick KM, Hartman JK, Mansouri K, Minto MS, Nicolas CI, Phillips MB, Slattery S, Yoon M, Clewell RA. Evaluating opportunities for advancing the use of alternative methods in risk assessment through the development of fit-for-purpose in vitro assays. Toxicol In Vitro. 2018;48:310–7.
22. Adami HO, Berry SC, Breckenridge CB, Smith LL, Swenberg JA, Trichopoulos D, Weiss NS, Pastoor TP. Toxicology and epidemiology: improving the science with a framework for combining toxicological and epidemiological evidence to establish causal inference. Toxicol Sci. 2011;122(2):223–34.
23. Hernández AF, Tsatsakis AM. Human exposure to chemical mixtures: challenges for the integration of toxicology with epidemiology data in risk assessment. Food Chem Toxicol. 2017;103:188–93. https://doi.org/10.1016/j.fct.2017.03.012.

24. EFSA. Guidance of the scientific committee on a request from EFSA on the use of the benchmark dose approach in risk assessment. The EFSA Journal. 2009;2009(1150):1–72.
25. Neumann HG. Risk assessment of chemical carcinogens and thresholds. Crit Rev Toxicol. 2009;39(6):449–61.
26. Adeleye Y, Andersen M, Clewell R, Davies M, Dent M, Edwards S, Fowler P, Malcomber S, Nicol B, Scott A, Scott S, Sun B, Westmoreland C, White A, Zhang Q, Carmichael PL. Implementing toxicity testing in the 21st century (TT21C): making safety decisions using toxicity pathways, and progress in a prototype risk assessment. Toxicology. 2015;5(332): 102–11.
27. McConnell ER, Bell SM, Cote I, Wang RL, Perkins EJ, Garcia-Reyero N, Gong P, Burgoon LD. Systematic omics analysis review (SOAR) tool to support risk assessment. PLoS One. 2014;9(12):e110379.
28. Dourson M, Becker RA, Haber LT, Pottenger LH, Bredfeldt T, Fenner-Crisp PA. Advancing human health risk assessment: integrating recent advisory committee recommendations. Crit Rev Toxicol. 2013;43(6):467–92.
29. Hartwig A, Arand M, Epe B, Guth S, Jahnke G, Lampen A, Martus HJ, Monien B, Rietjens IMCM, Schmitz-Spanke S, Schriever-Schwemmer G, Steinberg P, Eisenbrand G. Mode of action-based risk assessment of genotoxic carcinogens. Arch Toxicol. 2020;94(6):1787–877.
30. Thomas PC, Bicherel P, Bauer FJ. How in silico and QSAR approaches can increase confidence in environmental hazard and risk assessment. Integr Environ Assess Manag. 2019;15(1):40–50.
31. Gbeddy G, Egodawatta P, Goonetilleke A, Ayoko G, Chen L. Application of quantitative structure-activity relationship (QSAR) model in comprehensive human health risk assessment of PAHs, and alkyl-, nitro-, carbonyl-, and hydroxyl-PAHs laden in urban road dust. J Hazard Mater. 2020;5(383):121154.
32. Ågerstrand M, Beronius A. Weight of evidence evaluation and systematic review in EU chemical risk assessment: foundation is laid but guidance is needed. Environ Int. 2016;92-93: 590–6.
33. Barlow S, Renwick AG, Kleiner J, Bridges JW, Busk L, Dybing E, Edler L, Eisenbrand G, Fink-Gremmels J, Knaap A, Kroes R, Liem D, Müller DJ, Page S, Rolland V, Schlatter J, Tritscher A, Tueting W, Würtzen G. Risk assessment of substances that are both genotoxic and carcinogenic report of an International Conference organized by EFSA and WHO with support of ILSI Europe. Food Chem Toxicol. 2006;44(10):1636–50.
34. Embry MR, Bachman AN, Bell DR, Boobis AR, Cohen SM, Dellarco M, Dewhurst IC, Doerrer NG, Hines RN, Moretto A, Pastoor TP, Phillips RD, Rowlands JC, Tanir JY, Wolf DC, Doe JE. Risk assessment in the 21st century: roadmap and matrix. Crit Rev Toxicol. 2014;44(Suppl 3):6–16.
35. Stedeford T, Zhao QJ, Dourson ML, et al. The application of non-default uncertainty factors in the U.S. EPA's Integrated Risk Information System (IRIS). Part I: UF(L), UF(S), and "other uncertainty factors"[J]. J Environ Sci Health C. 2007;25(3):245–79.
36. Pohl HR, Chou CH, Ruiz P, Holler JS. Chemical risk assessment and uncertainty associated with extrapolation across exposure duration. Regul Toxicol Pharmacol. 2010;57(1):18–23.
37. Moretto A, Bachman A, Boobis A, Solomon KR, Pastoor TP, Wilks MF, Embry MR. A framework for cumulative risk assessment in the 21st century. Crit Rev Toxicol. 2017;47(2): 85–97.
38. Boobis AR, Ossendorp BC, Banasiak U, Hamey PY, Sebestyen I, Moretto A. Cumulative risk assessment of pesticide residues in food. Toxicol Lett. 2008;180(2):137–50.
39. Safe SH. Development validation and problems with the toxic equivalency factor approach for risk assessment of dioxins and related compounds. J Anim Sci. 1998;76(1):134–41.
40. Gallagher SS, Rice GE, Scarano LJ, Teuschler LK, Bollweg G, Martin L. Cumulative risk assessment lessons learned: a review of case studies and issue papers. Chemosphere. 2015;120: 697–705.

41. Cote I, Andersen ME, Ankley GT, Barone S, Birnbaum LS, Boekelheide K, Bois FY, Burgoon LD, Chiu WA, Crawford-Brown D, Crofton KM, DeVito M, Devlin RB, Edwards SW, Guyton KZ, Hattis D, Judson RS, Knight D, Krewski D, Lambert J, Maull EA, Mendrick D, Paoli GM, Patel CJ, Perkins EJ, Poje G, Portier CJ, Rusyn I, Schulte PA, Simeonov A, Smith MT, Thayer KA, Thomas RS, Thomas R, Tice RR, Vandenberg JJ, Villeneuve DL, Wesselkamper S, Whelan M, Whittaker C, White R, Xia M, Yauk C, Zeise L, Zhao J, DeWoskin RS. The next generation of risk assessment multi-year study-highlights of findings, applications to risk assessment, and future directions. Environ Health Perspect. 2016;124(11):1671–82.
42. Munro IC, Renwick AG, Danielewska-Nikiel B. The threshold of toxicological concern (TTC) in risk assessment. Toxicol Lett. 2008;180(2):151–6.
43. Lachenmeier DW, Rehm J. Comparative risk assessment of alcohol, tobacco, cannabis and other illicit drugs using the margin of exposure approach. Sci Rep. 2015;30(5):8126.
44. Tijhuis MJ, de Jong N, Pohjola MV, Gunnlaugsdóttir H, Hendriksen M, Hoekstra J, Holm F, Kalogeras N, Leino O, van Leeuwen FX, Luteijn JM, Magnússon SH, Odekerken G, Rompelberg C, Tuomisto JT, Ueland WBC, Verhagen H. State of the art in benefit-risk analysis: food and nutrition. Food Chem Toxicol. 2012;50(1):5–25.
45. Rietjens IM, Alink GM. Future of toxicology--low-dose toxicology and risk--benefit analysis. Chem Res Toxicol. 2006 Aug;19(8):977–81.
46. Fransen H, de Jong N, Hendriksen M, Mengelers M, Castenmiller J, Hoekstra J, van Leeuwen R, Verhagen H. A tiered approach for risk–benefit assessment of foods. Risk Anal. 2010;30:808–16.
47. Verhagen H, Andersen R, Antoine JM, Finglas P, Hoekstra J, Kardinaal A, Nordmann H, Pekcan G, Pentieva K, Sanders TA, van den Berg H, van Kranen H, Chiodini A. Application of the BRAFO tiered approach for benefit-risk assessment to case studies on dietary interventions. Food Chem Toxicol. 2012;50(Suppl 4):S710–23. https://doi.org/10.1016/j.fct.2011.06.068.
48. Gold MR, Stevenson D, Fryback DG. HALYS and QALYS and DALYS, oh my: similarities and differences in summary measures of population health. Annu Rev Public Health. 2002;2002 (23):115–34.
49. Hoekstra J, Fransen HP, van Eijkeren JC, Verkaik-Kloosterman J, de Jong N, Owen H, Kennedy M, Verhagen H, Hart A. Benefit-risk assessment of plant sterols in margarine: a QALIBRA case study. Food Chem Toxicol. 2013;54:35–42.
50. Hart A, Hoekstra J, Owen H, Kennedy M, Zeilmaker MJ, de Jong N, Gunnlaugsdottir H. Qalibra: A general model for food risk-benefit assessment that quantifies variability and uncertainty. Food Chem Toxicol. 2013;54:4–17.
51. Hoekstra J, Verkaik-Kloosterman J, Rompelberg C, van Kranen H, Zeilmaker M, Verhagen H, de Jong N. Integrated risk-benefit analyses: method development with folic acid as example. Food Chem Toxicol. 2008;46:893–909.
52. Krul L, Kremer BHA, Luijckx NBL, Leeman WR. Quantifiable risk-benefit assessment of micronutrients: from theory to practice. Crit Rev Food Sci Nutr. 2017;57(17):3729–46.
53. Cohen JT, Bellinger DC, Connor WE, Kris-Etherton PM, Lawrence RS, Savitz DA, Shaywitz BA, Teutsch SM, Gray GM. A quantitative risk-benefit analysis of changes in population fish consumption. Am J Prev Med. 2005;29:325–34.
54. Institute of Medicine (IoM). Sea food choices. In: Balancing benefits and risks. Washington, D. C: National Academy Press; 2007.
55. Ginsberg GL, Toal BF. Quantitative approach for incorporating methylmercury risks and omega-3 fatty acid benefits in developing species specific fish consumption advice. Environ Health Perspect. 2009;117:267–75.
56. Gao YX, Zhang HX, Li JG, Zhang L, Yu XW, He JL, Shang XH, Zhao YF, Wu YN. The benefit risk assessment of consumption of marine species based on benefit-risk analysis for foods (BRAFO)-tiered approach. Biomed Environ Sci. 2015;28(4):243–52.
57. Hoekstra J, Hart A, Boobis A, Claupein E, Cockburn A, Hunt A, Knudsen I, Richardson D, Schilter B, Schutte K, Torgerson PR, Verhagen H, Watzl B, Chiodini A. BRAFO tiered

approach for benefit–risk assessment of foods. Food Chem Toxicol. 2012;50(Suppl 4): S684–98.

58. van den Berg M, Kypke K, Kotz A, Tritscher A, Lee SY, Magulova K, Fiedler H, Malisch R. WHO/UNEP global surveys of PCDDs, PCDFs, PCBs and DDTs in human milk and benefit-risk evaluation of breastfeeding. Arch Toxicol. 2017;91(1):83–96.

# Chapter 11
# Nutrients/Nutrition and Drug Interaction

**Yan Zhao, Jie Shen, Lingyu Ma, and Li Wang**

**Abstract** Drug-nutrient interactions are the reactions between drugs and nutrients, which are involved in the chemical, physical, physiological, or pathophysiological process. If certain foods or nutrients are ingested in company with some drugs, drug absorption and metabolism may be affected. An individual's nutritional status may also influence the treatment effect of some drugs. Therefore, this chapter describes some common interactions between nutrients/nutrition and drugs, including the influence of nutrition on drug absorption, distribution, exposure and response, effects of specific foods or food components, specific nutrients, or other dietary ingredients on drug-nutrient interaction.

**Keywords** Drug-nutrient interactions · Drug absorption · Specific food components · Specific nutrients

## 11.1 Introduction

Drug-nutrient interactions (DNI) are the reactions between drugs and nutrients, which are involved in the chemical, physical, physiological, or pathophysiological process. Most significant drug-nutrient interactions are affected by many factors [1, 2]. It was overlooked for a long time that some unconscious and unmanaged drug-nutrient interactions may have an impact, which may lead to very serious outcomes. A drug-nutrient interaction may reduce the expected therapeutic effect to a therapeutic drug, influence an individual's nutritional status, or cause an acute or chronic drug toxicity. If certain foods or nutrients are ingested in company with some drugs, the overall bioavailability, pharmacokinetics, pharmacodynamics, and therapeutic efficacy of the medications may be affected [3]. An individual's nutritional status may also influence the therapeutic efficacy of the drugs. Therefore, the sufficient or lack of some food ingredients or nutrients in the gut or in the body's

Y. Zhao (✉) · J. Shen · L. Ma · L. Wang
Department of Nutrition and Food Hygiene, School of Public Health, Harbin Medical University, Harbin, China
e-mail: yanzhao@hrbmu.edu.cn


L. Zhang (ed.), *Nutritional Toxicology*,
https://doi.org/10.1007/978-981-19-0872-9_11

physiological system may have an influence on the absorption and distribution of the drug. On the other hand, the presence of some drugs can significantly affect the metabolism and bioavailability of food and nutrient in humans. Some drugs can alter appetite and taste, change the absorption and metabolism of nutrients, resulting in impaired nutritional status. A great deal of evidence shows that drugs affect the therapeutic effect of other drugs by regulating the nutritional level of the body [4–6]. This chapter describes some common interactions between nutrients/nutrition and drugs.

Chan, a PharmD, proposed four types of the drug-nutrient interactions according to the metabolic mechanisms of the food, nutrient, and drugs [2].

Type I interactions refer to in vitro inactivation, meaning the interaction between the drug and the nutritional element or formulation through biochemical or physical reactions, such as hydrolysis, oxidation, neutralization, precipitation, and complexation. These interactions typically take place with enteral tube feeding in the delivery device by the ingredients contacting directly. These interactions usually occur before the drug and nutrient metabolize in the body.

Type II interactions can influence the absorption of drugs and nutrients through oral administration or enteral delivery devices. The absorption efficiency of the drugs or nutrients can be increased or decreased by these interactions. Type II interactions include three subtypes. The drugs or the nutritional elements that cause the interaction may regulate the activity of an enzyme (type A interaction) or a transport mechanism (type B interaction) involved in the biotransformation or metabolize the drugs or nutrients before reaching the systemic circulation. Type C interaction involves complexation, chelation, or other inactivating processes in the gastrointestinal tract. The absorption of the drugs or nutrients may be affected.

Type III interactions may influence the systemic and physiologic disposition of the drugs or the nutrients. The cellular or tissue dispersion, systemic metabolism, transport, or entrance to particular organs and tissues of the drugs or the nutritional elements can be changed, when absorbed from the gastrointestinal tract and entering into the systemic circulation. Such interactions may also modify the functions of other cofactors and hormones in some conditions.

Type IV interactions affect the removement of drugs or nutritional materials. The mechanisms of the interactions may involve the antagonism, impairment, or modulation of elimination in the liver, intestinal tract, and kidney.

In many cases, there are two categories in drug-nutrition interactions. The induction and inhibition are the processes to affect the reaction of the drug metabolism, which may lead to the drug dose decreasing or the toxicity. One example is that oral calcium supplements may have an impact on the activity of the ciprofloxacin by chelation and complexation. Dietary carnitine needs to change fat into energy by valproic acid and is poorly absorbed in susceptible patients. In this condition, valproic acid binds competitively with the intestinal SLC22A transporter to inhibit carnitine absorption [7–10]. The cases of inducer and inhibitor would occur also in Type IIC and Type IIB, respectively.

Generally speaking, some food may affect the gastrointestinal secretions and the speed of peristalsis, which may reduce the metabolism of the drugs. The metabolism of the drugs may turn into affecting the nutrition level of the body. As a result, patients can become malnourished and fail to respond to medication.

Drug-nutrient interactions can be influenced by the two most critical factors, that is, the host and the drug or nutrient itself. The host factor involves age, sex, body size, body composition, genetics, lifestyle, underlying diseases, and medical conditions. The individual's response to a drug or nutrient varies greatly. Elderly, malnourished, and pregnant patients, or patients with acquired immunodeficiency syndrome, cancer, gastrointestinal tract dysfunctions or surgery, and enteral nutrition are more likely to have adverse events associated with drug-nutrient interactions. Genetics play important roles for the suitable dose and response to a specific drug or nutritional element. The dose of pyridoxine, cobalamin, folic acid, and riboflavin requirement may be modified by the polymorphisms of the methylenetetrahydrofolate reductase (*MTHFR*), which may be a critical determinant for thc threshold intake of the nutrients in preventing certain drug-nutrient interactions. Some genetic polymorphisms may also protect against clinically significant drug-nutrient interactions.

The amount, time, or route of administration can usually influence the efficiency of the drugs or nutrients. For example, the administration of the nutrient and the drug via oral or enteral route may lead to type II drug-nutrient interaction. Whereas, the administration via the intravenous route may avoid the interaction. Various type IIC interactions can be eradicated by truly regulating the administration time. Type I interactions are more frequent in intravenous therapy.

The possibility for drug-nutrient interactions significantly increases when every drug or dietary supplement is added.

## 11.2 Effects of Nutrients/Nutrition on Efficacy of Drugs

The amount of drug in the circulation to generate pharmacological actions is referred to as bioavailability, which depends on the amount of the drug absorbed in the gastrointestinal (GI) tract and the first-pass elimination metabolized by the hepatic and intestinal enzymes. The administration route of the drug can also obviously influence the bioavailability. The food in the GI tract can affect the activity of the drugs by slowing the gastrointestinal secretions and the speed of peristalsis. After oral administration, drug absorption is the result of a series of complex interactions between the drug and the gastrointestinal tract. Food not only affects drug solubility, transport curve, rate and/or degree of absorption, but also affects intestinal pH, which leads to changes in pre-systemic metabolism and systemic drug clearance. Such alterations can have an influence on the activity and toxicity of drugs. Historically, food in the gut may form a barrier that interferes with drug absorption and metabolism. So it was suggested that a drug should be taken on an empty stomach. However, the potential impact of nutrition should also be taken into account. In fact, the nature and extent of the effect of food on the function of a drug is not only related to the formulation and dosage of the drug, but also to the quantity and composition of the food.

The potential clinical significance of drug-nutrient interactions has reached consensus among regulatory agencies around the world. The recommended test meal for the design and conduct of food effect and fasted/fed state studies contains about 800–1000 kcal with a fat energy ratio of 50% using the concept of "worst-case scenario". Nutrients or nutrition in diet may influence drug response through several mechanisms, including the alteration of gastric acid and gastrin secretion, gastrointestinal transit time, dissolution of drugs in solid dosage forms and bile flow as well as binding or complexation of micro- or macronutrient with food contents.

### *11.2.1 Influence of Nutrition on Drug Dissolution and Dissociation*

Foods or nutritional materials may impact the pH in the gastrointestinal tract as well as drug dissolution and dissociation. So the effectiveness and solubility of certain drugs can be modified by these foods or nutrients [11, 12]. For example, the pH of the gastrointestinal tract may be altered by high levels of vitamin C, also called ascorbic acid, and then the solubility of some drugs can be changed. Gastric pH can influence saquinavir, a type of protease inhibitor for HIV treatment. The solubilization increases its bioavailability. What's more, foods can also inhibit dissolution of some drugs such as isoniazid through increasing gastric pH disintegration.

### *11.2.2 Effects of Nutrition on Drug Exposure and Response*

Drug absorption mainly occurs in the intestine, so the intestine is the site of drug-nutrition interactions. It is a multifactorial process from the drug absorption to the target site. The variety of drug-metabolizing enzymes and transport proteins is one of the limits to drug absorption in the enterocyte to detoxify, bioactivate, and shuttle xenobiotics [13–17]. Diet is also a challenge for therapeutic efficacy while avoiding toxicity. Drug absorption, distribution, metabolism, and/or excretion (ADME) can be changed by dietary components via physiologic/mechanical, physicochemical, and biochemical mechanisms.

Physiologic mechanisms are associated with the delay of gastric emptying, stimulation in bile or blood flow, and alterations of gastrointestinal pH or intestinal flora. All of these changes may decrease the absorption of certain drugs, such as angiotensin-converting enzyme inhibitors. Ingestion of the food composition can significantly affect the rate of gastric emptying and therefore the activity of the drugs [18]. Dietary fiber and fat are generally considered nutrients that can prolong gastric emptied time. On the other hand, low PH status in the stomach may also have an impact on gastric emptying. Some drugs, such as nitrofurantoin and hydralazine, are better absorbed, while others (such as penicillin and digoxin) are degraded and therefore inactive in the condition of the low pH for a long time.

Individuals with high blood pressure are suggested to restrict sodium from diet, which may also enhance renal tubular absorption of certain drugs, such as the antipsychotic lithium, leading to toxic blood levels. On the other hand, sodium restriction might disturb the blood pressure-regulating effects of diuretics and angiotensin-converting enzyme inhibitors, while enhance the pharmacological effects on calcium antagonists, nifedipine, and verapamil.

The binding of food with the drug is involved in the physicochemical mechanisms. For instance, phenytoin, an antiepileptic agent, can interact with proteins and salts in enteral nutrition formulations and then reduce the absorption of phenytoin and potentially inadequate control of seizure [19]. Therefore, phenytoin cannot be administered with enteral nutrition formulas simultaneously. Divalent cation-containing products (e.g., calcium in dairy) can bind to some tetracyclines and fluoroquinolones, decreasing the drug metabolism and causing the adverse outcome of the therapy [20, 21]. The fat in the food may enhance the absorption of some fat-soluble drugs like some antiretroviral protease inhibitors, such as with some antiretroviral protease inhibitors.

The processes that dietary components interfere with the formation and function of cofactors, potentiation of drug pharmacodynamics, and functional alteration of drug-metabolizing enzymes or transporters are related to biochemical interactions. Warfarin, used as an anticoagulant, may influence vitamin K absorption and enhance the risk of bleeding or clotting. So warfarin should be cautiously taken with vitamin K-rich foods [22, 23]. Vitamin K-rich foods can also disrupt the cofactor function. Isoniazid used for treatment with tuberculosis and monoamine oxidase inhibitors with depression inhibit the decomposition of endogenous and dietary amines; a tyramine-rich diet can cause a hypertensive crisis. Certain botanical supplements have some health benefits and therefore are used as complements to drug therapy. The consumption of fruit juices, teas, and alcoholic beverages are not only their taste and nutritional value, but more importantly their pharmacological effects.

However, the intestine is an important organ for drug metabolism and dietary nutrient absorption. The enzymes and transporters in the intestine have been shown to inhibit drug disposal. Systemic drug exposure might be increased by inhibition of metabolism and active efflux, whereas decreased by inhibition of active uptake.

## 11.3 Effects of Food Intake on Drug in the Body

### *11.3.1 The Significance of Food Intake for Drugs*

Some drugs can reduce irritation of the gastrointestinal tract when taken with food or drink. For some drugs, however, this way of taking drugs can also affect the absorption and distribution of certain drugs, causing the failure of the therapy. In the case of concurrent meal intake and drug administration, food intake generally stimulates gastric and intestinal secretions, and then facilitates drug dissolution and absorption. Higher fat in the meals tend to promote more complete absorption of certain drugs and nutrients, since they stimulate the release of bile salts, and also facilitate the intestinal uptake of lipophilic drugs. High-fat meals also contribute to slowing down the gastrointestinal tract motility and raising up the contact between the drugs and the intestinal epithelial tissues via the release of cholecystokinin [24, 25]. Albendazole and griseofulvin are such examples, as a fatty meal dramatically increases the oral bioavailability of these drugs. However, food intakes might make the drug absorption rather unregular and unpredictable through the underlying physicochemical mechanisms, the capacity of the drug-nutrient interactions, the dosage of the drug, and the composition of the meals themselves. For instance, the argument whether the food can affect the bioavailability of verapamil, a calcium channel blocker used to ease hypertension and cardiac arrhythmias, is still inconclusive.

A small reduction of oral bioavailability of verapamil was observed owing to meal intake in some studies, whereas no clinically relevant changes were shown in verapamil bioavailability in other studies. In general, no significant difference in the areas under the curve (AUCs) of verapamil was shown with or without food. But it was found that the presence of food could reduce the absorption of verapamil, suggesting that verapamil can be taken without the necessity to consider the effects of food.

In many cases, the drug absorption rate is measured by the time reached the highest serum drug concentration (or $t_{\max}$), while the AUC measured the total amount of the absorption. In many cases, food may reduce the drug absorption rate, which may cause the phenomenon that there is no significant difference in the area under the curve.

The individual's physiological response to food and the difference in the type of food content between two meals may delay the absorption of drugs. Food intake may increase $t_{\max}$ and therefore delay the absorption of some drugs, such as famciclovir, methotrexate, verapamil, and levodopa. But food has no significant impact on the overall amount of drug absorption. In these cases, food can slow down the onset of the drug action, but not affect the efficacy. However, food intake may lead to the reduction in the total amount of drug absorption upon the apparent modification in the AUC of a drug. Such food-drug interaction can result in detectable pharmacodynamic changes with more significance in clinic. So patients should be cautious for drug administration with food intake under specific instructions. Drug compliance should be monitored to keep the responses stable and consistent. Unwanted or side effects should also be cautious to be prevented [26].

Food-triggered dilatation of the stomach may change gastrointestinal internal environment by stimulating the autonomic nervous system. Food induces the endocrinologic changes, such as the release of insulin, cholecystokinin, and gastrin, and increases splanchnic blood flow. The increase of blood delivery to the liver can enhance presystemic metabolism, whereas rarely leads to significant drug-nutrient interaction, due to the fact that most drugs metabolize using the intrinsic clearance as well as the amount and the activity of the enzymes. But the metabolism of ethanol seems to be influenced by hepatic blood flow. Concurrent food intake with 530 kcal may increase the presystemic metabolism of ethanol by up to 49%.

### *11.3.2 Impact of Meal Intake on First-Pass Elimination*

The venous drainage system of the stomach and intestines is associated with drug-nutrient interactions. Active biotransformation (drug metabolism) occurs in the liver, resulting in the conversion of the venous drainage of most organs that goes directly to the heart, whereas that of the GI tract brings blood into the portal circulation and liver and then goes to the heart. An inactive parent substance (a prodrug) turns to its active metabolite(s) in some cases, but the production of less active metabolites than the parent substance in most circumstances. Drugs and their metabolites are drained to the heart through the vein, and enter the systemic circulation through the liver. So the portal circulation has a special impact on distribution during the "first-pass" into the circulation. Oral medication has the greatest first-pass effect [27]. The changes in GI and liver function or nutrients and food components may induce the alteration of biotransformation.

Food composition can directly influence the drug first-pass elimination. For example, grapefruit juice has significant clinical impact on the oral bioavailability of several drugs. The inhibitory can result in the risk of the toxicity in oral bioavailability of some drugs. Meal influences drug first-pass elimination elements through saturation and inhibition. The solubilization of an oral drug in meal lipid may enhance lipophilic drug concentration in the GI lumen. Therefore, food composition may affect the first-pass function in the intestines and liver [28]. Nutrients are more likely to inhibit the phase I metabolic pathways of drugs than phase II pathways. Grapefruit juice has been the hotspot studies on drug-nutrient interaction, since it suppresses the activities of the cytochrome P450-3A4 isoenzyme (CYP3A4). CYP3A4 is involved in the metabolism of the greatest number of drugs and drug candidates in the gastrointestine. Grapefruit juice may also inhibit membrane influx transporters (e.g., OATP) and efflux transporters (e.g., P-glycoprotein) [29–31].

Nutrient intake may affect other drug oxidases in the intestine and liver. In animal studies, methionine and cysteine inhibited cimetidine sulfoxidation mediated by flavin monooxygenase (FMO). In contrast, this interaction does not play a major role in humans. Meal intake has no impact on the absorption of a narrow therapeutic index drug with FMO-mediated sulfoxidation. However, as for the screening of metabolic enzymes in new drugs and the clarification of their drug metabolism mechanism, it can be found that the influence of nutrients on drugs is not just through CYP3A4.

In summary, according to the characteristics of the drug, choose whether to take it with food or on an empty stomach.

### 11.3.3 Effect of Meal Intake on Drug Bioavailability

A meal may also influence the bioavailability of drugs. The bioavailability (~70%) of deferasirox, an iron-chelating agent, would be increased with a meal. The administration of deferasirox at 20 mg/kg was assessed at different time points, including one-half hour before a meal with high calorie (1000 kcal with 50% from fat), one-half hour before or with a standard breakfast meal (450 kcal), or in fasting. Drug taken with food enhanced the drug bioavailability and even more at higher content of fat. The standard breakfast caused the greatest bioavailability (1580 μmol h $L^{-1}$) and fasted state resulted in the lowest bioavailability (1060 μmol h $L^{-1}$) of the drug. With proper pH, fat content, and surfactant levels, food is more easily dissolved and absorbed in the gastrointestinal tract. The current recommendation is to administer deferasirox 30 min before a meal. The magnitude of change in bioavailability would determine clinical significance to choose the fed or fasted states.

### 11.3.4 Effects of Meal Intake on Drug Absorption

Diet can affect the transport of the oral drug from the gut to the target site, which contains a complex reaction process. When ingested, the foodstuff is mechanically and chemically broken down into the nutrients that can participate in the body metabolism. Digestion begins with the GI tract consisting of the mouth, esophagus, stomach, small intestine, and large intestine [32]. The absorption means the digested nutrients are transported by mucosa to the blood, which are carried by the bloodstream to the target organ. The absorption process is regulated via mechanisms including passive and active transports, simple diffusion, endocytosis, and paracellular movement. The intestine is a barrier to drug absorption. The drug-metabolizing enzymes and transport proteins in the enterocyte vary greatly for the detoxification, bioactivation, and shuttle of xenobiotics. Some CYP enzymes and conjunctional enzymes have an impact on drugs metabolization. Diet may also affect the drugs by other pathways, such as transport by P-glycoprotein. It is actually an argument that specific enzyme isoform is influenced by diet. Meanwhile, it is difficult to speculate drug substrate from one to another. The possible mechanisms that a food or nutrient affects the absorption of the drug mainly include: (1) Delayed gastric emptying permits dissolution and absorption of the drugs, such as dicumarol, hydrochlorothiazide, nitrofurantoin, phenytoin, propoxyphene, spironolactone, and penicillin G; (2) food may reduce stomach fluid volume such as ampicillin and amoxicillin; (3) food reduces first-pass extraction and metabolism, blocks enzymatic transformation in GI tract, such as hydralazine, labetalol, and metaprolol; (4) food enhances enterohepatic recycling of drug and increases dissolution secondary to gastric acid secretion including diazepam; (5) calcium of iron in foods may form insoluble chelates with the drugs, such as penicillamine and tetracyclines; (6) foods may change gastric pH, resulting in the alteration of absorption of the drugs such as aspirin, isoniazid and nafcillin; (7) high-fiber and high-pectin foods act as absorbent and protectant of the drugs involving acetaminophen and digoxin.

## *11.3.5 Phase I Metabolism*

### 11.3.5.1 Cytochrome P450 3A

The cytochromes P450 (CYPs), the predominant phase I enzymes, are expressed in the metabolism of the drugs. The CYP3A subfamily is the most abundant CYPs in the intestine, which are identified to affect drug distribution. CYP3A is expressed by the oxidative metabolism of more than half of pharmaceutical agents. The impacts of several fruit juices on CYP3A activities and functions have been substantially studied in vitro and in human populations [33–35]. Specific repressive components in some fruit juices have been reported. On the contrary, the inhibitory effects of tea and alcoholic beverage on CYP3A activities are less known and the clinical significance remains to be identified.

### 11.3.5.2 Esterase

Carboxylesterase, acetylcholinesterase, arylesterase, and butyrylcholinesterase are major esterases. Esterases, vital to prodrugs (e.g., enalapril and lovastatin), should be activated via hydrolytic cleavage of the ester bond to form the active species. It was shown that grapefruit juice inhibited the enteric esterase activities in rats and enhanced the stability of the ester in the lumen and enterocytes. Therefore, a large amount of the ester was absorbed and rapidly hydrolyzed in plasma with more chances to be exposed to active metabolite.

## *11.3.6 Phase II Conjugation*

### 11.3.6.1 Uridine Diphosphate Glucuronosyl Transferase

Human glucuronosyl transferases (UGTs) increase the hydrophilicity via the formation of glucuronide conjugates and therefore facilitate the elimination of endogenous substrates and xenobiotics. Generally speaking, UGT can bind to the endoplasmic reticulum, exposing the substrate-binding site in the cavity. Intestinal UGTs can limit the oral bioavailability of many botanically derived products.

UGT1A1 glucuronidates bilirubin, estrogens, and several dietary carcinogens. Dietary substances exert cancer chemopreventive effects partly through up-regulation of UGTs. Cruciferous vegetables, citrus fruits, and soy foods are known to induce UGT activity. As the marker of UGT1A1 activty, serum bilirubin drops in the individuals homozygous for the *UGT1A1*28* variant alleles (7/7) with a high vegetable and fruit diet. A significant interaction between UGT1A1 genotype and citrus consumption was found in women in a follow-up study. Women with the 7/7 genotype had about 30% lower serum bilirubin when consumed $\geq$0.5 daily servings of citrus fruit than those consumed less. Therefore, studies have shown that the consumption of citrus fruits may increase the UGT1A1 activity in women with 7/7 genotypes, may increase the ability to remove certain carcinogens, and affect cancer susceptibility.

Isothiocyanates from cruciferous vegetables play critical roles in the chemoprevention partly through UGTs induction. In a randomized, controlled, crossover feeding trial in humans, the response of the three cruciferous diets to total bilirubin was lower than that of the basal diet. For the *UGT1A1*28/*28* genotype, all cruciferous-containing diets led to lower bilirubin concentrations compared to baseline, implicating that dietary intervention may influence the carcinogens and drugs metabolism, particularly among *UGT1A1*28/*28* individuals.

#### 11.3.6.2 Sulfotransferase

Sulfotransferases (SULTs) catalyze the sulfation of a multitude of xenobiotics, hormones, and neurotransmitters via conjugation with 3′-phosphoadenosine 5-′-phosphosulfate. It is identified that there are three human SULT subfamilies in the liver, brain, intestine, lung, kidney, and other tissues. SULT1As can prevent the reaction of xenobiotics and ingested catecholamine precursors. Grapefruit and orange juices and green tea inhibit SULT1A1 and SULT1A3. SULT1A3 mainly sulfates tyrosine and dopamine and is expressed only in extrahepatic tissues, including the intestine. SULT1A inhibits normal catecholamine inactivation. Ingestion of SULT1A inhibitors, such as coffee, tea, chocolate, bananas, and citrus fruits, can cause increased catecholamines, changes in blood pressure, migraine, and/or atrial fibrillation in susceptible people. However, the effects of diet-induced SULT1A inhibition on drug is still unknown. Controlled clinical studies with appropriate SULT substrates are needed.

### *11.3.7 Transporter-Mediated Efflux and Uptake*

#### 11.3.7.1 P-Glycoprotein

Inhibition of efflux transporters can influence systemic and local drug concentrations [36]. The substrates for P-glycoprotein (P-gp), an efflux transporter, are extruded back into the intestinal lumen and reduce systemic drug concentrations [37, 38]. Thus, inhibition of enteric P-gp might increase systemic drug exposure.

Citrus juices can suppress P-gp activity in vitro, but the clinical relevance of enteric P-gp needs to be further investigated.

#### 11.3.7.2 Organic Anion Transporting Polypeptide

In the initial clinical study, fexofenadine had the function of exploring the effects of fruit juices on enteric P-gp activity. An unexpected 63% decrease in fexofenadine AUC relative to water was observed. Mean elimination half-life was stable. The inhibition of an intestinal uptake transporter might lead to this interaction. Organic anion transport peptides (OATPs), as a transmembrane transporter, can promote the absorption of many endogenous compounds (such as bile acids and hormones) and drugs [39]. Grapefruit juice significantly decreased mean aliskiren AUC while not changed half-life relative to water in healthy individuals, implicating the inhibition of intestinal but not hepatic OATPs.

Grapefruit juice can also interact with celiprolol, a cardioselective β-adrenergic receptor blocker. Grapefruit juice inhibits celiprolol absorption in the intestine, resulting in an 85% decrease in mean AUC.

### *11.3.8 Effects of Nutritional Status on Drug Disposition*

Although the influence of nutritional status (such as protein-calorie malnutrition, obesity, micronutrient deficiency) on drug metabolism has been recognized, the effect of nutritional status on drug distribution is unclear. Malnutrition may influence drug pharmacokinetic parameters, such as drug distribution and clearance. For example, protein-calories malnutrition (PCM) may influence the level of body functions, changing the concentration of the drug. Severe PCM may lead to the decrease in drug absorption, the limitation of protein carriers, and the reduction of metabolism. The degree of malnutrition depends on the efficiency of the therapy and leads to increasing toxicity.

Much attention has been paid to antimicrobials in obesity. During therapy for cellulitis with piperacillin-tazobactam 3.375 g q4h intravenously in a morbidly obese patient (BMI 50 kg/m$^2$), pharmacokinetic monitoring data showed that the volume of distribution (Vd) (0.33 L/kg) and clearance (CL) (27 L/h) for piperacillin were changed, implicating that the dose of piperacillin can vary according to total body weight.

The preoperative administration of antimicrobials for the prevention of postoperative infection should be given particular concern in obese patients with surgery. Serum drug concentrations were below the minimum inhibitory concentration (MIC) for several organisms in patients with BMI $> 40$ kg/m$^2$ when 1 g of cefazolin was used as antibiotic prophylaxis for surgery, and the rates of surgical site infection were significantly decreased from 16.5 to 5.6% when the dose was raised to 2 g. Inadequate antimicrobial usage may also be associated with mediastinitis after cardiac surgery in obese patients. Obesity may enhance the clearance of cephalosporin and repeated dosing may be necessary during an operation lasting for more than 3 h.

## 11.4 Effects of Specific Foods or Food Components on Drug-Nutrient Interaction

Specific foods may have a unique impact on drug disposition. Dietary products containing divalent and trivalent cations (including dairy products) may be chelated with fluoroquinolone antibiotics and change their activity. Milk containing a high level of xanthine oxidase may reduce the bioavailability of the drugs, such as mercaptopurine.

### *11.4.1 Cruciferous Vegetables*

Cruciferous vegetables, such as cabbage and Brussels sprouts, and alfalfa meal markedly induce chemical oxidations. They are dietary sources of glucosinolates that are metabolized to isothiocyanates and indoles. The cruciferous vegetables significantly increase the oxidative metabolism of antipyrine and phenacetin (Fig. 11.1) and the conjugation of acetaminophen. The function of long-term anticoagulation with warfarin can be significantly influenced by cruciferous vegetables and vitamin supplements containing vitamin K. In addition, the elimination rate

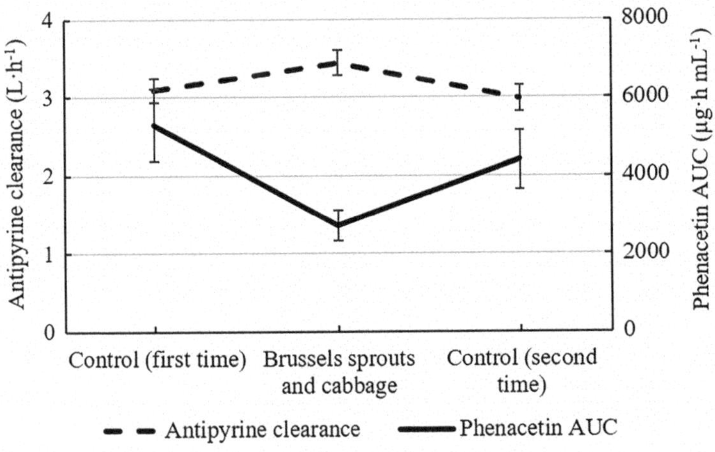


**Fig. 11.1** Effects of dietary Brussels sprouts and cabbage on antipyrine and phenacetin metabolism in healthy subjects

of warfarin can be increased by a Brussels sprouts-rich diet. Patients with low and unregular intakes of vitamin K may be at risk for unstable control of anticoagulation. Coumarin's effects in vivo anticoagulants may be enhanced by coumarins and suppressed by vitamin K. The intake of high-vitamin K food and supplements should be maintained by a reasonable constance, since they have an impact on anticoagulant drugs metabolism and make the prothrombin time within the desired range during long-term anticoagulating therapy.

Humans can appreciably absorb a glucosinolate precursor of phenethyl-isothiocyanate in watercress, which may influence the CYP2E1 activity and even alter the metabolism of chlorzoxazone. The ingestion of a watercress homogenate (50 g) showed similar effects as isoniazid, a CYP2E1 inhibitor. The watercress homogenate increased chlorzoxazone AUC by 56% and prolonged the chlorzoxazone elimination half-life by 53%. It significantly decreased the peak plasma concentration and AUC of acetaminophen, but had no effects on the formation and excretion of the major glucuronide and sulfate conjugates of acetaminophen.

It is argued for the interaction between isothiocyanates, the inductor of glutathione-*S*-transferase (GST) enzymes, and drugs.

### *11.4.2 Grapefruit Juice*

The interaction between fruit juice and drugs appears to affect transporters and metabolic enzymes. Studies have shown that these juices can affect drugs containing furanocoumarins and flavonoids. For some juices, the evidence is only circumstantial. Prospective studies may not find the pharmacokinetic mechanism of the effect of cranberry juice on warfarin. On the other hand, the effect of grapefruit juice on the metabolism of CYP3A4-metabolizing drugs is the reaction recognized by the public.

The grapefruit juice was first used to investigate the activity of felodipine with alcohol in 1989. It was found that grapefruit juice decreased the oral clearance and enhanced AUC of felodipine, resulting in the increase in the bioavailability of the drug and its systemic exposure and pharmacodynamic effect. Then research on the interaction of various drugs with grapefruit juice began to emerge.

Grapefruit juice, which contains CYP3A inhibitors, can affect the intestinal first-pass metabolism to enhance systemic drug exposure. The interaction might result in adverse events, such as muscle pain in some statins and severe hypotension with some calcium channel blockers [40]. Furanocoumarins have been established as major mediators of the "grapefruit juice effect" in humans, through intestinal CYP3A reversible inhibition and degradation of the protein. Naringin in grapefruit juice is easily combined with metabolic enzymes, which may affect the p-glycoprotein drug transporters in some intestinal cells. Grapefruit juice can inhibit organic anion transporting polypeptides as well and might decrease oral drug bioavailability. Drugs including calcium channel blockers, HMG-CoA reductase inhibitors (statins), immunosuppressants, phosphodiesterase type-5 inhibitors, cytotoxic agents, and some anxiolytics/hypnotics have the greatest interaction. Caffeine, nicotine, dextromethorphan, hyoscine, simvastatin, and omeprazole are also affected.

It is difficult to clarify the relationship between intake and response, even normal intake of grapefruit or juice (a whole grapefruit or 200 mL of juice) may cause interaction. Compared with water, the reduction was by 42 and 64%, a normal quantity of grapefruit juice (300 mL) and a large quantity (1200 mL) in fexofenadine exposure.

### 11.4.3 *Methylxanthines*

Methylxanthines include a series of compounds and are extensively metabolized by CYP enzymes. For example, caffeine (1,3,7-trimethylxanthine) is a component of coffee and tea, and is used as a supplement for many carbonated beverages. Theophylline (1,3-dimethylxanthine) is clinically used as a bronchodilator to treat asthma and other pulmonary diseases. When regularly administered, they can accumulate and have an influence on drug metabolism, which are complex and may involve saturation, inhibition, or induction of methylxanthines-metabolizing enzymes in the liver. The metabolism of theobromine (3,7-dimethylxanthine), a major methylxanthine in chocolate, is lowered by saturating or inhibiting hepatic enzymes with repeated uptake; however, the hepatic metabolism of theobromine is induced several days after the last administration. Theophylline can promote the body's own metabolism and the methylxanthine in the diet affects the metabolism of healthy subjects by competing with theophylline for saturable metabolic pathways.

In Africa and elsewhere, people ingest cola nuts with 2.3% caffeine to get the excitement effect. Cola nut chewing can prolong the half-life of antipyrine in Gambian villagers.

The interaction between caffeine and clozapine, both of which are substrates of CYP1A2, has been confirmed in patients with schizophrenia. On monotherapy, the concentration of clozapine on a 5-day decaffeinated diet was reduced. Therefore, the metabolism of this drug can be altered via habitual caffeine intakes, suggesting that caffeine intake and clozapine levels should be kept under medical surveillance in schizophrenic patients.

### 11.4.4 *Food Preparation*

Cooking may induce chemical changes in foods, particularly at high temperatures. When consumed, the chemical products enter into the body and are absorbed, and then affect drug metabolism. Polycyclic aromatic hydrocarbons, for example, may be formed in charcoal-broiled meats and cigarettes. Drug oxidation rates in smokers may probably be increased by polycyclic aromatic hydrocarbons in cigarette smoke.

Char-roasted beef has also been reported to have a significant effect on the metabolism of drugs in healthy subjects such as phenacetin, theophylline, and antipyrine in healthy subjects. Pharmacokinetic results showed that consumption of charcoal-broiled beef obviously suppressed plasma phenacetin concentrations and enhanced the ratio of *N*-acetyl-*p*-aminophenol (acetaminophen), the major metabolite of phenacetin, to phenacetin. So phenacetin *O*-dealkylation is increased by both charcoal-broiled beef and cigarette smoking in humans. Now acetaminophen, metabolized primarily by conjugation, has been largely used as a substitute for phenacetin, since the consumption of charcoal-broiled beef has no influence on the metabolism of acetaminophen. Ingestion of charcoal-broiled beef also enhanced the clearance of antipyrine and theophylline.

### 11.4.5 Green Tea

It was found that consuming green tea prevented tumor cell deaths with bortezomib and reduced the antihypertensive effects of nadolol, suggesting that green tea or catechin interacts with OATP1A2 substrate drugs to affect its efficacy, which is similar to the interaction of grapefruit juice with many drugs. However, there is no research showing that green tea has an effect on CYP3A4 drug metabolism [41–43].

### 11.4.6 Other Food or Components

Other food or components may also influence drug metabolism. For example, the expression and activity of CYP1A1 may be inhibited by soy protein isolates, most probably through a posttranslational decrease in the transcription factor aryl hydrocarbon receptor. Two drug transporters, three phase I and two phase II enzymes are markedly upregulated by soy isoflavones on the basis of gene array screening. On the other hand, losartan pharmacokinetics are not affected by a soy extract in healthy subjects.

Enteral nutrition may also influence drug disposition and drug effect. In the presence of enteral nutrition, drug bioavailability may be altered by several mechanisms. Although it is directly supplemented by intravenous injection instead of gastrointestinal absorption, parenteral nutrition can also interact with medication, including individual nutrients administered parenterally.

## 11.5 Effects of Specific Nutrients or Other Dietary Ingredients on Drug-Nutrient Interaction

Individual nutrients and non-nutrient dietary supplement ingredients are associated with drug interactions. Divalent and trivalent cations administered in drug doses can affect the activity and function of many drugs, such as iron and mycophenolic acid For example, dietary calcium affects the bioavailability of ciprofloxacin through chelation and complexation, reducing its activity, leading to failure of drug therapy [44–47].

### *11.5.1 Dietary Protein, Fat, and Carbohydrate*

The influence of diet on human drug metabolism was first recognized in crossover studies in male subjects. Under the condition of constant total energy and fat supply, dietary protein and carbohydrate were sequentially exchanged. Protein supplement (sodium caseinate at 100 g/day) for 2 weeks to a well-balanced diet led to the acceleration of the metabolic rates for antipyrine and theophylline, while carbohydrate supplement (sucrose at 200 g/day) for the same time periods decreased the metabolic rates of the drugs. The enhancement of dietary protein could also promote the metabolism of propranolol as well as aminopyrine and caffeine. Such an effect of protein can be observed in all the people. High-protein diet makes the metabolic clearance of antipyrine and theophylline enhanced and the plasma concentration decreased. For example, the clearances of antipyrine and theophylline are $0.71 \pm 0.05$ and $0.98 \pm 0.06$ mL·min$^{-1}$ kg$^{-1}$, respectively, during high protein (50%) diet, while those are $0.57 \pm 0.02$ and $0.76 \pm 0.06$ mL·min$^{-1}$ kg$^{-1}$, respectively, during high carbohydrate (80%) diet. The effects of high fat (70%) diet are similar to the high carbohydrate diet. The clearances of these drugs rely on the metabolic transformations by liver CYP1A2.

The effects of deficiency in carbohydrate, saturated fats, and unsaturated fats were confirmed in males. In the condition of constant dietary protein at 15% of total calories, it was found that isocaloric substitution of carbohydrate for either butter (rich in saturated fat) or corn oil (rich in unsaturated fat) resulted in no significant effects on the metabolism of antipyrine and theophylline. On the other hand, the metabolism of antipyrine was unchanged when substituting saturated and unsaturated fat in the diet of normal subjects. Taken together, there was no response for the metabolism of some substrates for CYP enzymes in the case of isocaloric exchanges of saturated fat, unsaturated fat, and carbohydrate in humans. However, hepatic drug oxidation can be affected by the changes in dietary fat of animals, since dietary fat may influence some CYP enzymes or other vital enzymes in drug metabolism.

Therefore, dietary protein seems more important for drug metabolism and oxidation rates than fat or carbohydrate in humans. Protein content of the diet may influence drug metabolism in patients with cirrhosis as well as other liver diseases.

For hospitalized children with asthma, a high-protein diet led to a higher clearance and serum levels of theophylline and less frequent wheezing episodes than a lower protein diet. Meanwhile, a high-protein diet caused lower levels of theophylline than a high-carbohydrate diet in adults with obstructive pulmonary disease. High-protein diets might increase the requirements for warfarin in some patients. However, the mechanisms are still unknown whether it was related to the changes in warfarin absorption or metabolism or the changes in vitamin K intake.

High-protein intakes enhance the hepatic microsomal CYP contents, weight, and mitotic indices. Certain amino acids, such as tryptophan and sulfur oxide amino acids, not only increase liver protein synthesis but also accelerate the induction of mixed-function oxidase systems. Under some conditions, high-carbohydrate and fat diets may lead to fat accumulation in hepatocytes and therefore affect the activities of drug-metabolizing enzymes.

The steroid hormones are mainly metabolized through CYP enzymes, microsomal reductases, and conjugating enzymes in the liver. High dietary protein and low carbohydrate diet can promote estrogen 2-hydroxylation in healthy subjects and reduce androgen 5-α reduction. It also changes the plasma concentrations of testosterone and cortisol and the globulins binding for these steroids. These events are similar to those induced by phenobarbital in humans.

Protein in diet can affect blood flow, creatinine clearance, and tubular transport, leading to the changes of the drugs disposition, especially reducing tubular transport of basic drugs or drug metabolites in the kidney. Obviously, protein can also affect the process of the absorption, distribution, and metabolism of the drugs For example, a high-protein diet makes theophylline absorption faster than a high-carbohydrate or high-fat diet, since protein has greater buffering capacity than lipid and carbohydrate. Therefore, a high-protein meal is preferred to enhance the bioavailability of acid-resistant drugs.

The delivery of a drug to its reaction center may also be affected by protein in the diet. Protein restriction does not affect levodopa absorption and blood levels. So using levodopa to treat patients with Parkinson's disease can uptake a low-protein diet to decrease unpredictable fluctuations, due to the fact that protein-decomposed amino acids inhibit the transport of levodopa through the blood-brain barrier through aromatic amino acid transporters, reducing its drug activity. Therefore, a high-protein intake can reduce brain dopamine formation from exogenous levodopa. A protein alternative diet between protein restriction and unrestriction was beneficial for patients. But before the dietary change, marginal intakes of protein or other nutrients may result in deficiency. A diet with balanced protein and carbohydrate is suggested.

### *11.5.2 Tyramine and Related Substances*

Tyramine, as an indirect sympathomimetic amine, exerts sympathomimetic effects by releasing norepinephrine stored in sympathetic nerve tissue and the adrenal medulla. Strong or aged cheeses and smoked meats are rich in Tyramine, which have an impact on the drug metabolism. Other fermented high-protein foods may also contain much tyramine or other phenylethylamines, including pickled herring, yeast preparations, broad beans, certain wines, and beers. Ingestion of foods containing tyramine may lead to hypertension crisis in patients using monoamine oxidase (MAO) inhibitors [48, 49]. The "tyramine reactions" or "cheese reactions" may include hypertension with palpation, nausea, vomiting, and headache. The patients are suggested to ingest less than 100 mg tyramine. But the tyramine contents are inconsistent in foods. Tyramine at as little as 6 mg may have a pressor effect. The potentially life-threatening hypertensive crises may occur within 1 h of ingestion of the tyramine-containing food. The "tyramine reactions" is a typical adverse reaction caused by food-drug interaction.

Tyramine is widely found in plants and animals and is metabolized by various enzymes. It is usually produced by decarboxylated tyrosine during fermentation or decay. In general, phenylethylamines absorbed from diet are oxidatively deaminated by MAO in the intestine and liver. When a drug inhibits the activities of MAO, phenylethylamines can be systemically absorbed and make neurotransmitter-norepinephrine replace from storage vesicles in the nervous system. Then acute hypertension and other symptoms, such as headache, myocardial infarction, and hemorrhagic stroke, may occur when large amounts of norepinephrine are released into synapses. Therefore, patients with depression or Parkinson's disease improve diet to reduce tyramine intake, although MAO inhibitors are no longer first-line medications.

Other nutrients may also affect the metabolism of tyramine by affecting the absorption of phenethylamine in the diet and the rate of transport to the systemic circulation. Iron deficiency may increase susceptibility to tyramine reactions. Sympathomimetic drugs may also enhance the reactions in the meantime. Ingestion of broad beans, containing dopa or dopamine, may cause similar reactions. Other MAO inhibitors also have the potential implicated in tyramine reactions, such as furazolidone, meperidine, linezolid, procarbazine, and isoniazid.

Strategies to avoid tyramine reactions include dietary intervention and development of new pharmaceutical products. Dietary plan should be made before drug therapy, including tyramine intake below 5 mg and increasing the intake of fresh food. Foods rich in aromatic amino acids will denature amino acids to produce tyramine due to prolonged exposure and microbial contamination. Dietary restrictions should continue 4 weeks after completion of drug therapy. New routes of administration have been discovered.

### *11.5.3 Minerals*

Sodium is interchangeable with lithium, a constant element in the body and also used as a drug for mood stabilization and mania, via the sodium channel and the sodium proton exchanger. Patients on a low-sodium diet or taking diuretics may suffer from lithium poisoning, due to the fact that sodium affects the reabsorption of the renal tubules and reduces the level of lithium. Therefore, it is necessary to advise patients to adjust the intake of dietary sodium and diuretics to keep the lithium level in a normal state.

Patients taking potassium-sparing diuretics must be aware of foods rich in potassium in their diets. High potassium levels can block aldosterone production, which is critical for the management of patients with chronic kidney disease. Potassium in the diet is mainly found in fruits and vegetables, dairy products, and animal organs. Potassium participates in the formation of sodium-potassium pump. When hypokalemia occurs in the body, it will affect the function of sodium-potassium ATPase, which may affect the metabolism of some drugs, such as digoxin. Foods that affect the degree of ionization and solubility or chelating reaction (i.e., forming an inactive complex) significantly inhibit the drug metabolism. For example, tetracycline may combine with divalent minerals, such as calcium in milk or antacids. Calcium may adversely affect the absorption of quinolones [50, 51]. Therefore, the ingestion of foods rich in calcium should be avoided at the same time as the administration of antimicrobial drugs. Food may react between iron and tetracycline, fluoroquinolone antibiotics, and floxacins. Iron might suppress the bioavailability of ciprofloxacin and ofloxacin by 52% and 64%, respectively. Similarly, zinc may decrease the absorption of fluoroquinolones.

### *11.5.4 Vitamins*

Large doses of vitamins may have the potential to alter drug metabolism. The role of vitamin C in the human body has been extensively studied. Vitamin C has an antioxidant effect. In guinea pigs and rats lacking vitamin C, there will be drug metabolism disorders and impaired functions of CYP and various metabolic enzymes. Vitamin C deficiency also does harm to drug metabolism in humans. For example, the half-life of antipyrine was longer in liver disease patients with low leukocyte ascorbate levels than in those with high ascorbate levels. Ascorbic acid supplementation caused shortened half-life of antipyrine in elderly or diabetic patients with low leukocyte or serum ascorbate levels. There are reasons to believe that malnutrition will lead to impaired drug metabolism. A substantial effect of vitamin C deficiency is not found in healthy subjects. Severe vitamin C deficiency possibly influences drug metabolism in humans.

Mono-oxygenase activities can be reduced by large doses of vitamin C in animals, but not in humans. Administration of vitamin C in large doses might have no significant influence on warfarin disposition, but increase antipyrine clearance and affect nonoxidative pathways of drug metabolism. For example, vitamin C can compete to inhibit the sulfate binding of drugs such as salicylamide and paracetamol. Large doses of vitamin C may reduce plasma concentrations of indinavir in steady state. However, daily intake of 1000 mg vitamin C or 800 IU of vitamin E cannot change the pharmacokinetics of linezolid in healthy subjects.

Vitamin D promotes the expression of several phase I and II metabolizing enzymes. Vitamin D receptor is a nuclear receptor, so vitamin D has a positive effect on drug metabolism such as the use of pyridoxine in the prevention or treatment of isoniazid toxicity. Studies have shown that vitamin B6 can promote the peripheral conversion of levodopa to dopamine through dopa decarboxylase. It is evidenced that fortification of folate in the US diet to prevent neural tube defects might contribute to higher methotrexate dosing in patients with rheumatoid arthritis.

### 11.5.5 Phytochemicals

Interaction may occur in polyphenols and other phytochemicals including the flavonoids, phenolic acids, stilbenes, and lignans. Bioactive peptides from plants and non-plant food sources such as herbs and spices may also affect drug absorption and metabolism [52]. The interaction should be considered on the risk assessment and safety evaluation of dietary intake of flavonoids. It has been confirmed in cells, tissues, and animal models that dietary polyphenols can inhibit phase I and phase II metabolic enzymes, thereby affecting drug metabolism. Flavonoids may influence the expression and activity of several CYP, GST, *N*-acetyltransferase (NAT), sulfotransferase (SULT), and uridine diphosphate glucuronosyltransferase (UGT) enzyme isoforms.

Daidzein, an isoflavone, may diminish the CYP1A2 activity and therefore increase the bioavailability of theophylline and reduce its elimination. The bioavailability of metronidazole may be enhanced by diosmin (a flavone) and reduced by silymarin (a flavonoid). Diosmin can increase the bioavailability of diclofenac possibly through CYP2C9 inhibition.

Drug and nutrition interactions should be recognized and understood in clinical practice. Obviously, nutrients in food can have a major impact on the metabolism and effects of certain drugs, but the evidence for these interactions is not complete. Many specific effects of dietary components on drug metabolism and actions need further study. The evaluation of drug-nutrient interactions is beneficial to drug development, efficacy evaluation, and risk assessment. On the other hand, diet may partly explain the intraindividual variability in drug response. Further studies on the effects of drug metabolism and dietary variations are needed.

## References


1. Chan LN. Drug-nutrient interactions. JPEN J Parenter Enteral Nutr. 2013;37(4):450–9.
2. Péter S, Navis G, de Borst MH, von Schacky C, van Orten-Luiten ACB, Zhernakova A, Witkamp RF, Janse A, Weber P, Bakker SJL, Eggersdorfer M. Public health relevance of drug-nutrition interactions. Eur J Nutr. 2017;56(Suppl 2):23–36.
3. Genser D. Food and drug interaction: consequences for the nutrition/health status. Ann Nutr Metab. 2008;52:29–32.

4. Li H, Wei Y, Zhang S, Xu L, Jiang J, Qiu Y, Mangin E, Zhao XM, Xie S. Pharmacokinetics and safety of posaconazole administered by intravenous solution and oral tablet in healthy Chinese subjects and effect of food on tablet bioavailability. Clin Drug Investig. 2019;39(11):1109–16.
5. Kvasnicka T, Malikova I, Zenahlikova Z, Kettnerova K, Brzezkova R, Zima T, Ulrych J, Briza J, Netuka I, Kvasnicka J. Rivaroxaban-metabolism, pharmacologic properties and drug interactions. Curr Drug Metab. 2017;18(7):636–42.
6. Biesdorf C, Martins FS, Sy SKB, Diniz A. Physiologically-based pharmacokinetics of ziprasidone in pregnant women. Br J Clin Pharmacol. 2019;85(5):914–23.
7. Boullata JI. Drug-nutrition interactions and the brain: it's not all in your head. Curr Nutr Rep. 2019;8(2):92–8.
8. Chopra A, Kolla BP, Mansukhani MP, Netzel P, Frye MA. Valproate-induced hyperammonemic encephalopathy: an update on risk factors, clinical correlates and management. Gen Hosp Psychiatry. 2012;34(3):290–8.
9. Coppola G, Epifanio G, Auricchio G, Federico RR, Resicato G, Pascotto A. Plasma free carnitine in epilepsy children, adolescents and young adults treated with old and new antiepileptic drugs with or without ketogenic diet. Brain and Development. 2006;28(6):358–65.
10. Boullata JI, Armenti VT. Handbook of drug-nutrient interactions. 2nd ed. New York: Springer Science+Business Media, LLC; 2010.
11. Moss DM, Siccardi M, Murphy M, Piperakis MM, Khoo SH, Back DJ, Owen A. Divalent metals and pH alter raltegravir disposition in vitro. Antimicrob Agents Chemother. 2012;56(6): 3020–6.
12. Bendich A, Deckelbaum RJ. Preventive nutrition: the comprehensive guide for health professionals. 5th ed. New York: Humana Press; 2016.
13. Boudry G, David ES, Douard V, Monteiro IM, Le Huërou-Luron I, Ferraris RP. Role of intestinal transporters in neonatal nutrition: carbohydrates, proteins, lipids, minerals, and vitamins. J Pediatr Gastroenterol Nutr. 2010;51(4):380–401.
14. Grandvuinet AS, Vestergaard HT, Rapin N, Steffansen B. Intestinal transporters for endogenic and pharmaceutical organic anions: the challenges of deriving in-vitro kinetic parameters for the prediction of clinically relevant drug-drug interactions. J Pharm Pharmacol. 2012;64(11): 1523–48.
15. Hediger MA, Romero MF, Peng JB, Rolfs AB, Takanaga A, Bruford EA. The ABCs of solute carriers: physiological, pathological and therapeutic implications of human membrane transport proteins. Pflugers Arch. 2004;447(5):465–8.
16. Sai Y. Biochemical and molecular pharmacological aspects of transporters as determinants of drug disposition. Drug Metab Pharmacokinet. 2005;20(2):91–9.
17. Shugarts S, Benet LZ. The role of transporters in the pharmacokinetics of orally administered drugs. Pharm Res. 2009;26(9):2039–54.
18. Sugano K, Kansy M, Artursson P, Avdeef A, Bendels S, Di L, Ecker GF, Faller B, Fischer H, Gerebtzoff G, Lennernaes H, Senner F. Coexistence of passive and carrier-mediated processes in drug transport. Nat Rev Drug Discov. 2010;9(8):597–614.
19. Ferreira Silva R, Rita Carvalho Garbi Novaes M. Interactions between drugs and drug-nutrient in enteral nutrition: a review based on evidences. Nutr Hosp. 2014;30(3):514–8.
20. Sánchez Navarro A, Martínez Cabarga M, Dominguez-Gil HA. Comparative study of the influence of $Ca^{2+}$ on absorption parameters of ciprofloxacin and ofloxacin. J Antimicrob Chemother. 1994;34(1):119–25.
21. Elkhatib WF, Haynes VL, Noreddin AM. Unexpected induction of resistant Pseudomonas aeruginosa biofilm to fluoroquinolones by diltiazem: a new perspective of microbiological drug-drug interaction. J Infect Public Health. 2008;1(2):105–12.
22. Holbrook AM, Pereira JA, Labiris R, McDonald H, Douketis JD, Crowther M, Wells PS. Systematic overview of warfarin and its drug and food interactions. Arch Intern Med. 2005;165(10):1095–106.

23. Hirsh J, Fuster V, Ansell J, Halperin JL, American Heart Association/American College of Cardiology Foundation. American Heart Association/American College of Cardiology foundation guide to warfarin therapy. J Am Coll Cardiol. 2003;41(9):1633–52.
24. Yeap YY, Trevaskis NL, Quach T, Tso P, Charman WN, Porter CJ. Intestinal bile secretion promotes drug absorption from lipid colloidal phases via induction of supersaturation. Mol Pharm. 2013;10(5):1874–89.
25. Moghimipour E, Ameri A, Handali S. Absorption-enhancing effects of bile salts. Molecules. 2015;20(8):14451–73.
26. Choi JH, Ko CM. Food and drug interactions. J Lifestyle Med. 2017;7(1):1–9.
27. van den Anker J, Reed MD, Allegaert K, Kearns GL. Developmental changes in pharmacokinetics and pharmacodynamics. J Clin Pharmacol. 2018;58(Suppl 10):S10–25.
28. Tozer TN, Rowland M. Introduction to pharmacokinetics and pharmacodynamics: the quantitative basis of drug therapy. Philadelphia: Lippincott Williams & Wilkins; 2006.
29. Manikandan P, Nagini S. Cytochrome P450 structure, function and clinical significance: a review. Curr Drug Targets. 2018;19(1):38–54.
30. Sridharan K, Sivaramakrishnan G. Interaction of citrus juices with cyclosporine: systematic review and meta-analysis. Eur J Drug Metab Pharmacokinet. 2016;41(6):665–73.
31. Huang SM, Lesko LJ. Drug-drug, drug-dietary supplement, and drug-citrus fruit and other food interactions: what have we learned? J Clin Pharmacol. 2004;44(6):559–69.
32. Koziolek M, Carrière F, Porter CJH. Lipids in the stomach-implications for the evaluation of food effects on oral drug absorption. Pharm Res. 2018;35(3):55.
33. Jeon H, Jang IJ, Lee S, Ohashi K, Kotegawa T, Ieiri I, Cho JY, Yoon SH, Shin SG, Yu KS, Lim KS. Apple juice greatly reduces systemic exposure to atenolol. Br J Clin Pharmacol. 2013;75 (1):172–9.
34. Bailey DG. Fruit juice inhibition of uptake transport: a new type of food-drug interaction. Br J Clin Pharmacol. 2010;70(5):645–55.
35. Hanley MJ, Cancalon P, Widmer WW, Greenblatt DJ. The effect of grapefruit juice on drug disposition. Expert Opin Drug Metab Toxicol. 2011;7(3):267–86.
36. Englund G, Rorsman F, Ronnblom A, Karlbom U, Lazorova L, Grasjo J, Kindmark A, Artursson P. Regional levels of drug transporters along the human intestinal tract: co-expression of ABC and SLC transporters and comparison with Caco-2 cells. Eur J Pharm Sci. 2006;29(3–4):269–77.
37. Estudante M, Morais JG, Soveral G, Benet LZ. Intestinal drug transporters: an overview. Adv Drug Deliv Rev. 2013;65(10):1340–56.
38. Niemi M, Pasanen MK, Neuvonen PJ. Organic anion transporting polypeptide 1B1: a genetically polymorphic transporter of major importance for hepatic drug uptake. Pharmacol Rev. 2011;63(1):157–81.
39. Evers R, Chu XY. Role of the murine organic anion-transporting polypeptide 1b2 (Oatp1b2) in drug disposition and hepatotoxicity. Mol Pharmacol. 2008;74(2):309–11.
40. Gjestad C, Hole K, Haslemo T, Diczfalusy U, Molden E. Effect of grapefruit juice intake on serum level of the endogenous CYP3A4 metabolite 4β-hydroxycholesterol-an interaction study in healthy volunteers. AAPS J. 2019;21(4):58.
41. An G, Mukker JK, Derendorf H, Frye RF. Enzyme- and transporter-mediated beverage-drug interactions: an update on fruit juices and green tea. J Clin Pharmacol. 2015;55(12):1313–31.
42. Roth M, Timmermann BN, Hagenbuch B. Interactions of green tea catechins with organic anion-transporting polypeptides. Drug Metab Dispos. 2011;39(5):920–6.
43. Knop J, Misaka S, Singer K, Hoier E, Müller F, Glaeser H, König J, Fromm MF. Inhibitory effects of green tea and (−)-epigallocatechin galate on transport by OATP1B1, OATP1B3, OCT1, OCT2, MATE1, MATE2-K and P-glycoprotein. PLoS One. 2015;10:e0139370.
44. Mueller BA, Brierton DG, Abel SR, Bowman L. Effect of enteral feeding with ensure on oral bioavailabilities of ofloxacin and ciprofloxacin. Antimicrob Agents Chemother. 1994;38(9): 2101–5.

45. Williams NT. Medication administration through enteral feeding tubes. Am J Health Syst Pharm. 2008;65(24):2347–57.
46. Mimoz O, Binter V, Jacolot A, Edouard A, Tod M, Petitjean O, Samii K. Pharmacokinetics and absolute bioavailability of ciprofloxacin administered through a nasogastric tube with continuous enteral feeding to critically ill patients. Intensive Care Med. 1998;24(10):1047–51.
47. Podilsky G, Berger-Gryllaki M, Testa B, Buclin T, Roulet M, Pannatier A. The bioavailability of bromazepam, omeprazole and paracetamol given by nasogastric feeding tube. Eur J Clin Pharmacol. 2009;65(5):435–42.
48. Patkar AA, Pae CU. Atypical antipsychotic augmentation strategies in the context of guideline-based care for the treatment of major depressive disorder. CNS Drugs. 2013;27(Suppl 1): S29–37.
49. Gillman K. "Much ado about nothing": monoamine oxidase inhibitors, drug interactions, and dietary tyramine. CNS Spectr. 2017;22(5):385–7.
50. Karadima V, Kraniotou C, Bellos G, Tsangaris GT. Drug-micronutrient interactions: food for thought and thought for action. EPMA J. 2016;7(1):10.
51. Boullata JI. Drug and nutrition interactions: not just food for thought. J Clin Pharm Ther. 2013;38(4):269–71.
52. Rodríguez-Fragoso L, Martínez-Arismendi JL, Orozco-Bustos D, Reyes-Esparza J, Torres E, Burchiel SW. Potential risks resulting from fruit/vegetable-drug interactions: effects on drug-metabolizing enzymes and drug transporters. J Food Sci. 2011;76(4):R112–24.

# Chapter 12
# Phytochemicals and Health

**Yan Yang**

**Abstract** Phytochemicals refer to a large group of non-nutritious but bioactive compounds derived from plant foods including fruits, vegetables, and grains. Due to the considerable diversity in their chemical structures, phytochemicals can be divided into phenolic compounds, carotenoids, terpenoids, organosulfur compounds, glucosinolates, saponins, phytoestrogens, phytic acid, phytosterols, etc. Phytochemicals are not only important for plant growth, but also helpful for plants to survive under various environmental stresses and resist infections of viruses, bacteria, yeasts, and fungi. At the same time, the beneficial roles of phytochemicals in improving human health, such as protecting against coronary heart disease, diabetes, cancers, hypertension, inflammation, and other chronic diseases, have gained increasing interest in recent years. In this chapter, we summarized the definition, classification, physicochemical property, pharmacokinetics, biological effects, and health functions of phytochemicals.

**Keywords** Phytochemicals · Plant foods · Bioactive compounds · Health

## 12.1 Overview of Phytochemicals

Phytochemicals are plant-derived chemicals. The *phyto-* of the word *phytochemical* is derived from the Greek word *phyto*, which means plant. There has been mounting evidence supporting the beneficial roles of phytochemicals in improving human health. They were shown to protect against cancers, coronary heart disease, diabetes, hypertension, inflammation, infections, psychotic diseases, spasmodic conditions, ulcers, and other chronic diseases based on clinical and preclinical studies. Although the underlying mechanisms have not been fully elucidated, it was suggested that phytochemicals could interfere with cellular functions, alter the activation of

Y. Yang (✉)
Department of Nutrition and Food Safety, School of Public Health (Shenzhen), Shenzhen Campus of Sun Yat-sen University, Sun Yat-Sen University, Shenzhen, People's Republic of China
e-mail: yangyan3@mail.sysu.edu.cn



L. Zhang (ed.), *Nutritional Toxicology*,
https://doi.org/10.1007/978-981-19-0872-9_12

transcription factors, modulate intracellular metabolism via various signaling pathways, and exert health benefits as substrates for biochemical reactions, cofactors of enzymatic reactions, inhibitors of enzymatic reactions, scavengers of reactive or toxic chemicals, ligands that agonize or antagonize cell surface or intracellular receptors. To date, over 5000 phytochemicals have been identified with more kinds undiscovered yet. In addition, the health benefits of phytochemicals could be even greater than that people currently know.

### 12.1.1 *Definition of Phytochemicals*

Phytochemicals, also known as plant secondary metabolites, anti-nutritional factors, or plant xenobiotics, are defined as bioactive non-nutrient plant compounds in fruits, vegetables, grains, and other plant foods [1]. Phytochemicals participate in various phytophysiological processes such as defense against potential threats which may include bacteria, viruses, and fungi [2].

Phytochemicals are present at fairly low concentrations and they are restricted to specific groups of organisms and are produced in response to particular conditions [3]. Thus, it was until recent centuries that we got to investigate the mechanisms underlying the health benefits of phytochemicals. Although phytochemicals are not essential nutrients, epidemiological evidence has shown that consumptions of fruits, vegetables and whole grains containing plentiful phytochemicals were strongly related with lower risks of cardiovascular diseases (CVDs), cancers, diabetes, and cognitive dysfunction. Moreover, these associations remained significant after adjusting the intake of the essential nutrients (proteins, carbohydrates, fats and vitamins) from plant foods [4]. Therefore, it indicated that other bioactive constituents in the plant foods account for their beneficial effects on life and health. Many laboratories started to seek and identify phytochemicals in plants since 1980s followed by extensive studies concerning the roles of phytochemicals in human health.

### 12.1.2 *Classification of Phytochemicals*

According to the chemical structures and functional characteristics, phytochemicals can be classified into several categories (see Table 12.1), including polyphenols, carotenoids, terpenoids, organosulfur compounds, saponins, phytic acids, and phytosterols. Besides, some phytochemicals, including curcumin, capsaicin, and chlorophyll, also have certain bioactivities [5, 6].

**Table 12.1** Classification of phytochemicals

| Category | Representative compounds | Food sources | References |
|---|---|---|---|
| Polyphenols | Catechins | Tea | [3] |
| | Anthocyanins | Black rice, blueberry | [5] |
| | Resveratrol | Grapes, red wine, berries | [7] |
| | Flavonoids | Fruits, vegetables | [8] |
| Carotenoids | Carotene | Yellow vegetables, leaves | [9] |
| | Lycopene | Watermelons, pink grapefruits, apricots | [10] |
| | Zeaxanthin | Corn, eggs | [11] |
| Terpenoids | Monoterpenoids | Citrus | [12] |
| | Sesquiterpenoids | | |
| | Diterpenoids | | |
| | Triterpenoids | | |
| Organosulfur compounds | Glucosinolates | Garlic, onions | [13] |
| Saponins | Steroidal saponins | Loquat, beans | [14] |
| | Triterpenoid saponins | Grain | [15] |
| Phytoestrogens | Isoflavone | Soybeans, tofu | [16] |
| | Lignans | Flaxseed, sesame seeds, whole-grain cereals | [17] |
| Phytic acid | Inositol hexaphosphate | Cereals, vegetables, nuts, natural oil | [18] |
| Phytosterols | Beta-sitosterol | Vegetable oil, fruits | [19] |
| | Stigmasterol | Beans, nuts | [20] |

## *12.1.3 The Physicochemical Property of Phytochemicals*

Phytochemicals are structurally diverse, so it results in different physicochemical properties. The chemical structures and physical properties are closely related to their functions. In this part, we focus on the physicochemical properties of polyphenols, carotenoids, terpenoids, organosulfur compounds, glucosinolates, saponins, phytoestrogens, phytic acid and phytosterols.

### 12.1.3.1 Phenolic Compounds

Phenolic compounds constitute a great portion of phytochemicals in fruits and vegetables. To date, more than 800 phenolic structures have been studied. Phenolic compounds can be classified as flavonoids and phenolic acids based on their chemical structures [21].

Flavonoids contain a 15-carbon skeleton consisting of two benzene rings attached via a heterocyclic pyrane ring, labeled as rings A, B, and C, in a C6–C3–C6

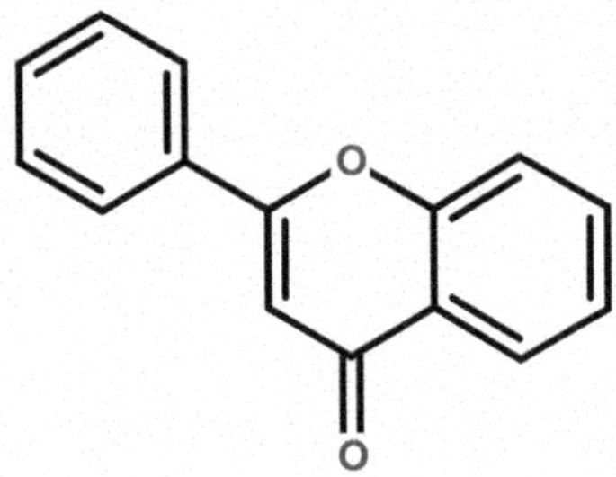

**Fig. 12.1** Flavonoid basic structure(aglycone) [23]

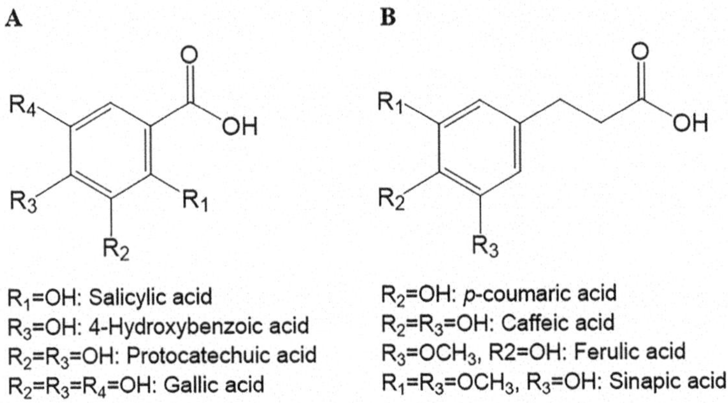


**Fig. 12.2** Backbones of hydroxybenzoic acid (**A**) and hydroxycinnamic acid (**B**) derivatives [24]

arrangement, as shown in Fig. 12.1. They can be divided into a variety of classes such as flavones (e.g., flavone, apigenin and luteolin), flavonols (e.g., quercetin, kaempferol, myricetin and fisetin), flavanones (e.g., flavanone, hesperetin and naringenin), and others. The chemical nature of flavonoids depends on their structural properties including the degree of hydroxylation, different substitutions and conjugations, and the degree of polymerization. Most flavonoids are crystalline solid and poorly soluble in water, while a few are soluble in orgainc solvents such as methanol, ethanol and ethyl acetate. Flavonoids are widely used as antioxidants because their phenolic hydroxyl groups can donate active hydrogen atoms and stop/delay the automatic oxidation of lipids [22].

Phenolic acids can be divided into two major groups, namely hydroxybenzoic acids and hydroxycinnamic acids, which are derived from non-phenolic molecules of benzoic and cinnamic acid, respectively (Fig. 12.2) [24]. Immobilized *Candida antarctica* lipase can be used to catalyze the direct acetylation of flavonoids with phenolic acids [25]. Phenolic acids behave as antioxidants, due to the reactivity of the phenol moiety (hydroxyl substituent on the aromatic ring). Although there are several mechanisms, the predominant mode of antioxidant activity is believed to be radical scavenging via hydrogen atom donation. Other established antioxidants, radical-quenching mechanisms are through electron donation and singlet oxygen quenching. Substituents on the aromatic ring affect the stabilization and therefore affect the radical-quenching ability of these phenolic acids. The antioxidant behavior of the free, esterified, glycosylated, and nonglycosylated phenolics has been reported [26].

#### 12.1.3.2 Carotenoids

Carotenoids are composed of polyene chains through conjugating double bonds. Because of the different conjugated double bonds, the colors of the carotenoids are various. The more conjugated double bonds they have, the darker color they present, and the stronger light absorption capacity they show. Carotenoids are tetraterpenoids, which are composed of eight condensed isoprene precursors that generate a linear backbone of 40 carbons (Fig. 12.3) [27]. They are dissolvable in most organic reagents and insoluble in water. They have a strong absorption at 400–500 nm, and present as red, orange, and yellow. Carotenoids are unstable in oxygen, acid, intense light, and high temperature, but relatively stable in alkaline conditions.

#### 12.1.3.3 Terpenoids

Terpenoids belong to a large and diverse class of naturally occurring organic chemicals similar to terpenes. They are derived from five-carbon isoprene units, and most of them have multi-cyclic structures that differ from each other in their functional groups and basic carbon skeletons [21]. The terpenoids can be classified according to the number of isoprene units or the number of cyclic structures they contain. Monoterpenes are terpenes that consist of two isoprene units (Fig. 12.4) [28]. Most of them are highly volatile and contain non-polar (lipophilic) structures that conferring the property of high membrane-penetration [29]. They possess variable basic skeletons and exhibit stereoisomerism. Moreover, there is a wide range of oxygenated derivatives (alcohols, aldehydes, ketones, and carboxylic acids) that can be derived from native basic skeletons [28].

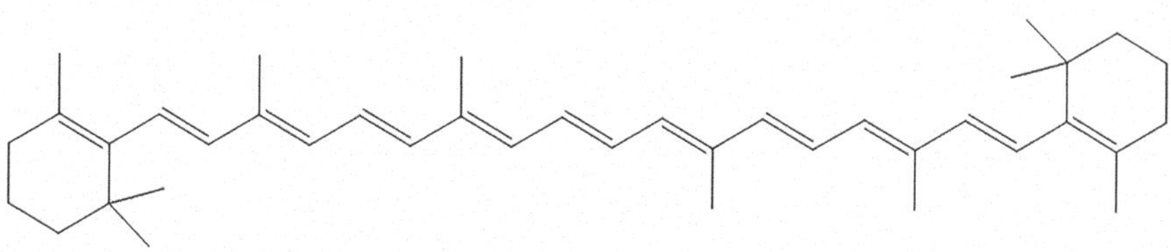

**Fig. 12.3** Chemical structure of β-carotene [27]

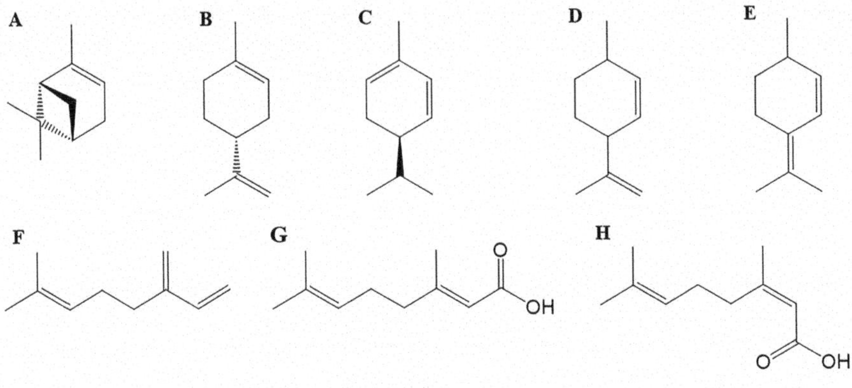


**Fig. 12.4** Chemical structures of (−)-α-pinene (**a**), (+)-limonene (**b**), R-(−)-α-phellandrene (**c**), isolimonene (**d**), isoterpinolene (**e**), myrcene (**f**), geranic acid (**g**), and neric acid (**h**) [28]

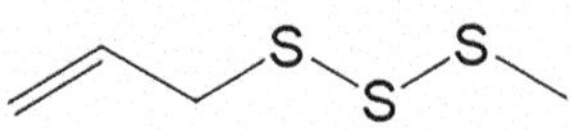


**Fig. 12.5** Chemical structure of methyl allyl trisulfide (MATS), a type of representative organosulfur compounds [30]

**Fig. 12.6** Chemical structure of glucosinolate [31]

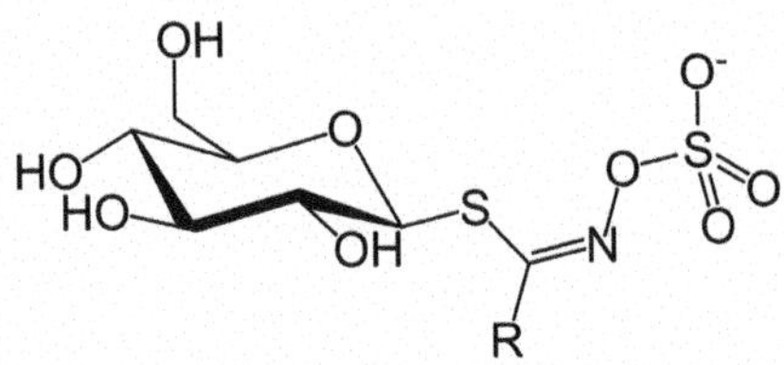


#### 12.1.3.4 Organosulfur Compounds

Organosulfur compounds are organic compounds that contain sulfur (Fig. 12.5). They are often associated with foul odors. The most characteristic feature of sulfur is that, because it has a larger atomic radius and a lower ionization strength than oxygen, it can bind together and make chains of 2–5 sulfurs. Sulfur shares the chalcogen group with oxygen, selenium, and tellurium, thus it is suggested that organosulfur compounds share some similarities in their physiological properties with carbon-oxygen, carbon-selenium, and carbon-tellurium compounds.

#### 12.1.3.5 Glucosinolates

Glucosinolates are β-thioglucoside *N*-hydroxysulfates, with a side chain and a sulfur-linked β-d-glucopyranose moiety (Fig. 12.6). Glucosinolates can be degraded under enzyme addition, high temperature, and high pressure. In intact plant tissues, glucosinolates are stored in compartments that are physically separated from compartments containing myrosinase enzyme. Upon tissue damage, glucosinolates could be hydrolyzed by myrosinase, and produce a range of breakdown products including isothiocyanates.

#### 12.1.3.6 Saponins

Saponins are amphipathic surface-active glycosides mainly produced by plants [32]. More specifically, saponins contain one or more hydrophilic glycoside moieties combined with a lipophilic triterpene derivative leading to their soap-like foaming property when shaken in aqueous solutions (Fig. 12.7). The lipophilic aglycone can be a wide variety of polycyclic organic structures ranging from a 10-carbon terpene to a 30-carbon triterpene skeleton, more often, a 27-carbon steroidal skeleton.

#### 12.1.3.7 Phytoestrogens

Phytoestrogens are a group of biologically active plant-derived compounds with estrogen-like functions. According to their chemical structures, phytoestrogens can

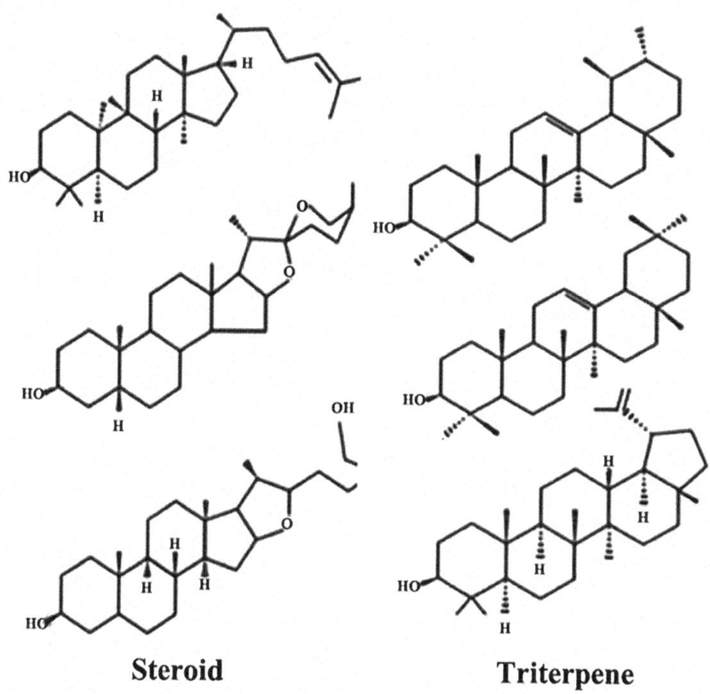


**Fig. 12.7** Chemical structures of saponins [32]

be further divided into subclasses of isoflavones, lignans, and coumestans. In foods, isoflavones are in the biologically inactive glucoside form but are converted to active aglycone form by gut bacteria in the intestine (Fig. 12.8). It is suggested that the estrogenic effects of soy can be attributed to the aglycones genistein, daidzein, and glycitein, though levels of these molecules vary considerably across soy food products [33].

#### 12.1.3.8 Phytic Acid

Phytic acid is synthesized from myo-inositol via a series of phosphorylation steps, thus it consists of an inositol ring with six phosphate ester bonds (Fig. 12.9) [34]. Phytic acid is light yellow or light brown slurry liquid, which is soluble in water, ethanol, acetone but insoluble in anhydrous ether, chloroform, benzene, and hexane. Phytic acid is a kind of strong acid, which has a strong ability to chelate, in addition to combining with metal cation [35]. It can also bind with protein molecules, thus reducing protein digestibility in animals.

#### 12.1.3.9 Phytosterols

Phytosterols refer to plant-derived sterols and stanols which are similar in chemical structures with animal-derived cholesterol. Stanols are saturated sterols without double bonds in their sterol rings (Fig. 12.10). The relative density of phytosterols is slightly higher than water. They are insoluble in water, acid, and alkali, while soluble in many organic solvents, such as ether, benzene, chloroform, ethyl acetate,

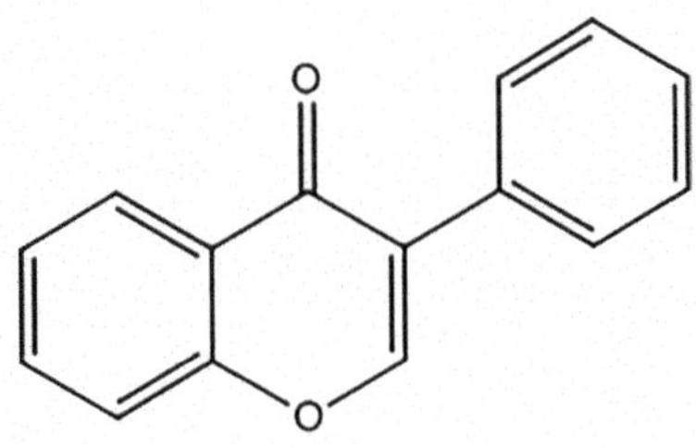


**Fig. 12.8** Chemical structure of isoflavone, a subclass of phytoestrogens [33]

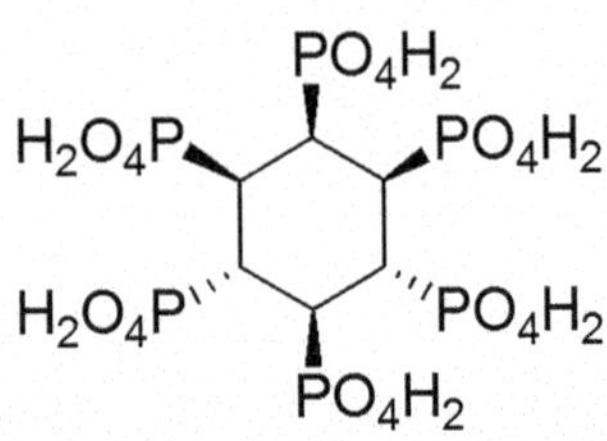


**Fig. 12.9** Chemical structure of phytic acid [34, 36]

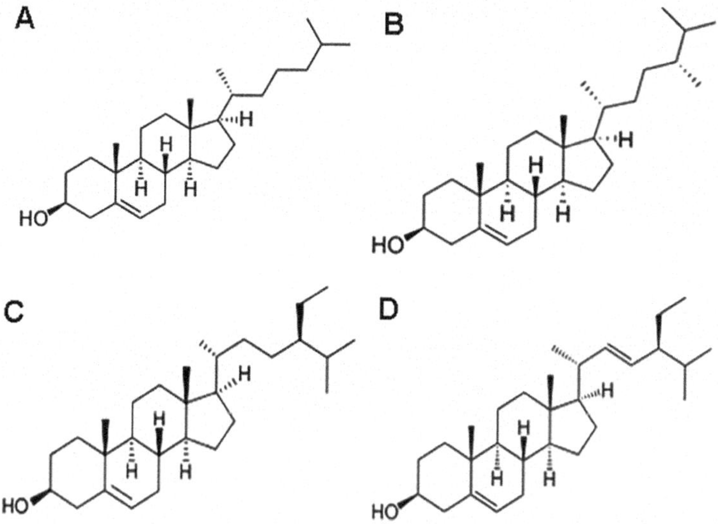


**Fig. 12.10** Chemical structures of cholesterol (**A**), campesterol (**B**), β-sitosterol (**C**), and stigmasterol (**D**) [37]

and petroleum ether. Due to their amphiphilic properties, phytosterols could act as emulsifiers during small intestinal absorption of lipids and lipid-soluble nutrients.

### *12.1.4 Pharmacokinetics of Phytochemicals*

Phytochemicals are metabolized in the human body by biotransformation enzymes similar to other xenobiotics. Phytochemicals undergo conjugation with glutathione, glucuronide, and sulfate moieties quickly and excretion in urine and bile. Metabolism of phytochemicals could be affected by polymorphisms of biotransformation enzymes, such as the glutathione *S*-transferases (GST), UDP-glucuronosyltransferases (UGT), and sulfotransferases (SULT) [38].

The metabolism of phytochemicals takes place throughout the digestive process, and is affected by gastrointestinal micro-environment, such as: (1) microorganisms and salivary enzymes in the oral cavity; (2) acidic condition in the stomach; (3) pancreatic and microbial enzymes in the small and large intestines; (4) endogenous phase I and phase II enzymes during the transmembrane transport; and (5) human gut microbiota. Thus, the metabolism of phytochemicals differs along the digestive tract, and the biological characteristics of phytochemicals are also changing simultaneously. The metabolic process of phytochemicals can be divided into the following steps: the change of active functional groups; partial or total degradation of the compounds; the combination of the compounds with other molecules. We will take flavonoids as an example to enhance the understanding of the metabolism of phytochemicals [39].

#### 12.1.4.1 Absorption

Most dietary flavonoids exist in food as their *O*-glycosides. The most common glycosidic unit is glucose, and other glycosidic units include glucorhamnose, galactose, arabinose, and rhamnose [40]. The biological fate of the dietary flavonoid glycosides was still controversial over the years. It was long believed that these fairly large and highly polar molecules could not be absorbed after oral ingestion but were hydrolyzed to their aglycones by bacterial enzymes in the gut. The aglycones might then be partially absorbed or further utilized by gut bacteria [41, 42].

#### 12.1.4.2 Metabolism

Flavonoid compounds and their metabolites bind to albumin in the blood. The affinity of flavonoids with albumin varies based on their chemical structures. The binding of flavonoid compounds and their metabolites with albumin will affect their clearance rate and the rate of transport into the cells and tissues. It is suggested that maintaining a constant supply of flavonoid glucuronides in the circulation may be of benefit because tissues can secret β-glucuronidases to release the bioactive aglycone when needed. The pressing challenge is to know the local concentration of flavonoids and their metabolites and utilize this as the foundation for designing in vitro bioactivity assays to determine the bioactive concentration of flavonoids, rather than counting on the concentration of flavonoids found in the blood as the maximum cut-off [40].

The metabolism of flavonoids is mediated by phase I and phase II enzymes in the liver after absorption in the intestine [43–45]. Hydroxylation and demethylation are the main acting forms of cytochrome P450 monooxygenase system, and both may participate in the metabolism of flavonoids. Demethylation occurs when methyl is at the 4′ but not the 3′ position. Researches have shown that phase I metabolism of dietary flavonoids in the body can hardly be detected because the hydroxyls on the reactive sites of flavonoids are more prone to the phase II conjugative reaction.

After absorption, the flavonoids are prone to conjugation, such as sulfation, methylation, and glucuronidation. Conjugation can reduce the active hydroxyls and increase the solubility of flavonoids, making flavonoids easier to be excreted into bile and urine. Additionally, the difference in conjugative sites can affect the bioactivity of conjugative products [46].

#### 12.1.4.3 Excretion

Most flavonoids and their derivatives were excreted out through urine and bile (Fig. 12.11). Most of the conjugated metabolites are excreted through bile, while other metabolites with a simpler structure such as monosulfate are mainly excreted through urine. The total amounts of metabolites excreted by urine correlate with the peak concentration of metabolites in plasma. But consumption of phytochemical or its precursor does not necessarily equate with exposure at the tissue level. Most of the phytochemicals excreted were in the form of their intestinal metabolites via gut fermentation [47]. An enhanced understanding of the factors that contribute to interindividual differences in the metabolism and disposition of phytochemicals

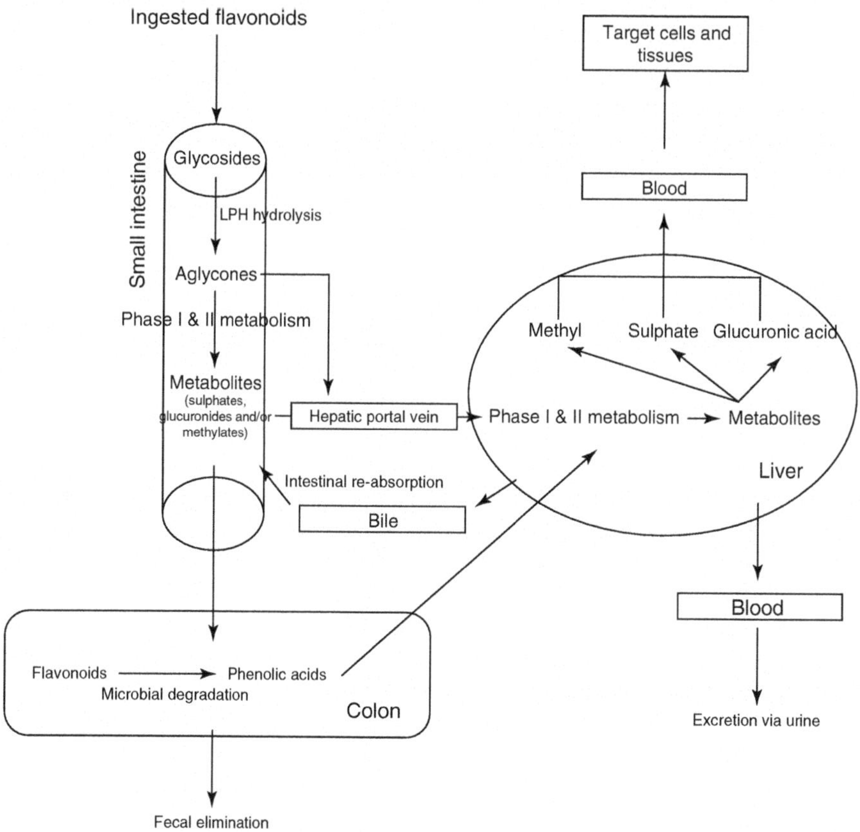


**Fig. 12.11** A simplified schematic of human flavonoid metabolism. Ingested flavonoids undergo extensive intestinal metabolism. Metabolites are then transported to the liver via hepatic portal vein and undergo further metabolism. The liver metabolites can be transported to targeted cells and tissues, excreted to bile and undergo enterohepatic re-circulation, or eliminated via urine and/or feces. The aglycones or flavonoid metabolites that reach the colon can undergo microbial degradation and reabsorption [49]

may allow for a more comprehensive evaluation of the role of these dietary constituents in chronic diseases prevention [48].

## 12.2 Biological Effects and Health Function of Phytochemicals

The following sections will focus on the physiological functions of major phytochemicals.

### *12.2.1 Anti-inflammation Effects*

Inflammation is a pathophysiological phenomenon associated with many diseases. It has been clearly demonstrated that low-grade chronic inflammatory status is considerably involved in the early onset and progression of metabolic syndrome, diabetes, cardiometabolic diseases, and oncology [50]. Phytochemicals could inhibit inflammation through various mechanisms.

#### 12.2.1.1 Phenolic Compounds

The anti-inflammatory activities of natural polyphenols have been well demonstrated before [51, 52]. Dietary anthocyanins and catechins could suppress the production of inflammatory cytokines via targeting nuclear transcription factors such as factor-κappa B (NF-κB), IRF-3, and AP-1, inhibiting their translocation into the nucleus and binding to the promoter region of target genes [53–55]. Tea catechins could decrease IL-1β production while not affecting the production of IL-6 and TNF-α in peripheral blood mononuclear cells [56, 57].

The anti-inflammation property of protocatechuic acid (PCA), the gut metabolite of anthocyanin [58], has been observed in vitro and in vivo. The anti-inflammation effects of PCA were likely to be mediated by modulating PI3K/Akt, MAPK, and NF-κB signaling pathways [59–61]. In addition, oral administration of PCA at 50 mg/kg body weight for 45 days significantly suppressed NO and TNF-α levels in the pancreas, cerebrum, and cerebellum of streptozotozin (STZ)-induced diabetic rats [62].

A large number of studies have demonstrated that resveratrol can inhibit inflammation. It has been revealed that the anti-inflammatory effects of resveratrol in cardiac tissue are mediated by inhibiting the mRNA expressions of intercellular adhesion molecule 1 (ICAM-1), inducible nitric oxide synthase (iNOS), and IL-1β [63]. Resveratrol has been also reported to attenuate inflammatory response by attenuating CD40 associated with sirtuin-1 and sirtuin-5 pathways in TNF-α induced endothelial cells injury [64]. The MAPKs and NF-κB signaling pathways have been reported to be involved in the inhibitory effects of resveratrol on TNF-α or IL-13-induced human pulmonary artery endothelial cells [65]. In a clinical trial, it has been revealed that resveratrol plays an important role in inflammatory responses in many diseases, including in Alzheimer's disease (AD) and angina [66]. However, results from clinical trials were not always consistent, the anti-inflammation effects of resveratrol may depend on the dosages, study duration, and the disease type [67–69].

It was summarized that the mechanisms underlying the anti-inflammation effects of polyphenols include: (1) inhibition of key enzymes in the arachidonic acid pathway, such as cyclooxygenases (COX) and lipoxygenases (LOX); (2) inhibition of nitric oxide synthase (NOS); (3) inhibition of nuclear NF-κB; (4) regulation of mitogen-activated protein kinase (MAPK); (5) activation of nonsteroidal anti-inflammatory drugs (NSAID)-activated gene-1 (NAG-1) and (6) regulation of cytokines production [52, 70].

#### 12.2.1.2 Carotenoids

Carotenoids have been shown to have anti-inflammatory effects in many studies. Previous studies have shown that retinoids and carotenoids could prevent Alzheimer's disease progression through inhibiting the production and accumulation of Aβ peptide and pro-inflammatory mediators [71]. NF-κB and MAPKs signaling are important pathways for the anti-inflammatory action of monoterpenes. Lycopene has been reported to inhibit inflammatory mediators via reducing ROS generation, suppressing the synthesis and release of proinflammatory cytokines, and modifying the arachidic acid metabolism, modulating the NF-kB, activated protein 1 (AP), and MAPK signaling pathways [72].

#### 12.2.1.3 Terpenoids

The terpenoid-rich essential oils could exert anti-inflammatory activities in vitro and in vivo [73]. Growing evidences have revealed that artemisinin and its derivatives also possess potent anti-inflammatory properties [74]. It has been demonstrated that the mechanisms of labdane diterpenoids underlying its anti-inflammatory activity were attributed to inhibition of NF-κB activity, modulation of arachidonic acid metabolism, and reduction of nitric oxide production [75]. In addition, it was found that triterpenoids isolated from *Euphorbia maculate* exhibited significant anti-inflammatory activities in TPA-induced inflammation in mice [76].

#### 12.2.1.4 Organosulfur Compounds

Numerous studies have proven the anti-inflammatory effects of organosulfur compounds [77, 78]. Allicin could inhibit the expression of intracellular cell adhesion molecule (ICAM) through down-regulating intracellular AP-1 and cJNK signaling pathways in gamma-irradiated HUVECs [79]. In addition, allicin and ajoene could reduce the expression of iNOS. Furthermore, 1,2-vinyldithiin could decrease lipid accumulation in preadipocytes and reduce the secretion of inflammatory molecules such as IL-6 and monocyte chemoattractant protein-1 (MCP-1) induced by macrophage-secreted factors [80].

Sulforaphane (SFN) is an isothiocyanate compound produced during myrosinase-catalyzed conversion of glucosinolates in cruciferous vegetables. SFN also shows an anti-inflammatory potency in both Nrf2-dependent and -independent manners. Sulforaphane could act as an anti-inflammatory molecule by preventing IL-1R-associated kinase-1 degradation, activation of NF-κB and interferon regulatory factor 3, and COX-2 expression [81]. SFN could also suppress the production of immunological factors such as IL-1β [82]. Moreover, several studies have shown that SFN reduces inflammation through the TLR4 signaling pathway. SFN suppressed TLR4 oligomerization, inhibited the TLR4/MyD88 pathway, and reduced the TNF-α and IL-6 levels by inhibiting PI3K/Akt and HIF-1α activation [81, 83, 84].

#### 12.2.1.5 Saponins

The furotrilliumoside from the roots and rhizomes of *Trillium Tschonoskii Maxim, a new steroidal saponin,* was reported to inhibit the LPS-induced inflammation in RAW264.7 cells by targeting PI3K/Akt, MARK, and Nrf2/HO-1 pathways [85]. The steroidal saponin monomer DT-13 inhibited endothelium vascular inflammation through regulating nitric oxide production and the expression of ROS, TNF-α, IL-8, and MCP-1 [86]. In addition, a C21 steroidal saponin isolated from a Chinese medicine *Cynanchi Stauntonii* was found to suppress macrophage activation induced by LPS in RAW264.7 cells [87].

*P.notoginseng* (*Panax Notoginoside*, PNS) is an important resource of saponins. Water extracts of PNS could inhibit neutrophil hyperactivity, and suppress overproduction of nitric oxide and prostaglandin E2 (PGE2) through decreasing the iNOS and COX-2 expression in mouse peritoneal macrophages stimulated with lipopolysaccharide (LPS) [88]. Besides, PNS could reduce the expression of pro-inflammatory factors by inhibiting the expression of RAGE and MAPK signaling pathways in the lesions of $ApoE^{-/-}$ mice [89]. Meanwhile, methanol extracts of PNS flower could suppress the expression of the inflammation-associated genes by inhibiting the NF-κB activation in LPS-stimulated mouse macrophage cells [90] and decrease TNF-α mRNA expression regulated by the PKC-NF-κB signaling pathway in scalded mice [91]. In addition, saponins from Ginsenoside could also inhibit the pro-inflammatory mediators, iNOS and COX-2, and pro-inflammatory cytokines, IL-1β, IL-6, and TNF-α, and increase the expression of the anti-inflammatory cytokine IL-10 [92].

#### 12.2.1.6 Phytosterols

Phytosterols consumption could decrease proinflammatory biomarkers secretion [93]. A recent study showed that stigmasterol could inhibit pro-inflammatory as well as matrix degradation mediators possibly via inhibiting the NF-κB signaling pathway in vitro [94]. Overexpression of serum vascular cell adhesion molecule-1 (VCAM-1) and ovalbumin-specific immunoglobulin E elicited by ovalbumin sensitization and challenge was significantly controlled with stigmasterol in guinea pig [95]. Beta-sitosterol also markedly reduces TNF-α, IL-1β, and IL-6 levels. In addition, recent research also found that Beta-sitosterol could protect against acute liver injury induced by LPS/GalN via suppressing the pro-inflammatory TLR4 and Nrf2 signaling pathways [96]. Overall, the anti-inflammatory roles and mechanisms of phytosterols are still ambiguous and need more studies.

### *12.2.2 Cancer Chemoprevention*

Tumorigenesis or carcinogenesis is a multi-step process that initiates with cellular transformation and progresses to hyperproliferation leading to the development of cancer. Epidemiological studies in the last decades have shown that regular consumption of fruits and vegetables could reduce the risk of cancers. Phytochemicals have been shown to suppress cancer by inhibiting cell proliferation, transformation, and angiogenesis. The anti-cancer activity of the major phytochemicals is summarized below.

#### 12.2.2.1 Phenolic Compounds

There is considerable evidence that polyphenols could inhibit tumor development and carcinogenesis inhibition at the cellular level. The mechanisms are different among the phenolic compounds. Anthocyanins, curcumin, resveratrol, and the gut metabolite of anthocyanins, namely PCA, will be discussed below in detail to demonstrate the anti-cancer activities of phenolic compounds.

It has been demonstrated that anthocyanins could inhibit trastuzumab-resistant breast cancer both in vitro and in vivo [97]. Moreover, recent studies have found the activity of anthocyanins on anti-cancer was mediated by its pro-apoptotic effect via increasing phosphatidylserine exposure and enhancing caspase-3 activity, and modulating MAPK, PI3K, and NF-κB signaling pathways [98].

Previous research has demonstrated that curcumin could inhibit tumor stem cell phenotypes in *ex vivo* models of colorectal liver metastases [99]. Curcumin could increase the activity of detoxifying enzymes, such as p21 and glutathione-*S*-transferases, leading to the inhibition of cell proliferation [100]. Moreover, curcumin has shown anti-cancer activity through inhibiting the expression of vascular endothelial growth factor (VEGF), epithelial-mesenchymal transition (EMT), migration, and invasion of thyroid cancer cells by regulating HIF-1α, PI3K/Akt, and TGF-β pathways, and NF-κB [101]. In addition, curcumin could increase the therapeutic efficacy of drugs for the treatment of thyroid cancer including sorafenib and docetaxel [102–104].

The anti-cancer activity of PCA has been investigated in various cancer cell lines, for example, in ovarian, breast, lung, liver, cervix, and prostate cancer cell lines [105]. The underlying mechanisms of anti-cancer activities of PCA may be different and depend on the cell types and animal models [106, 107].

Resveratrol has been reported to exert anti-cancer activities in different types of human cancers, including breast, cervical, uterine, blood, kidney, liver, eye, bladder, thyroid, esophageal, prostate, brain, and lung cancer [108–110]. Accordingly, owing to the preventive and therapeutic effects of resveratrol against different types of cancer, it is conceivable that the underlying mechanisms may also differ [108, 110].

#### 12.2.2.2 Carotenoids

Alpha-carotene, beta-carotene, and lycopene have received great interest because of their potential activity in reducing the risk of tumors [111, 112]. Meta-analyses of epidemiological studies have shown that alpha-carotene and beta-carotene could help reduce the risk of gastric cancer and breast cancer [113]. The molecular mechanisms underlying the association between carotenoids and cancer prevention include their strong antioxidant activity and modulation of immune function [114–116]. However, several studies have come to contrast findings with carotenoids. Results from the beta-Carotene and Retinol Efficacy Trial (CARET) have identified an increased risk of death from lung cancer in the group of individuals who received the supplement [117, 118].

#### 12.2.2.3 Terpenoids

Current research has shown that oleuropein acts as an anti-cancer agent by several mechanisms, including targeting HER2, epigenetic modifications, interfering with MAPK pathway, modulation of apoptosis and PI3K/Akt signaling axis as well as by reducing ROS production in different cell lines. Moreover, other studies also found that oleuropein effectively induced complete regressed tumors in mice experimental model [119]. A recent study also found 14 diterpenoids isolated from *E. fischeriana* could inhibit the proliferation of several cancer cells [120]. In in vitro study, the cytotoxicity activities of Lanostane triterpenoids isolated from *Ganoderma luteomarginatum* against human cancer cell lines (HGC-27, HeLa, A549, and SMMC-7721) were assayed and found that all isolates could suppress the cancer cells and the (5α,24E)-3β-acetoxyl-26-hydroxylanosta-8,24-dien-7-one exhibited the highest cytotoxicity against HeLa and A549 cell lines [121].

#### 12.2.2.4 Organosulfur Compounds

Adopting a diet with vegetables containing high levels of organosulfur compounds might reduce the risk of gastric cancer, breast cancer, and colorectal cancer [122, 123]. Organosulfur compounds have shown activity through multiple signaling pathways, such as carcinogen detoxification pathways as well as PI3K/Akt/mTOR signaling pathways. It has been demonstrated that organosulfur compounds could accelerate the clearance of carcinogens from the body by blocking the carcinogen activation in phase I metabolism, and thus inhibiting enzymes of the cytochrome P450 family and inducing phase II enzymes, such as glutathione S-transferases [123–125].

#### 12.2.2.5 Saponins

Saponins could also help reduce the risk of various types of cancer, such as gastric cancer, breast cancer, and so on [126]. Numerous studies have shown the anti-cancer activity of saponins through mechanisms including anti-proliferation, anti-angiogenesis, and anti-metastasis by induction of apoptosis and promotion of cell differentiation [127]. Ginsenoside, one of the triterpenoid saponins, is the main active component of *Ginseng radix* (Renshen in Chinese), which is widespread in Northeast China, Korea, and Japan. Ginsenoside Rg3 and Rh2 are the most well-studied active PPD-type anti-cancer congeners. The anti-cancer activity of Ginsenoside Rg3 was reported to affect a broad range of signaling pathways, including inactivation of NF-κB by reducing phosphorylation of Erk and Akt [128], inhibition of CXC receptor 4 (CXCR4) [129], down-regulation of PI3K/Akt family proteins and inhibition of apoptosis protein (IAP) family proteins [130].

### *12.2.3 Antioxidative Effects*

The antioxidant effects of phytochemicals act as important roles in the prevention of non-communicable chronic diseases [131]. The mechanisms include scavenging free radicals, accelerating the activities of antioxidant enzymes, defending intracellular DNA oxidative stress damage, and attenuating lipid peroxidation [132–134]. However, it is noteworthy that some phytochemicals, including glucosinolates and carotenoids, might promote oxidation at high doses [135, 136]. The antioxidant effects of plant chemicals still have much to be established, which probably can be profound for the prevention or treatment of oxidative stress-related diseases. Here, the antioxidant effects and mechanisms of several phytochemicals are summarized below.

#### 12.2.3.1 Phenolic Compounds

The antioxidant effects of phenolic compounds have gained increasing interest in recent years. Phenolic compounds, including curcumin, resveratrol, quercetin, protocatechuic acid, tannic acid, and chlorogenic acid, could exert their antioxidant properties by regulating the transcription of the antioxidant enzyme genes through PKC signaling in an isoform-dependent way [137, 138].

Anthocyanins are well known for their antioxidant activities. A human intervention study showed that intake of anthocyanin-rich juices could significantly decrease the malondialdehyde (MDA) level in plasma and DNA oxidative damage in lymphocytes [139]. The ability to remove free radicals of anthocyanins is greater than most common antioxidants, including butyl hydroxy acid (BHA), vitamin E, catechins, and quercetin [140, 141]. Due to the molecular structure with multiple phenolic hydroxyl groups, anthocyanins can directly obliterate all kinds of free radicals through liberating electrons of self-redox and make the balance between oxidation-reduction system and free radicals [142]. In addition, anthocyanins could increase the activities of intracellular superoxide dismutase (SOD) and glutathione transferase (GST) to reduce oxidative stress [143].

PCA has been reported to possess a strong antioxidant activity. The antioxidant property of PCA has been well clarified in vivo. For example, administration of PCA significantly increased the cerebral glutathione (GSH) level in diabetic rats [62]. In Wistar albino rats, dietary supplementation of PCA attenuated dermal wounds via up-regulating enzymatic antioxidant activity, such as increased SOD, CAT, and glutathione peroxidase (GPx) [144]. Moreover, it has been indicated that the effective chemoprotective role of PCA in colitis and the associated hepatotoxicity is related to its antioxidative properties in rats [145].

Chlorogenic acid could attenuate glucotoxicity in H9c2 cells via inhibition of glycation and PKCα up-regulation and safeguarding innate antioxidant status [146]. Dietary chlorogenic acid supplementation affected gut morphology, increased antioxidant capacity, and regulated intestinal selected bacterial populations in weaned piglets [147]. Moreover, it has been also shown that chlorogenic acid ameliorated intestinal mitochondrial injury by increasing antioxidant effects and activity of respiratory complexes in the intestinal mitochondrial injury induced by hydrogen peroxide [148].

The antioxidant property of resveratrol has been also well clarified. It has been shown that resveratrol alleviates inflammatory hyperalgesia by modulating ROS generation, antioxidant enzymes, and Erk1/2 activation in hyperalgesic rats [149]. In the brain of streptozotocin-induced diabetic rats, resveratrol can also decrease malondialdehyde and oxidized glutathione levels, and increase the action of all antioxidant enzymes [150]. In general, the antioxidant properties of resveratrol in vivo are more likely to be attributable to its effect as a gene regulator. Resveratrol inhibits NADPH oxidase-mediated production of ROS by down-regulating the expression and activity of the oxidase [151]. The antioxidant effect of resveratrol is mainly through chelating metal ions or eliminating free radicals in order to achieve LDL oxidation inhibition [152].

Curcumin could inhibit the generation of ROS, thus scavenging free radicals and peroxide by increasing the activities of intracellular SOD, glutathione peroxidase (GSH-Px) [153]. Another clinical study in which 106 healthy humans have administrated curcumin capsules orally for 158 days showed that the SOD activity in serum of the curcumin group was significantly increased compared with the control group.

#### 12.2.3.2 Carotenoids

Carotenoids are one of the most powerful antioxidants. The antioxidant activity of carotenoids depends on their structure and the presence of other co-antioxidants [154]. In general, the quenching activity of carotenoids is affected by the number of conjugated double bonds, the nature of substituents in carotenoids containing cyclic end groups, and the end groups (cyclic or acyclic) [155]. However, it is still controversial whether carotenoids are antioxidants or pro-oxidants, the circumstances that define each kind of activity are very specific and particular, which makes difficult a clear separation between the two concepts. Indeed, carotenoids act as the chain-breaking antioxidants at low oxygen, while they are readily autoxidized and exhibit pro-oxidant activity at high oxygen pressure [156]. Thus, carotenoids have a bidirectional effect on health effects.

The natural antioxidants oleuropein and quercetin could counteract the Cyclo-induced hepatotoxicity through activation of Nrf2/HO-1 signaling pathway with subsequent suppression of oxidative stress and inflammation [157]. The antioxidant activity of the fractions was due to oxygenated monoterpenes, specifically α-terpineol and *cis*-sabinene hydrate. Researches have showed the residues (R1 and R4) and butylated hydroxytoluene had greater antioxidant activity than either the distillate fractions or original rosemary essential oil [158].

Eight new sesquiterpenoids, two new diterpenoids, and 17 known sesqui- and diterpenoids were isolated from the radix of curcuma aromatica. Among these 27 compounds, 12 of them exhibited notable antioxidant effects on oxidative injury induced by $H_2O_2$ [159]. Moreover, in vitro study also revealed two triterpenoid glucosides markedly increased SOD activity and reduced MDA level [160].

#### 12.2.3.3 Organosulfur Compounds

Several studies have shown that isothiocyanates had both direct and indirect antioxidant effects [136]. By inhibiting phase I enzyme, isothiocyanates could decrease the formation of toxic intermediates through oxidation, and as the inducer of phase II enzyme. Therefore, isothiocyanates could sweep or neutralize oxide to improve the efficiency of eliminating toxic intermediates. Allicin, diallyl disulfide, and diallyl trisulfide appeared to be the main antioxidative compounds in *Allium* spp. [161, 162]. Garlic has been shown to suppress nitric oxide production through the inhibition of NF-κB, down-regulating expression of inducible nitric oxide synthase (iNOS), and blocking nuclear translocation of NF-κB [163]. Garlic has also been found to inhibit oxidative injury in the liver through adenosine monophosphate (AMP)-activated protein kinase [164].

#### 12.2.3.4 Phytoestrogens

Previous clinical research has proven that 6-month intervention of soybean isoflavone in healthy women could attenuate DNA damage and enhance the activities of amino glucoside enzyme. Both in vivo and in vitro studies have shown that the antioxidant effects of soybean isoflavones were mediated by inhibiting active oxygen free radicals, hydrogen peroxide, oxidative damage of DNA, and lipid peroxidation [165]. In addition, soy isoflavones could decrease the susceptibility of oxidative stress to LDL and DNA [166].

Sesaminol has been shown to increase the availability of tocopherols in biological systems by raising liver and plasma concentrations of vitamin E. Besides, sesaminol glucosides from defatted sesame flour were proved to decrease susceptibility to oxidative stress and inhibit allergen absorption in vitro [167]. Due to this high antioxidant activity, sesamol and pinoresinol which are in minor amounts in sesame were reported to protect cell membrane against lipid peroxidation, prevent LDL oxidation and microsomal peroxidation.

### *12.2.4 Immunomodulatory Effects*

The immune system is a very sophisticated defensive mechanism of vertebrates in response to severe perturbation of homeostasis, pathogens, injury, external contaminants, and infections. And the immune responses are activated and mediated by innate and adaptive immune systems to compact against undesirable and extraneous agents. A growing number of studies have shown that phytochemicals have powerful immunomodulatory effects.

#### 12.2.4.1 Phenolic Compounds

It has been found that enhanced immune function was observed in healthy volunteers consuming polyphenol-rich foods, such as resveratrol, anthocyanin, epigallocatechin gallate, and so on. Cyanidin-3-glucoside, delphinidin-3-glucoside, and petunidin-3-glucoside could inhibit NF-κB activities via mitogen-activated protein kinase (MAPK) pathways, and cyanidins can inhibit cyclooxygenase enzyme activities. The immunomodulatory activities of anthocyanins could partly account for their cardioprotective effects [53].

Several previous reports have suggested the immunomodulatory effects of PCA. An in vitro study showed that PCA inhibited human dendritic cell functional activation via up-regulation of PPARγ [168]. It has been reported that PCA suppressed ovalbumin-induced airway inflammation in a mouse allergic asthma model via regulation of Th2-mediated inflammatory responses, including decreasing IL-4, IL-5, IL-13, and immunoglobulin E levels [169]. Moreover, Guo et al. have revealed that PCA induced a better antiviral effect by immune enhancement in specific pathogen-free chickens [170].

It has been reported that resveratrol has powerful antagonistic actions on the inhibition of cellular immunity induced by cyclophosphamide via regaining the phagocytic index. Moreover, the concomitant immune effects of resveratrol on human peripheral blood mononuclear cell proliferation and releasing of IFN-γ and TNF-α in vitro are mediated by inhibition of the transcription factor NF-κB [171]. Resveratrol could induce dendritic cell-associated tolerance particularly during DC differentiation. Resveratrol-treated DCs lost the ability to produce IL-12p70 after activation, but have an enhanced ability to produce IL-10 [172]. A clinical trial showed that resveratrol-treated AD subjects showed an enhancement of adaptive immunity and demonstrated that sirtuin-1 activation may be a viable target for treatment or prevention of neurodegenerative disorders [173].

Curcumin was shown to reduce the expression of inflammatory cytokines including TNF and IL-1, adhesion molecules including ICAM-1 and VCAM-1 in HUVECs, inflammatory mediators including prostaglandins, and enzymes involved in inflammation like cyclooxygenase and lipoxygenase in mice. Moreover, curcumin could down-regulate NF-κB and STAT3 and reduce the expression of TLR-2 and TLR-4 [174].

Epigallocatechin gallate (EGCG), the most abundant catechin in tea, could increase the percentage of CD3, T-cell, CD19, B-cell, and Macrophage-3 antigen (Mac-3), and macrophages, but reduce the cell surface expression of CD11b in monocytes [175]. What's more, EGCG significantly suppressed IFN-γ production and the proliferation of peripheral blood mononuclear cells in vitro [176].

#### 12.2.4.2 Terpenoids

Terpenoids may possess potential therapeutic effects on Th1-mediated immune disorders, via suppressing IFN-γ excretion both in vivo and in vitro, which is the key cytokine secreted by Th-1 cells and plays a pivotal role in Th1 immune responses. The molecular mechanism by which diterpenoids exert immunomodulatory activity could be attributed to four signaling pathways, including TCR signaling pathway, TLR signaling pathway, VEGF signaling pathway, and osteoclast differentiation pathway [177].

It was reported that Paeoniflorin, one of the major bioactive components of *Paeony* root, inhibited T and B cells proliferation in arthritis animal model, up-regulated the regulatory T cells (Treg) function in systemic lupus erythematosus patients and Sjogren's syndrome animal model, decreased maturation of dendritical cells in collagen-induced arthritis mice, reduced the number of F4/80$^{+}$CD68$^{+}$ macrophages in imiquimod-induced psoriasis-like mice, and inhibited the functions of B cells stimulated by lipopolysaccharide [178].

Perillyl alcohol (POH), a naturally occurring monoterpene, could suppress lymphocyte proliferation, inhibite T-cell receptor-mediated calcium ion signaling and production of autoantibodies in vitro. Perillic acid, a metabolite of POH, could reduce IL-2 production in T lymphocytes by inhibiting Ras/MAPK signaling pathway [179].

#### 12.2.4.3 Saponins

Previous studies have revealed that saponins had the immuno-modulatory effects in animals, and could enhance macrophage phagocytosis, antibody secretions, and the production of cytotoxic T-lymphocytes (CTLs) againsts exogenous antigens. In addition, saponins have been found to enhance phagocytosis, promote IL-1β production by peritoneal macrophages, and stimulate secretion of cytokines such as IL-2, IL-4, IL-6, IL-10, IFN-γ, and TNF-α [180].

*Asparagus racemosus willd.* Roots contain steroidal saponins Shatavarin IV which exhibited immunomodulatory activity on Th1/Th2 immunity both in vivo and in vitro [181, 182]. Gypenoside can significantly increase the number of white blood cells as well as enhance the activity of natural killer (NK) cells. Soyasaponins could increase the secretion of IL-2 and promote the production of lymphocyte factor in T lymphocyte. Furthermore, soyasaponins can significantly improve the proliferation of B lymphocytes as well as enhance humoral immune function. Soyasaponins could also significantly improve the activity of NK cells.

#### 12.2.4.4 Phytoestrogens

Genistein, an abundant isoflavone in soybeans, could reduce inflammatory cell counts and enhance Th1/Th2 cytokine-mediated immunoreactions against rheumatoid arthritis and allergic asthma [183]. Puerarin, a secondary metabolite of the plant *Pueraria lobata*, has been used for the treatment of autoimmune disorders in traditional Chinese medicine.

A recent study suggested that supplementation of soy isoflavones effectively improved immunological status of pigs and produced mild improvements of growth performance under certain disease challenges including porcine reproductive and respiratory syndrome viruses [184].

Sesamin can significantly decrease expression levels of IL-4, IL-5, IL-13, serum IgE and the numbers of total inflammatory cells and eosinophils in BALF, thus attenuating OVA-induced eosinophil infiltration, airway goblet cell hyperplasia, mucus occlusion, and MUC5AC expression in the lung tissue [185].

#### 12.2.4.5 Phytosterols

It has been found that phytosterols could influence autoimmune diseases and virus infection by affecting the Th1–Th2 balance. A recent study reported that acute intake of plant sterols and stanols down-regulated the signaling pathway involved in T-cell functions and reduced the number of CD3, CD4, and Foxp3$^{+}$ cells in the jejunum [186]. Beta-sitosterol, the major phytosterol, which was found in various plant oils, nuts, and seeds, and its glycoside, β-sitosterol glycoside (BSSG) plays a key role in immunomodulation. Both lactose-beta-sitosterol (L-BSS) and beta-sitosterol (BSS) mitigated immunomodulation by eosinophil infiltration and mucus hypersecretion by goblet hyperplasia. The mRNA and protein expression of IL-4 and IL-5 were significantly inhibited by L-BSS and BSS [187]. A mixture of BSS and BSSG could influence the cellular proliferation of T lymphocytes when activated by mitogens in vitro [188]. Additionally, Bouic and Lamprecht suggested that phytosterols could regulate the immune system by increasing the activity of T lymphocytes and NK cells which are phenotypically lymphocytes that contribute to innate immunity, adaptive immunity, and placental reproduction. In addition, β-sitosterol was able to increase the effect of vitamin D on the immune function of macrophages, and repress mitogen-induced IL-2 production in cells of human Jurkat T in vitro [189].

### *12.2.5 Hypolipidemic Effect*

Hyperlipidemia is a major risk factor for CVDs. Phytochemicals with triacylglycerol- and cholesterol-lowering effects could be protective against metabolic disturbance and dyslipidemia.

#### 12.2.5.1 Phenolic Compounds

Anthocyanidins could reduce the plasma levels of low-density lipoprotein-cholesterol (LDL-C) and increase high-density lipoprotein-cholesterol (HDL-C) [190]. A double-blind, randomized, placebo-controlled trial suggested that anthocyanin supplementation increased HDL-C, decreased LDL-C concentrations, and enhanced cellular cholesterol efflux to serum, these benefits might be due to the inhibition of cholesteryl ester transfer protein (CETP). Besides, the cholesterol-lowering effects of anthocyanins were also mediated by their stimulatory effects on fecal sterol excretion. In a study by Liang et al., blueberry anthocyanins, mainly cyanidin-3-glucoside and petunidin-3-glucoside, at doses of 0.5 and 1% effectively lowered plasma cholesterol by increasing fecal excretion of acidic and neutral sterols in hamsters fed a cholesterol-enriched diet [190, 191].

Our previous research has demonstrated that protocatechuic acid, as the gut microbiota metabolite of cyanidin-3-glucoside, exerts the antiatherogenic effect partly through this newly defined miRNA-10b-ABCA1/ABCG1-cholesterol efflux signaling cascade. Another also reported that oral administration of PCA at 100 mg/kg body weight for 20 days significantly decreased the plasma levels of very low-density lipoprotein cholesterol (VLDL-C) and LDL-C, and increased HDL-C concentrations in D-galactosamine-induced hepatotoxic rats [192].

Despite that resveratrol could significantly reduce plasma lipids in experimental animals, the results of randomized clinical trials (RCTs) are contradictory. A recent meta-analysis of RCTs concluded that resveratrol does not significantly improve lipid profiles and the lipid-lowering effects of resveratrol require more studies exclusively on dyslipidemic patients [193].

Previous studies have found that grape seed procyanidin extract (GSPE) effectively improved blood lipid profiles in the fructose-fed rat model. This might be achieved via down-regulating of the hepatic expression of sterol regulatory element-binding protein 1c (SREBP1c) and stearoyl-CoA desaturase-1 (SCD1) and increasing endogenous hepatic cholesterol synthesis. The innovative findings obtained from this research implied that GSPE altered the conversion of endogenously synthesized cholesterol, directing it through TICE for export via the feces [194].

As major polyphenolic compounds, tea catechins could lower blood lipids via the following mechanisms: First, tea polyphenols could hinder emulsification, hydrolysis, and micellar solubilization of lipids thus interfering intestinal lipid absorption [195]; second, tea catechins could inhibit the biosynthesis of cholesterol and fatty acid in the liver [196]; third, tea catechin could enhance clearance of plasma LDL-C by elevating hepatic LDL receptor activity in rats [197].

#### 12.2.5.2 Carotenoids

Lycopene is a kind of carotenoid, with a strong capacity of antioxidation and regulating the blood lipids, and it can inhibit the elevation of serum total cholesterol (TC), Total triglycerides (TG), and LDL-C in hyperlipidemia rats with 25 mg/kg and 85 mg/kg lycopene added to the diet [198]. Lycopene could reduce blood cholesterol concentration via down-regulating the expression of hepatic proprotein convertase subtilisin/kexin type 9 (PCSK-9), HMGR and increasing LDL-R gene expression [199].

It was found that β-carotene could significantly decrease cholesterol absorption in the small intestine while increasing fecal sterol excretion thus improving hypocholesterolemia in rats fed a cholesterol-enriched diet. Besides, vitamin A, which can be transformed from β-carotene by the intestinal enzyme β, β-carotene 15,15′-monooxygenase, plays a major role in lipid metabolism. Chronic feeding of a vitamin A-enriched diet in hypercholesterolemic obese rats normalized the plasma HDL-C level and presumably improved reverse cholesterol transport, with an effective dose of 52 mg/kg diet, possibly through up-regulating hepatic SR-BI-mediated pathway [200].

#### 12.2.5.3 Organosulfur Compounds

Garlic consumption could lower blood total and LDL-cholesterol, increase HDL-C, and reduce LDL-C and triglycerides, thus protecting blood vessels and heart. Diallyl disulfide (DADS), another active principle of garlic (*Allium sativum*), is known for its antihyperlipidemic properties [201].

#### 12.2.5.4 Phytoestrogens

A meta-analysis of 11 RCTs incorporating 471 participants showed that soy protein that contained enriched soy isoflavones significantly decreased serum total cholesterol and LDL-C compared with the same amounts of isoflavone-depleted soy protein [202]. These results suggested that consumption of 102 mg soy isoflavones for 1–3 months could lower total cholesterol by 3.9 mg/dL and LDL-C by 5.0 mg/dL independent of the amount of soy protein ingested.

Sesamin could down-regulate the expression of peroxisome proliferator-activator receptor alpha (PPARα) and decrease the expression of sterol regulatory element-binding protein-1c, acetyl-CoA carboxylase, and fatty acid synthase, and inhibit cholesterol absorption and biosynthesis in human and rats leading to the reduction of LDL-C and increase the levels of HDL-C [167].

#### 12.2.5.5 Phytosterols

The hypolipidemic effect of phytosterols was found in animal experiments for the first time in 1951 [203]. In recent years, more than 200 clinical trials and several meta-analyses have confirmed the efficacy, effectiveness, and safety of their use as a cholesterol-lowering agent [204].

Although the effects of phytosterols in reducing blood cholesterol have been demonstrated for several decades, the precise mechanism has not been fully elucidated yet. The research from Hayes suggested that free phytosterols could promote the excretion of endogenous cholesterol in gerbils [205]. Other researchers reported that the cholesterol-lowering effects of phytosterols were partly attributed to their conversions into the liver X receptor agonist, thus activating the expression of ABC sterol efflux transporters, such as ABCG5 and ABCG8, which act as transporters for cholesterol and phytosterols [206]. But some researchers suggested that phytosterol inhibition of cholesterol absorption in mice could also be independent of ABCA1 [207].

Remarkably, a recent clinical research showed that combination of phytosterols and ezetimibe significantly enhanced the efficacy of ezetimibe on whole-body cholesterol metabolism [208]. The large cumulative action of combined dietaries and pharmacological treatments on cholesterol metabolism emphasized the potential importance of dietary phytosterols as adjunctive therapy for the treatment of hypercholesterolemia. The combination of phytosterols and ezetimibe caused 7% and 22% reduction of LDL-C compared with ezetimibe alone and placebo, respectively. It is obvious that this combined treatment could be a better clinical option for treating hypercholesterolemia.

## *12.2.6 Hypoglycemic Effect*

It is generally known that improving glucose metabolism is crucial for preventing and curing diabetes and other metabolic disorders. In recent years, a growing number of studies have focused on the hypoglycemic effects of phytochemicals.

### 12.2.6.1 Phenolic Compounds

Epidemiological studies have clearly shown that dietary intake of polyphenols was inversely related to risks of type 2 diabetes. Results from the clinical trials have shown that polyphenols could reduce blood glucose and increase insulin sensitivity.

Several studies implied that dietary polyphenols could modulate insulin signaling pathway by reducing inflammatory adipokines and modifying microRNA profiles [209, 210]. Polyphenols could also improve the small intestinal glucose absorption via inhibition of sodium glucose cotransporter. The benefits of dietary polyphenols for type 2 diabetes could be summarized as the protection of pancreatic β-cells against glucose toxicity, inhibition of α-amylases and α-glucosidases leading to decreased starch digestion, and inhibition of advanced glycation end products formation [211].

Dietary cyanidin-3-glucoside-rich bayberry fruit extracts could ameliorate impaired glucose tolerance in streptozotocin-induced diabetic ICR mice [212]. Besides, ethanolic extracts of cherry fruit, which were rich in polyphenols, could result in a significant reduction in blood glucose in alloxan-induced diabetic rats [213]. Findings mentioned above suggested that developing novel functional food containing anthocyanin extract would be a promising strategy to improve glycemic control in diabetes patients.

It has been demonstrated that PCA decreased blood glucose and oxidative stress, and increased plasma nitric oxide in diabetic rats induced by streptozotocin [214]. PCA isolated from the *Sansevieria roxburghiana* rhizomes decreased blood glucose in a rodent model of type 2 diabetes via stimulating glucose metabolism [215]. Moreover, oral administration of PCA at 100 mg/kg body weight per day to STZ-diabetic rats for 45 days could effectively improve glycemic profiles [216].

Although regular consumption of resveratrol has been known to improve glucose homeostasis and reverse insulin resistance in type 2 diabetes mellitus (T2DM), the reported results are inconsistent. A recent systematic review and meta-analysis includes nine randomized controlled trials involving 283 participants, and demonstrated that resveratrol supplementation at doses of $\geq$100 mg/day resulted in a significant decrement in fasting plasma glucose [217].

Results from animal research implied that quercetin potentiated insulin secretion in INS-1 cells induced by glucose, glibenclamide, or potassium chloride. And quercetin also improved the normal function of β-cell. Both effects might be mediated via phosphorylation of extracellular signal-regulated kinase 1 and 2 (Erk 1/2) [218]. In addition, clinical research indicates a single oral dose of quercetin at 400 mg/day could effectively suppress postprandial hyperglycemia in patients with type 2 diabetes [219, 220].

#### 12.2.6.2 Carotenoids

Studies have demonstrated that administration of 90 mg/kg lycopene to streptozotocin-induced hyperglycemic rats caused a decrease in blood glucose levels with an increase in insulin concentration, and the effect was mediated via lowering the free radical activity. However, there is still little evidence about the precise relationship between lycopene-rich diets and the incidence of diabetes yet [221].

#### 12.2.6.3 Organosulfur Compounds

The beneficial effects of garlic supplements on diabetes mellitus have been well established [222]. It was hypothesized that the hyperglycemic effects of garlic were partly through the potential release of insulin from β-cells of islets [223]. Scientific evidence illustrated that organosulfur compounds are the predominant bioactive compounds in garlic species and are responsible for the antidiabetic effect of garlic [224, 225]. Moreover, garlic oil, which contains abundant allyl methyl sulfide, was found to stimulate insulin secretion in pancreatic β-cells and might increase sensitivity to insulin [226].

Previous studies have found that sulforaphane could suppress hepatic glucose production by modulating nuclear translocation of nuclear factor erythroid 2-related factor 2 (NRF2) and decreasing the expression of key enzymes in gluconeogenesis [227]. Moreover, some researchers suggested that sulforaphane attenuated dysregulated glucose production and glucose intolerance in a similar way to metformin [227]. Sulforaphane also reduced fasting blood glucose and glycated hemoglobin (HbA1c) in obese type 2 diabetes patients [227]. Additionally, it was indicated that sulforaphane could act as a promising agent to improve glucose tolerance through up-regulating insulin signaling mainly involving the IRS-1/Akt/GLUT4 pathway in the skeletal muscle [228].

#### 12.2.6.4 Phytoestrogens

Systematic reviews of observational studies showed that intake of soy isoflavones was associated with lower risks of type 2 diabetes especially in women [229]. Ding et al. reported a significant inverse association of soy isoflavones with type 2 diabetes (multivariate-adjusted HR of 0.89 (95% CI: 0.83–0.96) in the highest compared to the lowest consumption of soy isoflavones [230].

Total lignans from *Fructus arctii* have significant hypoglycemic potential in GK rats through stimulating insulin secretion, promoting the release of GLP-1, and decreasing intestinal absorption of glucose [231]. Besides, lignan-rich extracts from *Fructus schisandrae* could improve glucose homeostasis by increasing glucose disposal rates and enhancing hepatic insulin sensitivity in type-2 diabetic rats [232].

#### 12.2.6.5 Phytosterols

Misawa et al. showed that the administration of lophenol and cycloartenol significantly decreased the levels of fasting blood glucose and reduced visceral fat weight in Zucker diabetic rats. In addition, the expression levels of hepatic genes which encode gluconeogenic enzymes (glucose-6-phosphatase and phosphopyruvate carboxykinase) and lipogenic enzymes (acetyl-CoA carboxylase and fatty acid synthase) were decreased significantly after the administration of lophenol and cycloartenol [233]. Moreover, Ponnulakshmi et al. indicated that beta-sitosterol could improve glycemic control through activating insulin receptor substrate (IRS) and glucose transporter 4 (GLUT4) in the adipose tissue of high fat- and high sucrose-induced type-2 diabetic rats [234]. Although phytosterols have been reported as compounds with antidiabetic effects, further clinical evidence is necessary to confirm this function.

### *12.2.7 Anti-thrombosis Effect*

Cardiovascular disease is the leading cause of death in developed countries. Approximately 50% of all deaths associated with malignant neoplasms are due to thrombotic events. Normal hemostasis may be damped by pathological factors leading to uncontrolled clot formation and vessel occlusion in the arteries and veins. Platelets, endothelial cells, and circulating coagulation proteins are crucial mediators of vascular hemostasis and thrombosis. Studies have reported phytochemicals could effectively inhibit thrombosis through various mechanisms. The sections below focus on the anti-thrombosis effect of phenolic compounds, carotenoids, and saponins.

#### 12.2.7.1 Phenolic Compounds

Anthocyanins have been found to significantly inhibit platelet activation, aggregation, and thrombus formation in vitro [197, 198] through inhibiting platelet PI3K/Akt activation, attenuating eNOS phosphorylation and cGMP production [199], attenuating fibrinogen binding to platelets following agonist treatment [197], and decreasing the chemokine secretion. Anthocyanins also have been shown to influence vascular endothelial cells to inhibit thrombosis. They could inhibit NF-κB secretion and increase NO production to decrease the expression of endothelial cell surface adhesion molecules. Anthocyanins could also regulate immune cells adhesion by decreasing the expression of vascular endothelial growth factor (VEGF) receptor and ICAM-1 in endothelial cells [235].

Resveratrol could reduce the incidence of portal vein system thrombosis after splenectomy in a rat fibrosis model [236]. Moreover, resveratrol preserves the function of human platelets stored for transfusion and inhibits thrombosis in an in vivo mouse model of transfusion [237]. It has been reported that resveratrol could also inhibit portal vein thrombogenesis in rats [238].

Many studies have reported curcumin can inhibit platelet activation, aggregation, and thrombus formation [239]. It is because that curcumin could modulate cycloxygenase activity, inhibit tyrosine phosphorylation of various proteins [204], suppress transcription factor Egr-1 [205], and interfere with the kinase activity of Syk with subsequent activation of PLCγ2 [207]. In addition, curcumin could also suppress angiogenesis by down-regulating the expression of VEGF [240].

#### 12.2.7.2 Carotenoids

Epidemiological studies supported the hypothesis that antioxidants might act as a cheap but effective means to prevent cardiovascular disease [241]. Platelets play an important role in thrombosis, and lycopene has been shown to inhibit platelet aggregation and thrombosis induced by various agonists [242].

#### 12.2.7.3 Saponins

Saponins could activate the fibrinolytic system, promote fibrin dissolution, enhance the anticoagulant effect, reduce the release of thromboxane, and inhibit platelet aggregation, ultimately inhibiting thrombus formation. Notoginsenoside could reduce thrombogenesis through up-regulating tissue-type PA (t-PA) and down-regulating the expression of PAI-1. In addition, it could inhibit TNF-α induced PAI-1 overexpression in vascular smooth muscle cells by suppressing the Erk and PKB signaling pathways. It could also decrease intimal thickness, promote endothelial regeneration, as well as down-regulate VEGF and MMP-2 expression of rabbit iliac arteries after balloon endothelial denudation injury and increase NO production via the PI3K/Akt/eNOS pathway [243].

### 12.2.8 Antimicrobial Effects

The current clinical antimicrobial therapies aim to inhibit the microbial growth. Undesirably it could impose a strong selective pressure on the cells and consequently induce the development of resistance. Unlike synthetic molecules, phytochemicals exhibit an unmatched structural diversity with intricate and novel multilayer mechanisms of action [244, 245]. Therefore, compounds that inhibit bacterial growth by different mechanisms provide promising approaches to control the drug-resistant infections. The antimicrobial activity of some major phytochemicals is summarized below.

#### 12.2.8.1 Polyphenols

Flavonoids have been suggested as potential antimicrobial agents. Their activity is partly due to the ability of forming a complex with extracellular proteins thus increasing their permeability through the cell membrane. Flavonoids with greater lipophilicity might also disrupt microbial membranes. Flavonoids with fewer hydroxyl groups on their $\beta$-rings are more active against microorganisms via targeting the hydroxyl groups on the surface of the cell membrane [246, 247].

It has been shown that PCA could inhibit hepatitis B virus replication by activating Erk1/2 pathway and down-regulating HNF4α and HNF1α in Huh7 and HepG2 cells [248]. Moreover, PCA could mediate bacterial lethality in vitro via attenuating oxidative stress [249]. PCA has been considered as an antiviral agent against avian influenza virus (AIV) and infectious bursal disease (IBD) virus via enhancing immune response in specific pathogen-free (SPF) chickens [170].

Resveratrol has been reported to show anti-vital activity, antifungal activity, antiparasitic activity, antibacterial activity in vitro and in vivo [250]. The antiviral effect of resveratrol has been reported in hepatitis C virus, respiratory syncytial virus, herpes simplex virus, varicella-zoster virus, Epstein-Barr virus, influenza virus, human immunodeficiency virus, human metapneumonia virus, African swine fever virus, enterovirus, and duck enteritis virus [251–253].

Curcumin and its analogs are well known as antimicrobial agents against both gram-positive and gram-negative bacteria, and also have been used as a preservative. The antimicrobial mechanism of curcumin was mediated via the interaction with protein FtsZ essential as cell division initiating agent in bacteria. Experimental data supported that the methoxy and hydroxyl groups are directly involved in the antimicrobial activity of curcumin and its analogs [254].

#### 12.2.8.2 Terpenoids

Due to their recognized antimicrobial potential, terpenoids have been the subject of several studies along the years. It was suggested that the antimicrobial effects of terpenoids were mediated by membrane disruption by the lipophilic compounds, and their antimicrobial activities depend on their structures. This antibacterial action can result in the increase of membrane fluidity/permeability, disruption of membrane-embedded proteins, and change of ion transport processes in both gram-positive and gram-negative bacteria [255, 256].

Sesquiterpenoids showed antimicrobial activity against gram-positive and gram-negative bacteria and inhibited the growth of *M. tuberculosis*. It was demonstrated that diterpenoids isolated from the bark of *Podocarpus nagi*, in which the most abundant compound was totarol, showed strong bactericidal activity against the gram-positive bacteria *propionebacterium acnes*, *S. mutans*, and *S. aureus*. Similarly, the antibacterial activity of diterpenoids isolated from roots of *Salvia sclarea L.* was also observed against *S. aureus* and *Staphylococcus epidermidis* [257, 258].

#### 12.2.8.3 Organosulfur Compounds

Allicin has antimicrobial effects on a variety of gram-positive and gram-negative bacteria, which are comparable with some antibiotics. Allicin could inhibit the activity of bacteria sulfhydryl protease leading to bacteriostasis. In addition to the antibacterial effect, garlic also has anti-fungal, anti-parasitic, and anti-viral effects [259, 260].

Intact glucosinolates (GLS) are biologically inactive, but their hydrolysis products such as isothiocyanates (ITCs), nitriles, thiocyanates, epithionitriles, and oxazolidinediones have various functions including antimicrobial effects. Among the glucosinolate hydrolysis products of GLS, ITCs are regarded as the most robust inhibitors of bacterial activity [261, 262]. ITCs from seeds of *Sinapis alba* L., which were comprised of phenethyl, benzyl, and benzoyl groups, exhibited good antimicrobial activity against intestinal bacteria, namely *Clostridium difficile*, *Clostridium perfringens*, and *E. coli.* Allyl isothiocyanate, an aliphatic ITC, has strong bactericidal activities against foodborne pathogens.

#### 12.2.8.4 Phytoestrogens

Isoflavone possesses excellent antifungal effects toward *Candida glabrata* and *Cryptococcus neoformans* with a minimum inhibitory concentration value of 1 μg/mL. The antifungal effects of isoflavone toward *Cryptococcus neoformans* were comparable with those of standard compounds terbinafine phenol, fluconazole, and amphotericin B [263].

The ethanoic extracts from the leaves of *P. amarus*, which were rich in lignans, showed antimicrobial activity against gram-negative strains, but inactive against gram-positive and yeast strains. Moreover, the extracts and its major compound Phyllanthin could potentiate the antibiotic activity of norfloxacin against SA1199-B that overproduces the NorA efflux pump. Modulation of fluroquinolone resistance could be related to the inhibition of proton-motive force-dependent efflux pump NorA by Phyllanthin [264].

### 12.2.9 Neuroprotective Effects

Polyphenols have been suggested to be effective in the prevention and treatment of cognitive diseases due to their anti-amyloidogenic features. Curcumin is a promising candidate for dementia therapy due to its neuroprotective activities in healthy rats. Resveratrol administration could increase the activity of detoxifying enzymes including SOD and catalase (CAT), while decreasing the activity of MDA in the mouse brain [229].

The protective effects of PCA on Parkinson's disease were also investigated. It was revealed that the neuroprotective effects of PCA were mediated by a combination of cellular mechanisms including antioxidant cytoprotection and anti-inflammation effects [61]. An in vivo study showed that PCA supplementation could protect against seizure-induced neuronal death in adult male rats [265]. It has been also reported that PCA exerted neuroprotective effects under conditions of nitrosative stress, which occurs during inflammation in the central nervous system [266]. Moreover, PCA could also protect against hydrogen peroxide-induced retinal ganglion injury in retinal ganglion cells by attenuating oxidative damage [267]. The neuroprotective effects of PCA on PC12 cells treated with 1-methyl-4-phenylpyridinium ion may be through its inhibitory effects of cytotoxicity, apoptotic morphology, reduction of tyrosine hydroxylase (TH) expression, and abnormal oligomeration of alpha-synuclein, indicating the protective roles of PCA in Parkinson's disease [268].

A meta-analysis of 10 RCTs concerning the effects of soy isoflavones on cognitive function in post-menopausal women showed that subjects receiving soy isoflavones had significantly better scores of cognitive function and visual memory. Their subgroup analyses showed that the neuroprotective effects of soy isoflavones were more pronounced in women aged above 60 years [269]. Results from RCTs conducted in men and premenopausal women also reported similar benefits as those in post-menopausal women [270]. Overall, soy isoflavones have the potential for improving cognitive function.

Daidzein and genistein are two major isoflavones in soybeans. Daidzein exhibited neuroprotective effects by inhibiting microglial activation and the release of pro-inflammatory factors in murine microglial cells stimulated with lipopolysaccharide [271]. In animal models of Parkinsonism (P) and Parkinsonism + ovariectomized (OP), genistein has been shown to protect dopaminergic neurons while failing to influence the kinetic problems [272]. The mechanism underlying the neuroprotective action of genistein may be attributed to its estrogenic characteristics [273]. Arbabi et al. also suggested that genistein could prevent 6-OHDA-induced neuronal loss when substituted for estrogen in ovariectomized rats [274].

### *12.2.10 Other Physiological Functions*

Except for the functions mentioned above, flavonoids exert some properties of antimutagen, anti-aging, and resistance to radiation; monoterpenes have good analgesic effect; phytoestrogens can prevent osteoporosis through activating estrogen receptor in the bone tissue and inhibiting bone resorption of osteoclast; saponins can improve liver function and also have an anti-fatigue effect. Numerous studies have shown that carotenoids have protective effects on the eyes. They have also been linked with reducing the risk of age-related macular degeneration and cataracts.

## References


1. Leitzmann C. Characteristics and health benefits of phytochemicals. Forsch Komplementmed. 2016;23:69–74.
2. Aharoni A, Galili G. Metabolic engineering of the plant primary-secondary metabolism interface. Curr Opin Biotechnol. 2011;22:239–44.
3. Hussain MS, Fareed S, Ansari S, Rahman MA, Ahmad IZ, et al. Current approaches toward production of secondary plant metabolites. J Pharm Bioallied Sci. 2012;4:10–20.
4. Tennant DR, Davidson J, Day AJ. Phytonutrient intakes in relation to European fruit and vegetable consumption patterns observed in different food surveys. Br J Nutr. 2014;112:1214–25.
5. Phan MAT, Paterson J, Bucknall M, Arcot J. Interactions between phytochemicals from fruits and vegetables: effects on bioactivities and bioavailability. Crit Rev Food Sci Nutr. 2018;58:1310–29.
6. Favero G, Franceschetti L, Buffoli B, Moghadasian MH, Reiter RJ, et al. Melatonin: protection against age-related cardiac pathology. Ageing Res Rev. 2016;35:336–49.
7. Nunes S, Danesi F, Del Rio D, Silva P. Resveratrol and inflammatory bowel disease: the evidence so far. Nutr Res Rev. 2018;31:85–97.
8. Afshari K, Haddadi NS, Haj-Mirzaian A, Farzaei MH, Rohani MM, et al. Natural flavonoids for the prevention of colon cancer: a comprehensive review of preclinical and clinical studies. J Cell Physiol. 2019;234(12):21519–46.
9. Tan BL, Norhaizan ME. Carotenoids: how effective are they to prevent age-related diseases? Molecules. 2019;24:1801.
10. Grabowska M, Wawrzyniak D, Rolle K, Chomczynski P, Oziewicz S, et al. Let food be your medicine: nutraceutical properties of lycopene. Food Funct. 2019;10(6):3090–102.
11. Eisenhauer B, Natoli S, Liew G, Flood VM. Lutein and zeaxanthin-food sources, bioavailability and dietary variety in age-related macular degeneration protection. Nutrients. 2017;9:120.
12. Yazaki K, Arimura GI, Ohnishi T. ‘Hidden’ terpenoids in plants: their biosynthesis, localization and ecological roles. Plant Cell Physiol. 2017;58:1615–21.

13. Melino S, Sabelli R, Paci M. Allyl sulfur compounds and cellular detoxification system: effects and perspectives in cancer therapy. Amino Acids. 2011;41:103–12.
14. Zhao YZ, Zhang YY, Han H, Fan RP, Hu Y, et al. Advances in the antitumor activities and mechanisms of action of steroidal saponins. Chin J Nat Med. 2018;16:732–48.
15. Han Q, Qian Y, Wang X, Zhang Q, Cui J, et al. Cytotoxic oleanane triterpenoid saponins from *Albizia julibrissin*. Fitoterapia. 2017;121:183–93.
16. Rienks J, Barbaresko J, Nothlings U. Association of isoflavone biomarkers with risk of chronic disease and mortality: a systematic review and meta-analysis of observational studies. Nutr Rev. 2017;75:616–41.
17. Durazzo A, Lucarini M, Camilli E, Marconi S, Gabrielli P, et al. Dietary lignans: definition, description and research trends in databases development. Molecules. 2018;23:3251.
18. Silva EO, Bracarense AP. Phytic acid: from antinutritional to multiple protection factor of organic systems. J Food Sci. 2016;81:R1357–62.
19. Bin Sayeed MS, Karim SMR, Sharmin T, Morshed MM. Critical analysis on characterization, systemic effect, and therapeutic potential of beta-sitosterol: a plant-derived orphan phytosterol. Medicines (Basel). 2016;3:29.
20. Wang S, Ye K, Shu T, Tang X, Wang XJ, et al. Enhancement of galloylation efficacy of stigmasterol and beta-sitosterol followed by evaluation of cholesterol-reducing activity. J Agric Food Chem. 2019;67:3179–87.
21. Barbieri R, Coppo E, Marchese A, Daglia M, Sobarzo-Sanchez E, et al. Phytochemicals for human disease: an update on plant-derived compounds antibacterial activity. Microbiol Res. 2017;196:44–68.
22. Kumar S, Pandey AK. Chemistry and biological activities of flavonoids: an overview. ScientificWorldJournal. 2013;2013:162750.
23. Seleem D, Pardi V, Murata RM. Review of flavonoids: a diverse group of natural compounds with anti-*Candida albicans* activity in vitro. Arch Oral Biol. 2017;76:76–83.
24. Heleno SA, Martins A, Queiroz MJ, Ferreira IC. Bioactivity of phenolic acids: metabolites versus parent compounds: a review. Food Chem. 2015;173:501–13.
25. Stevenson DE, Wibisono R, Jensen DJ, Stanley RA, Cooney JM. Direct acylation of flavonoid glycosides with phenolic acids catalysed by *Candida antarctica* lipase B (Novozym 435®). Enzym Microb Technol. 2006;39:1236–41.
26. Robbins RJ. Phenolic acids in foods: an overview of analytical methodology. J Agric Food Chem. 2003;51:2866–87.
27. Domonkos I, Kis M, Gombos Z, Ughy B. Carotenoids, versatile components of oxygenic photosynthesis. Prog Lipid Res. 2013;52:539–61.
28. Oz M, Lozon Y, Sultan A, Yang KH, Galadari S. Effects of monoterpenes on ion channels of excitable cells. Pharmacol Ther. 2015;152:83–97.
29. Clarke S. Essential chemistry for aromatherapy. Edinburgh: Churchill Livingstone; 2008. p. 41–77.
30. Ariga T, Seki T. Antithrombotic and anticancer effects of garlic-derived sulfur compounds: a review. Biofactors. 2006;26:93–103.
31. Fahey JW, Zalcmann AT, Talalay P. The chemical diversity and distribution of glucosinolates and isothiocyanates among plants. Phytochemistry. 2002;59:237.
32. Dong J, Liang W, Wang T, Sui J, Wang J, et al. Saponins regulate intestinal inflammation in colon cancer and IBD. Pharmacol Res. 2019;144:66–72.
33. Thaung Zaw JJ, Howe PRC, Wong RHX. Does phytoestrogen supplementation improve cognition in humans? A systematic review. Ann N Y Acad Sci. 2017;1403:150–63.
34. Dersjant-Li Y, Awati A, Schulze H, Partridge G. Phytase in non-ruminant animal nutrition: a critical review on phytase activities in the gastrointestinal tract and influencing factors. J Sci Food Agric. 2015;95:878–96.
35. Changhao S, Wenhua L, Guowei H. Nutrition and Food Hygiene. Beijing: People's Health Publishing House; 2017. p. 135–6.

36. Graf E, Empson KL, Eaton JW. Phytic acid. A natural antioxidant. J Biol Chem. 1987;262: 11647–50.
37. Miras-Moreno B, Sabater-Jara AB, Pedreno MA, Almagro L. Bioactivity of phytosterols and their production in plant in vitro cultures. J Agric Food Chem. 2016;64:7049–58.
38. Lampe JW, Chang JL. Interindividual differences in phytochemical metabolism and disposition. Semin Cancer Biol. 2007;17:347–53.
39. Koistinen VM, Hanhineva K. Microbial and endogenous metabolic conversions of rye phytochemicals. Mol Nutr Food Res. 2016;61(7):1600627.
40. Gonzales GB, Smagghe G, Grootaert C, Zotti M, Raes K, et al. Flavonoid interactions during digestion, absorption, distribution and metabolism: a sequential structure-activity/property relationship-based approach in the study of bioavailability and bioactivity. Drug Metab Rev. 2015;47:175–90.
41. Gonzales GB. In vitro bioavailability and cellular bioactivity studies of flavonoids and flavonoid-rich plant extracts: questions, considerations and future perspectives. Proc Nutr Soc. 2017;76:175–81.
42. Heim KE, Tagliaferro AR, Bobilya DJ. Flavonoid antioxidants: chemistry, metabolism and structure-activity relationships. J Nutr Biochem. 2002;13:572–84.
43. Chen Z, Zheng S, Li L, Jiang H. Metabolism of flavonoids in human: a comprehensive review. Curr Drug Metab. 2014;15:48–61.
44. Feng X, Li Y, Brobbey Oppong M, Qiu F. Insights into the intestinal bacterial metabolism of flavonoids and the bioactivities of their microbe-derived ring cleavage metabolites. Drug Metab Rev. 2018;50:343–56.
45. Hollman PC, Katan MB. Absorption, metabolism and health effects of dietary flavonoids in man. Biomed Pharmacother. 1997;51:305–10.
46. Williamson G. Common features in the pathways of absorption and metabolism of flavonoids. In: Meskin MS, et al., editors. Phytochemicals: mechanisms of action. Boca Raton, FL: CRC Press; 2004. p. 21–33.
47. Nemeth K, Piskula MK. Food content, processing, absorption and metabolism of onion flavonoids. Crit Rev Food Sci Nutr. 2007;47:397–409.
48. Walle T. Absorption and metabolism of flavonoids. Free Radic Biol Med. 2004;36:829–37.
49. Thilakarathna SH, Rupasinghe HPV. Flavonoid bioavailability and attempts for bioavailability enhancement. Nutrients. 2013;5:3367–87.
50. Murakami A, Ohigashi H. Targeting NOX, INOS and COX-2 in inflammatory cells: chemoprevention using food phytochemicals. Int J Cancer. 2007;121:2357–63.
51. Bungau S, Abdel-Daim MM, Tit DM, Ghanem E, Sato S, et al. Health benefits of polyphenols and carotenoids in age-related eye diseases. Oxidative Med Cell Longev. 2019;2019:9783429.
52. Santangelo C, Vari R, Scazzocchio B, Di Benedetto R, Filesi C, et al. Polyphenols, intracellular signalling and inflammation. Ann Ist Super Sanita. 2007;43:394–405.
53. Lee YM, Yoon Y, Yoon H, Park HM, Song S, et al. Dietary anthocyanins against obesity and inflammation. Nutrients. 2017;9:1089.
54. Chan MM-Y, Fong D, Ho C-T, Huang H-I. Inhibition of inducible nitric oxide synthase gene expression and enzyme activity by epigallocatechin gallate, a natural product from green tea. Biochem Pharmacol. 1997;54:1281–6.
55. Paquay JBG, Haenen G, Stender G, Wiseman SA, Tijburg LBM, et al. Protection against nitric oxide toxicity by tea. J Agric Food Chem. 2000;48:5768–72.
56. Wang Y, Ji S, Zang W, Wang N, Cao J, et al. Identification of phenolic compounds from a unique citrus species, finger lime (*Citrus australasica*) and their inhibition of LPS-induced NO-releasing in BV-2cell line. Food Chem Toxicol. 2019;129:54–63.
57. Crouvezier S, Powell B, Keir D, Yaqoob P. The effects of phenolic components of tea on the production of pro- and anti-inflammatory cytokines by human leukocytes in vitro. Cytokine. 2001;13:280–6.

58. Wang D, Xia M, Yan X, Li D, Wang L, et al. Gut microbiota metabolism of anthocyanin promotes reverse cholesterol transport in mice via repressing miRNA-10b. Circ Res. 2012;111:967–81.
59. Wongwichai T, Teeyakasem P, Pruksakorn D, Kongtawelert P, Pothacharoen P. Anthocyanins and metabolites from purple rice inhibit IL-1beta-induced matrix metalloproteinases expression in human articular chondrocytes through the NF-kappaB and ERK/MAPK pathway. Biomed Pharmacother. 2019;112:108610.
60. Ormazabal P, Scazzocchio B, Vari R, Santangelo C, D'Archivio M, et al. Effect of protocatechuic acid on insulin responsiveness and inflammation in visceral adipose tissue from obese individuals: possible role for PTP1B. Int J Obes. 2018;42:2012–21.
61. Zhang Z, Li G, Szeto SSW, Chong CM, Quan Q, et al. Examining the neuroprotective effects of protocatechuic acid and chrysin on in vitro and in vivo models of Parkinson disease. Free Radic Biol Med. 2015;84:331–43.
62. Adedara IA, Fasina OB, Ayeni MF, Ajayi OM, Farombi EO. Protocatechuic acid ameliorates neurobehavioral deficits via suppression of oxidative damage, inflammation, caspase-3 and acetylcholinesterase activities in diabetic rats. Food Chem Toxicol. 2019;125:170–81.
63. Huang FC, Kuo HC, Huang YH, Yu HR, Li SC, et al. Anti-inflammatory effect of resveratrol in human coronary arterial endothelial cells via induction of autophagy: implication for the treatment of Kawasaki disease. BMC Pharmacol Toxicol. 2017;18:3.
64. Yu H, Pan W, Huang H, Chen J, Sun B, et al. Screening analysis of sirtuins family expression on anti-inflammation of resveratrol in endothelial cells. Med Sci Monit. 2019;25:4137–48.
65. Wu JM, Hsieh TC, Yang CJ, Olson SC. Resveratrol and its metabolites modulate cytokine-mediated induction of eotaxin-1 in human pulmonary artery endothelial cells. Ann N Y Acad Sci. 2013;1290:30–6.
66. Militaru C, Donoiu I, Craciun A, Scorei ID, Bulearca AM, et al. Oral resveratrol and calcium fructoborate supplementation in subjects with stable angina pectoris: effects on lipid profiles, inflammation markers, and quality of life. Nutrition. 2013;29:178–83.
67. van der Made SM, Plat J, Mensink RP. Trans-resveratrol supplementation and endothelial function during the fasting and postprandial phase: a randomized placebo-controlled trial in overweight and slightly obese participants. Nutrients. 2017;9:596.
68. Khodabandehloo H, Seyyedebrahimi S, Esfahani EN, Razi F, Meshkani R. Resveratrol supplementation decreases blood glucose without changing the circulating CD14(+) CD16 (+) monocytes and inflammatory cytokines in patients with type 2 diabetes: a randomized, double-blind, placebo-controlled study. Nutr Res. 2018;54:40–51.
69. Vors C, Couillard C, Paradis ME, Gigleux I, Marin J, et al. Supplementation with resveratrol and curcumin does not affect the inflammatory response to a high-fat meal in older adults with abdominal obesity: a randomized, placebo-controlled crossover. Trial J Nutr. 2018;148:379–88.
70. Biesalski HK. Polyphenols and inflammation: basic interactions. Curr Opin Clin Nutr Metab Care. 2007;10:724–8.
71. Mohammadzadeh Honarvar N, Saedisomeolia A, Abdolahi M, Shayeganrad A, Taheri Sangsari G, et al. Molecular anti-inflammatory mechanisms of retinoids and carotenoids in Alzheimer's disease: a review of current evidence. J Mol Neurosci. 2017;61:289–304.
72. Palozza P, Parrone N, Catalano A, Simone R. Tomato lycopene and inflammatory cascade: basic interactions and clinical implications. Curr Med Chem. 2010;17:2547–63.
73. Gallily R, Yekhtin Z, Hanus LO. The anti-inflammatory properties of Terpenoids from cannabis. Cannabis Cannabinoid Res. 2018;3:282–90.
74. Shi C, Li H, Yang Y, Hou L. Anti-inflammatory and immunoregulatory functions of artemisinin and its derivatives. Mediat Inflamm. 2015;2015:435713.
75. Tran QTN, Wong WSF, Chai CLL. Labdane diterpenoids as potential anti-inflammatory agents. Pharmacol Res. 2017;124:43–63.
76. Sun Y, Gao LL, Tang MY, Feng BM, Pei YH, et al. Triterpenoids from *Euphorbia maculata* and their anti-inflammatory effects. Molecules. 2018;23:2112.

77. Chen W, Liang X, Syed AK, Jessup P, Church WR, et al. Inhibiting GPIb shedding preserves post-transfusion recovery and hemostatic function of platelets after prolonged storage. Arterioscler Thromb Vasc Biol. 2016;36:1821–8.
78. Lei Y-P, Chen H-W, Sheen L-Y, Lii C-K. Diallyl disulfide and diallyl trisulfide suppress oxidized LDL-induced vascular cell adhesion molecule and E-selectin expression through protein kinase A- and B-dependent signaling pathways. J Nutr. 2008;138:996–1003.
79. Son E-W, Mo S-J, Rhee D-K, Pyo S. Inhibition of ICAM-1 expression by garlic component, allicin, in gamma-irradiated human vascular endothelial cells via downregulation of the JNK signaling pathway. Int Immunopharmacol. 2006;6:1788–95.
80. Keophiphath M, Priem F, Jacquemond-Collet I, Clement K, Lacasa D. 1,2-Vinyidithiin from garlic inhibits differentiation and inflammation of human Preadipocytes. J Nutr. 2009;139: 2055–60.
81. Youn HS, Kim YS, Park ZY, Kim SY, Choi NY, et al. Sulforaphane suppresses oligomerization of TLR4 in a thiol-dependent manner. J Immunol. 2010;184:411–9.
82. Durham A, Jazrawi E, Rhodes JA, Williams C, Kilty I, et al. The anti-inflammatory effects of sulforaphane are not mediated by the Nrf2 pathway. Eur Respir J. 2014;44:P3332.
83. Zeng X, Liu X, Bao H, Zhang Y, Wang X, et al. [Effects of sulforaphane on Toll-like receptor 4/myeloid differentiation factor 88 pathway of monocyte-derived macrophages from patients with chronic obstructive pulmonary disease]. Zhonghua Jie He He Hu Xi Za Zhi. 2014;37: 250–4.
84. Kim SY, Jeong E, Joung SM, Lee JY. PI3K/Akt contributes to increased expression of Toll-like receptor 4 in macrophages exposed to hypoxic stress. Biochem Biophys Res Commun. 2012;419:466–71.
85. Yan T, Yu X, Sun X, Meng D, Jia JM. A new steroidal saponin, furotrilliumoside from Trillium tschonoskii inhibits lipopolysaccharide-induced inflammation in Raw264.7 cells by targeting PI3K/Akt, MARK and Nrf2/HO-1 pathways. Fitoterapia. 2016;115:37–45.
86. Fan R, Han Y, Han H, Chen Z, Yu B, et al. DT-13 ameliorates TNF-alpha-induced nitric oxide production in the endothelium in vivo and in vitro. Biochem Biophys Res Commun. 2018;495: 1175–81.
87. Liu J, Tang J, Zuo Y, Yu Y, Luo P, et al. Stauntoside B inhibits macrophage activation by inhibiting NF-kappaB and ERK MAPK signalling. Pharmacol Res. 2016;111:303–15.
88. Lee SM. Anti-inflammatory effects of ginsenosides Rg5, Rz1, and Rk1: inhibition of TNF-alpha-induced NF-kappaB, COX-2, and iNOS transcriptional expression. Phytother Res. 2014;28:1893–6.
89. Taborskaya KI, Frolova MY, Kuleva NV. Comparative analysis of serotonin levels in rat platelets, serum and brain on the aging. Tsitologiia. 2016;58:115–9.
90. Jung H-W, Seo U-K, Kim J-H, Leem K-H, Park Y-K. Flower extract of *Panax notoginseng* attenuates lipopolysaccharide-induced inflammatory response via blocking of NF-kappa B signaling pathway in murine macrophages. J Ethnopharmacol. 2009;122:313–9.
91. Wang Y, Peng D, Huang W, Zhou X, Liu J, et al. Mechanism of altered TNF-alpha expression by macrophage and the modulatory effect of *Panax notoginseng* saponins in scald mice. Burns. 2006;32:846–52.
92. Metwaly AM, Lianlian Z, Luqi H, Deqiang D. Black ginseng and its saponins: preparation, phytochemistry and pharmacological effects. Molecules. 2019;24:1856.
93. Alemany L, Barbera R, Alegria A, Laparra JM. Plant sterols from foods in inflammation and risk of cardiovascular disease: a real threat? Food Chem Toxicol. 2014;69:140–9.
94. Gabay O, Sanchez C, Salvat C, Chevy F, Breton M, et al. Stigmasterol: a phytosterol with potential anti-osteoarthritic properties. Osteoarthr Cartil. 2010;18:106–16.
95. Antwi AO, Obiri DD, Osafo N. Stigmasterol modulates allergic airway inflammation in guinea pig model of ovalbumin-induced asthma. Mediat Inflamm. 2017;2017:2953930.
96. Yin Y, Liu X, Liu J, Cai E, Zhu H, et al. Beta-sitosterol and its derivatives repress lipopolysaccharide/d-galactosamine-induced acute hepatic injury by inhibiting the oxidation and inflammation in mice. Bioorg Med Chem Lett. 2018;28:1525–33.

97. Li X, Xu J, Tang X, Liu Y, Yu X, et al. Anthocyanins inhibit trastuzumab-resistant breast cancer in vitro and in vivo. Mol Med Rep. 2016;13:4007–13.
98. Chen M, Zhao Z, Yu S. Cytotoxicity and apoptotic effects of polyphenols from sugar beet molasses on colon carcinoma cells in vitro. Int J Mol Sci. 2016;17:993.
99. James MI, Iwuji C, Irving G, Karmokar A, Higgins JA, et al. Curcumin inhibits cancer stem cell phenotypes in ex vivo models of colorectal liver metastases, and is clinically safe and tolerable in combination with FOLFOX chemotherapy. Cancer Lett. 2015;364:135–41.
100. Khan AQ, Siveen KS, Prabhu KS, Kuttikrishnan S, Akhtar S, et al. Curcumin-mediated degradation of S-phase kinase protein 2 induces cytotoxic effects in human papillomavirus-positive and negative squamous carcinoma cells. Front Oncol. 2018;8:399.
101. Durko L, Malecka-Panas E. Lifestyle modifications and colorectal cancer. Curr Colorectal Cancer Rep. 2014;10:45–54.
102. Perna A, De Luca A, Adelfi L, Pasquale T, Varriale B, et al. Effects of different extracts of curcumin on TPC1 papillary thyroid cancer cell line. BMC Complement Altern Med. 2018;18:63.
103. Shin HJ, Hwang KA, Choi KC. Antitumor effect of various phytochemicals on diverse types of thyroid cancers. Nutrients. 2019;11:125.
104. Zhang L, Cheng X, Gao Y, Bao J, Guan H, et al. Induction of ROS-independent DNA damage by curcumin leads to G2/M cell cycle arrest and apoptosis in human papillary thyroid carcinoma BCPAP cells. Food Funct. 2016;7:315–25.
105. Yin MC, Lin CC, Wu HC, Tsao SM, Hsu CK. Apoptotic effects of protocatechuic acid in human breast, lung, liver, cervix, and prostate cancer cells: potential mechanisms of action. J Agric Food Chem. 2009;57:6468–73.
106. Tsao SM, Hsia TC, Yin MC. Protocatechuic acid inhibits lung cancer cells by modulating FAK, MAPK, and NF-kappaB pathways. Nutr Cancer. 2014;66:1331–41.
107. Lin HH, Chen JH, Chou FP, Wang CJ. Protocatechuic acid inhibits cancer cell metastasis involving the down-regulation of Ras/Akt/NF-kappaB pathway and MMP-2 production by targeting RhoB activation. Br J Pharmacol. 2011;162:237–54.
108. Rauf A, Imran M, Butt MS, Nadeem M, Peters DG, et al. Resveratrol as an anti-cancer agent: a review. Crit Rev Food Sci Nutr. 2018;58:1428–47.
109. De Amicis F, Chimento A, Montalto FI, Casaburi I, Sirianni R, et al. Steroid receptor signallings as targets for resveratrol actions in breast and prostate cancer. Int J Mol Sci. 2019;20:1087.
110. Elshaer M, Chen Y, Wang XJ, Tang X. Resveratrol: an overview of its anti-cancer mechanisms. Life Sci. 2018;207:340–9.
111. Carini F, David S, Tomasello G, Mazzola M, Damiani P, et al. Colorectal cancer: an update on the effects of lycopene on tumor progression and cell proliferation. J Biol Regul Homeost Agents. 2017;31:769–74.
112. Nelson SM, Panagiotou OA, Anic GM, Mondul AM, Mannisto S, et al. Metabolomics analysis of serum 25-hydroxy-vitamin D in the alpha-tocopherol, beta-carotene cancer prevention (ATBC) study. Int J Epidemiol. 2016;45:1458–68.
113. Yang T, Yang X, Wang X, Wang Y, Song Z. The role of tomato products and lycopene in the prevention of gastric cancer: a meta-analysis of epidemiologic studies. Med Hypotheses. 2013;80:383–8.
114. Bolhassani A. Cancer chemoprevention by natural carotenoids as an efficient strategy. Anti Cancer Agents Med Chem. 2015;15:1026–31.
115. Satomi Y. Antitumor and cancer-preventative function of fucoxanthin: a marine carotenoid. Anticancer Res. 2017;37:1557–62.
116. Liu RH. Potential synergy of phytochemicals in cancer prevention: mechanism of action. J Nutr. 2004;134:3479S–85S.
117. Abar L, Vieira AR, Aune D, Stevens C, Vingeliene S, et al. Blood concentrations of carotenoids and retinol and lung cancer risk: an update of the WCRF-AICR systematic review of published prospective studies. Cancer Med. 2016;5:2069–83.

118. Omenn GS, Goodman GE, Thornquist MD, Balmes J, Cullen MR, et al. Risk factors for lung cancer and for intervention effects in CARET, the beta-carotene and retinol efficacy trial. J Natl Cancer Inst. 1996;88:1550–9.
119. Ahmad Farooqi A, Fayyaz S, Silva AS, Sureda A, Nabavi SF, et al. Oleuropein and cancer chemoprevention: the link is hot. Molecules. 2017;22:705.
120. Jian B, Zhang H, Han C, Liu J. Anti-cancer activities of diterpenoids derived from *Euphorbia fischeriana* Steud. Molecules. 2018;23:387.
121. Su HG, Zhou QM, Guo L, Huang YJ, Peng C, et al. Lanostane triterpenoids from *Ganoderma luteomarginatum* and their cytotoxicity against four human cancer cell lines. Phytochemistry. 2018;156:89–95.
122. Pan JH, Abernathy B, Kim YJ, Lee JH, Kim JH, et al. Cruciferous vegetables and colorectal cancer prevention through microRNA regulation: a review. Crit Rev Food Sci Nutr. 2018;58: 2026–38.
123. Costea T, Hudita A, Ciolac OA, Galateanu B, Ginghina O, et al. Chemoprevention of colorectal cancer by dietary compounds. Int J Mol Sci. 2018;19:3787.
124. Wang S, Li M, Wang X, Li X, Yin H, et al. Diallyl trisulfide attenuated n-hexane induced neurotoxicity in rats by modulating P450 enzymes. Chem Biol Interact. 2017;265:1–7.
125. Wu X, Zhou QH, Xu K. Are isothiocyanates potential anti-cancer drugs? Acta Pharmacol Sin. 2009;30:501–12.
126. Xie ZZ, Li MM, Deng PF, Wang S, Wang L, et al. Paris saponin-induced autophagy promotes breast cancer cell apoptosis via the Akt/mTOR signaling pathway. Chem Biol Interact. 2017;264:1–9.
127. Chen MF, Huang SJ, Huang CC, Liu PS, Lin KI, et al. Saikosaponin d induces cell death through caspase-3-dependent, caspase-3-independent and mitochondrial pathways in mammalian hepatic stellate cells. BMC Cancer. 2016;16:532.
128. Kim BM, Kim DH, Park JH, Surh YJ, Na HK. Ginsenoside Rg3 inhibits constitutive activation of NF-kappaB signaling in human breast cancer (MDA-MB-231) cells: ERK and Akt as potential upstream targets. J Cancer Prev. 2014;19:23–30.
129. Chen XP, Qian LL, Jiang H, Chen JH. Ginsenoside Rg3 inhibits CXCR4 expression and related migrations in a breast cancer cell line. Int J Clin Oncol. 2011;16:519–23.
130. Wang JH, Nao JF, Zhang M, He P. 20(s)-ginsenoside Rg3 promotes apoptosis in human ovarian cancer HO-8910 cells through PI3K/Akt and XIAP pathways. Tumour Biol. 2014;35: 11985–94.
131. Bjorklund G, Chirumbolo S. Role of oxidative stress and antioxidants in daily nutrition and human health. Nutrition. 2017;33:311–21.
132. Ma LY, Sun ZH, Zeng YW, Luo MC, Yang JZ. Molecular mechanism and health role of functional ingredients in blueberry for chronic disease in human beings. Int J Mol Sci. 2018;19:2785.
133. Loffredo L, Perri L, Nocella C, Violi F. Antioxidant and antiplatelet activity by polyphenol-rich nutrients: focus on extra virgin olive oil and cocoa. Br J Clin Pharmacol. 2017;83:96–102.
134. Wang H, Guo X, Hu X, Li T, Fu X, et al. Comparison of phytochemical profiles, antioxidant and cellular antioxidant activities of different varieties of blueberry (Vaccinium spp.). Food Chem. 2017;217:773–81.
135. Gammone MA, Pluchinotta FR, Bergante S, Tettamanti G, D'Orazio N. Prevention of cardiovascular diseases with carotenoids. Front Biosci (Schol Ed). 2017;9:165–71.
136. Guerrero-Beltran CE, Calderon-Oliver M, Pedraza-Chaverri J, Chirino YI. Protective effect of sulforaphane against oxidative stress: recent advances. Exp Toxicol Pathol. 2012;64:503–8.
137. Spormann TM, Albert FW, Rath T, Dietrich H, Will F, et al. Anthocyanin/polyphenolic-rich fruit juice reduces oxidative cell damage in an intervention study with patients on hemodialysis. Cancer Epidemiol Biomark Prev. 2008;17:3372–80.
138. Yin TF, Wang M, Qing Y, Lin YM, Wu D. Research progress on chemopreventive effects of phytochemicals on colorectal cancer and their mechanisms. World J Gastroenterol. 2016;22: 7058–68.

139. Riso P, Visioli F, Gardana C, Grande S, Brusamolino A, et al. Effects of blood orange juice intake on antioxidant bioavailability and on different markers related to oxidative stress. J Agric Food Chem. 2005;53:941–7.
140. Krga I, Milenkovic D. Anthocyanins: from sources and bioavailability to cardiovascular-health benefits and molecular mechanisms of action. J Agric Food Chem. 2019;67:1771–83.
141. Thompson K, Pederick W, Santhakumar AB. Anthocyanins in obesity-associated thrombogenesis: a review of the potential mechanism of action. Food Funct. 2016;7:2169–78.
142. Mazza GJ. Anthocyanins and heart health. Ann Ist Super Sanita. 2007;43:369–74.
143. Reis JF, Monteiro VV, de Souza Gomes R, do Carmo MM, da Costa GV, et al. Action mechanism and cardiovascular effect of anthocyanins: a systematic review of animal and human studies. J Transl Med. 2016;14:315.
144. Yadav E, Singh D, Yadav P, Verma A. Attenuation of dermal wounds via downregulating oxidative stress and inflammatory markers by protocatechuic acid rich n-butanol fraction of *Trianthema portulacastrum* Linn. in wistar albino rats. Biomed Pharmacother. 2017;96:86–97.
145. Farombi EO, Adedara IA, Awoyemi OV, Njoku CR, Micah GO, et al. Dietary protocatechuic acid ameliorates dextran sulphate sodium-induced ulcerative colitis and hepatotoxicity in rats. Food Funct. 2016;7:913–21.
146. Preetha Rani MR, Anupama N, Sreelekshmi M, Raghu KG. Chlorogenic acid attenuates glucotoxicity in H9c2 cells via inhibition of glycation and PKC alpha upregulation and safeguarding innate antioxidant status. Biomed Pharmacother. 2018;100:467–77.
147. Zhang Y, Wang Y, Chen D, Yu B, Zheng P, et al. Dietary chlorogenic acid supplementation affects gut morphology, antioxidant capacity and intestinal selected bacterial populations in weaned piglets. Food Funct. 2018;9:4968–78.
148. Zhou Y, Zhou L, Ruan Z, Mi S, Jiang M, et al. Chlorogenic acid ameliorates intestinal mitochondrial injury by increasing antioxidant effects and activity of respiratory complexes. Biosci Biotechnol Biochem. 2016;80:962–71.
149. Singh AK, Vinayak M. Resveratrol alleviates inflammatory hyperalgesia by modulation of reactive oxygen species (ROS), antioxidant enzymes and ERK activation. Inflamm Res. 2017;66:911–21.
150. Sadi G, Konat D. Resveratrol regulates oxidative biomarkers and antioxidant enzymes in the brain of streptozotocin-induced diabetic rats. Pharm Biol. 2016;54:1156–63.
151. Xia N, Daiber A, Forstermann U, Li H. Antioxidant effects of resveratrol in the cardiovascular system. Br J Pharmacol. 2017;174:1633–46.
152. Zou JG, Huang YZ, Chen Q, Wei EH, Hsieh TC, et al. Resveratrol inhibits copper ion-induced and azo compound-initiated oxidative modification of human low density lipoprotein. Biochem Mol Biol Int. 1999;47:1089–96.
153. Pungcharoenkul K, Thongnopnua P. Effect of different curcuminoid supplement dosages on total in vivo antioxidant capacity and cholesterol levels of healthy human subjects. Phytother Res. 2011;25:1721–6.
154. Schex R, Lieb VM, Jimenez VM, Esquivel P, Schweiggert RM, et al. HPLC-DAD-APCI/ESI-MS(n) analysis of carotenoids and alpha-tocopherol in Costa Rican Acrocomia aculeata fruits of varying maturity stages. Food Res Int. 2018;105:645–53.
155. Gammone MA, Riccioni G, D'Orazio N. Marine carotenoids against oxidative stress: effects on human health. Mar Drugs. 2015;13:6226–46.
156. Young AJ, Lowe GM. Antioxidant and prooxidant properties of carotenoids. Arch Biochem Biophys. 2001;385:20–7.
157. Sherif IO. The effect of natural antioxidants in cyclophosphamide-induced hepatotoxicity: role of Nrf2/HO-1 pathway. Int Immunopharmacol. 2018;61:29–36.
158. Mezza GN, Borgarello AV, Grosso NR, Fernandez H, Pramparo MC, et al. Antioxidant activity of rosemary essential oil fractions obtained by molecular distillation and their effect on oxidative stability of sunflower oil. Food Chem. 2018;242:9–15.

159. Dong S, Li B, Dai W, Wang D, Qin Y, et al. Sesqui- and diterpenoids from the radix of *Curcuma aromatica*. J Nat Prod. 2017;80:3093–102.
160. Yang HM, Yin ZQ, Zhao MG, Jiang CH, Zhang J, et al. Pentacyclic triterpenoids from *Cyclocarya paliurus* and their antioxidant activities in FFA-induced HepG2 steatosis cells. Phytochemistry. 2018;151:119–27.
161. Sun K, Min K. Antioxidative activity of sulfur-containing flavor compounds in garlic. Biosci Biotechnol Biochem. 1997;9:1482–5.
162. Rabinkov A, Miron T, Konstantinovski L, Wilchek M, Mirelman D, et al. The mode of action of allicin: trapping of radicals and interaction with thiol containing proteins. Biochim Biophys Acta. 1998;1379:233–44.
163. Ryu JH, Park HJ, Jeong YY, Han S, Shin JH, et al. Aged red garlic extract suppresses nitric oxide production in lipopolysaccharide-treated RAW 264.7 macrophages through inhibition of NF-κB. J Med Food. 2015;18:439–45.
164. Han CY, Ki SH, Kim YW, Noh K, Lee DY, et al. Ajoene, a stable garlic by-product, inhibits high fat diet-induced hepatic steatosis and oxidative injury through LKB1-dependent AMPK activation. Antioxid Redox Signal. 2011;14:187–202.
165. Kim SH, Choi KC. Anti-cancer effect and underlying mechanism(s) of kaempferol, a phyto-estrogen, on the regulation of apoptosis in diverse cancer cell models. Toxicol Res. 2013;29: 229–34.
166. Erba D, Casiraghi MC, Martinez-Conesa C, Goi G, Massaccesi L. Isoflavone supplementation reduces DNA oxidative damage and increases O-beta-N-acetyl-D-glucosaminidase activity in healthy women. Nutr Res. 2012;32:233–40.
167. Dar AA, Arumugam N. Lignans of sesame: purification methods, biological activities and biosynthesis—a review. Bioorg Chem. 2013;50:1–10.
168. Del Corno M, Varano B, Scazzocchio B, Filesi C, Masella R, et al. Protocatechuic acid inhibits human dendritic cell functional activation: role of PPARgamma up-modulation. Immunobiology. 2014;219:416–24.
169. Wei M, Chu X, Guan M, Yang X, Xie X, et al. Protocatechuic acid suppresses ovalbumin-induced airway inflammation in a mouse allergic asthma model. Int Immunopharmacol. 2013;15:780–8.
170. Guo Y, Zhang Q, Zuo Z, Chu J, Xiao H, et al. Protocatechuic acid (PCA) induced a better antiviral effect by immune enhancement in SPF chickens. Microb Pathog. 2018;114:233–8.
171. Lai X, Pei Q, Song X, Zhou X, Yin Z, et al. The enhancement of immune function and activation of NF-kappaB by resveratrol-treatment in immunosuppressive mice. Int Immunopharmacol. 2016;33:42–7.
172. Svajger U, Obermajer N, Jeras M. Dendritic cells treated with resveratrol during differentiation from monocytes gain substantial tolerogenic properties upon activation. Immunology. 2010;129:525–35.
173. Moussa C, Hebron M, Huang X, Ahn J, Rissman RA, et al. Resveratrol regulates neuro-inflammation and induces adaptive immunity in Alzheimer's disease. J Neuroinflammation. 2017;14:1.
174. Marchiani A, Rozzo C, Fadda A, Delogu G, Ruzza P. Curcumin and curcumin-like molecules: from spice to drugs. Curr Med Chem. 2014;21:204–22.
175. Huang A-C, Cheng H-Y, Lin T-S, Chen W-H, Lin J-H, et al. Epigallocatechin Gallate (EGCG), influences a murine WEHI-3 leukemia model in vivo through enhancing phagocytosis of macrophages and populations of T- and B-cells. In Vivo. 2013;27:627–34.
176. Saleh F, Raghupathy R, Asfar S, Oteifa M, Al-Saleh N. Analysis of the effect of the active compound of green tea (EGCG) on the proliferation of peripheral blood mononuclear cells. BMC Complement Altern Med. 2014;14:322.
177. Wang Y, Hu B, Peng Y, Xiong X, Jing W, et al. In silico exploration of the molecular mechanism of cassane diterpenoids on anti-inflammatory and immunomodulatory activity. J Chem Inf Model. 2019;59:2309–23.

178. Zhai T, Sun Y, Li H, Zhang J, Huo R, et al. Unique immunomodulatory effect of paeoniflorin on type I and II macrophages activities. J Pharmacol Sci. 2016;130:143–50.
179. Imamura M, Sasaki O, Okunishi K, Nakagome K, Harada H, et al. Perillyl alcohol suppresses antigen-induced immune responses in the lung. Biochem Biophys Res Commun. 2014;443: 266–71.
180. Sun T, Yan X, Guo W, Zhao D. Evaluation of cytotoxicity and immune modulatory activities of soyasaponin Ab: an in vitro and in vivo study. Phytomedicine. 2014;21:1759–66.
181. Gautam M, Saha S, Bani S, Kaul A, Mishra S, et al. Immunomodulatory activity of *Asparagus racemosus* on systemic Th1/Th2 immunity: implications for immunoadjuvant potential. J Ethnopharmacol. 2009;121:241–7.
182. Pise MV, Rudra JA, Upadhyay A. Immunomodulatory potential of shatavarins produced from *Asparagus racemosus* tissue cultures. J Nat Sci Biol Med. 2015;6:415–20.
183. Gao F, Wei D, Bian T, Xie P, Zou J, et al. Genistein attenuated allergic airway inflammation by modulating the transcription factors T-bet, GATA-3 and STAT-6 in a murine model of asthma. Pharmacology. 2012;89:229–36.
184. Smith BN, Dilger RN. Immunomodulatory potential of dietary soybean-derived isoflavones and saponins in pigs. J Anim Sci. 2018;96:1288–304.
185. Lin CH, Shen ML, Zhou N, Lee CC, Kao ST, et al. Protective effects of the polyphenol sesamin on allergen-induced T(H)2 responses and airway inflammation in mice. PLoS One. 2014;9:e96091.
186. De Smet E, Mensink RP, Boekschoten MV, de Ridder R, Germeraad WT, et al. An acute intake of plant stanol esters alters immune-related pathways in the jejunum of healthy volunteers. Br J Nutr. 2015;113:794–802.
187. Yuk JE, Woo JS, Yun CY, Lee JS, Kim JH, et al. Effects of lactose-beta-sitosterol and beta-sitosterol on ovalbumin-induced lung inflammation in actively sensitized mice. Int Immunopharmacol. 2007;7:1517–27.
188. Bouic PJ. The role of phytosterols and phytosterolins in immune modulation: a review of the past 10 years. Curr Opin Clin Nutr Metab Care. 2001;4:471–5.
189. Bouic PJ, Lamprecht JH. Plant sterols and sterolins: a review of their immune-modulating properties. Altern Med Rev. 1999;4:170–7.
190. Liang YT, Chen JN, Zuo YY, Ma KY, Jiang Y, et al. Blueberry anthocyanins at doses of 0.5 and 1% lowered plasma cholesterol by increasing fecal excretion of acidic and neutral sterols in hamsters fed a cholesterol-enriched diet. Eur J Nutr. 2013;52:869–75.
191. Du C, Shi Y, Ren Y, Wu H, Yao F, et al. Anthocyanins inhibit high-glucose-induced cholesterol accumulation and inflammation by activating LXRalpha pathway in HK-2 cells. Drug Des Devel Ther. 2015;9:5099–113.
192. Radhiga T, Sundaresan A, Viswanathan P, Pugalendi KV. Effect of protocatechuic acid on lipid profile and DNA damage in D-galactosamine-induced hepatotoxic rats. J Basic Clin Physiol Pharmacol. 2016;27:505–14.
193. Haghighatdoost F, Hariri M. Effect of resveratrol on lipid profile: an updated systematic review and meta-analysis on randomized clinical trials. Pharmacol Res. 2018;129:141–50.
194. Zang Y, Zhang L, Igarashi K, Yu C. The anti-obesity and anti-diabetic effects of kaempferol glycosides from unripe soybean leaves in high-fat-diet mice. Food Funct. 2015;6:834–41.
195. Shishikura Y, Khokhar S, Murray BS. Effects of tea polyphenols on emulsification of olive oil in a small intestine model system. J Agric Food Chem. 2006;54:1906–13.
196. Lu CH, Hwang LS. Polyphenol contents of Pu-Erh teas and their abilities to inhibit cholesterol biosynthesis in Hep G2 cell line. Food Chem. 2008;111:67–71.
197. Bursill CA, Roach PD. A green tea catechin extract upregulates the hepatic low-density lipoprotein receptor in rats. Lipids. 2007;42:621–7.
198. Yang W, Shen Z, Wen S, Wang W, Hu M. Mechanisms of multiple neurotransmitters in the effects of Lycopene on brain injury induced by Hyperlipidemia. Lipids Health Dis. 2018;17: 13.

199. Sultan Alvi S, Ansari IA, Khan I, Iqbal J, Khan MS. Potential role of lycopene in targeting proprotein convertase subtilisin/kexin type-9 to combat hypercholesterolemia. Free Radic Biol Med. 2017;108:394–403.
200. Prashanth A, Jeyakumar SM, Giridharan NV, Vajreswari A. Vitamin A-enriched diet modulates reverse cholesterol transport in hypercholesterolemic obese rats of the WNIN/Ob strain. J Atheroscler Thromb. 2014;21:1197–207.
201. Rai SK, Sharma M, Tiwari M. Inhibitory effect of novel diallyldisulfide analogs on HMG-CoA reductase expression in hypercholesterolemic rats: CREB as a potential upstream target. Life Sci. 2009;85:211–9.
202. Taku K, Umegaki K, Sato Y, Taki Y, Endoh K, et al. Soy isoflavones lower serum total and LDL cholesterol in humans: a meta-analysis of 11 randomized controlled trials (vol 85, pg 1148, 2007). Am J Clin Nutr. 2007;86:809.
203. Peterson DW. Effect of soybean sterols in the diet on plasma and liver cholesterol in chicks. Proc Soc Exp Biol Med. 1951;78:143–7.
204. Plat J, Mackay D, Baumgartner S, Clifton PM, Gylling H, et al. Progress and prospective of plant sterol and plant stanol research: report of the Maastricht meeting. Atherosclerosis. 2012;225:521–33.
205. Hayes KC, Pronczuk A, Wijendran V, Beer M. Free phytosterols facilitate excretion of endogenous cholesterol in gerbils. J Nutr Biochem. 2005;16:305–11.
206. Brauner R, Johannes C, Ploessl F, Bracher F, Lorenz RL. Phytosterols reduce cholesterol absorption by inhibition of 27-hydroxycholesterol generation, liver X receptor alpha activation, and expression of the basolateral sterol exporter ATP-binding cassette A1 in Caco-2 enterocytes. J Nutr. 2012;142:981–9.
207. Calpe-Berdiel L, Escola-Gil JC, Blanco-Vaca F. New insights into the molecular actions of plant sterols and stanols in cholesterol metabolism. Atherosclerosis. 2009;203:18–31.
208. Lin X, Racette SB, Lefevre M, Ma L, Spearie CA, et al. Combined effects of ezetimibe and phytosterols on cholesterol metabolism: a randomized, controlled feeding study in humans. Circulation. 2011;124:596–601.
209. Otton R, Bolin AP, Ferreira LT, Marinovic MP, Rocha ALS, et al. Polyphenol-rich green tea extract improves adipose tissue metabolism by down-regulating miR-335 expression and mitigating insulin resistance and inflammation. J Nutr Biochem. 2018;57:170–9.
210. Santangelo C, Zicari A, Mandosi E, Scazzocchio B, Mari E, et al. Could gestational diabetes mellitus be managed through dietary bioactive compounds? Current knowledge and future perspectives. Br J Nutr. 2016;115:1129–44.
211. Xiao JB, Hogger P. Dietary polyphenols and type 2 diabetes: current insights and future perspectives. Curr Med Chem. 2015;22:23–38.
212. Sun CD, Zhang B, Zhang JK, Xu CJ, Wu YL, et al. Cyanidin-3-glucoside-rich extract from Chinese bayberry fruit protects pancreatic beta cells and ameliorates hyperglycemia in streptozotocin-induced diabetic mice. J Med Food. 2012;15:288–98.
213. Asgary S, Rafieian-Kopaei M, Shamsi F, Najafi S, Sahebkar A. Biochemical and histopathological study of the anti-hyperglycemic and anti-hyperlipidemic effects of cornelian cherry (*Cornus mas* L.) in alloxan-induced diabetic rats. J Complement Integr Med. 2014;11:63–9.
214. Semaming Y, Kukongviriyapan U, Kongyingyoes B, Thukhammee W, Pannangpetch P. Protocatechuic acid restores vascular responses in rats with chronic diabetes induced by streptozotocin. Phytother Res. 2016;30:227–33.
215. Bhattacharjee N, Dua TK, Khanra R, Joardar S, Nandy A, et al. Protocatechuic acid, a phenolic from *Sansevieria roxburghiana* leaves, suppresses diabetic cardiomyopathy via stimulating glucose metabolism, ameliorating oxidative stress, and inhibiting inflammation. Front Pharmacol. 2017;8:251.
216. Harini R, Pugalendi KV. Antihyperglycemic effect of protocatechuic acid on streptozotocin-diabetic rats. J Basic Clin Physiol Pharmacol. 2010;21:79–91.

217. Zhu X, Wu C, Qiu S, Yuan X, Li L. Effects of resveratrol on glucose control and insulin sensitivity in subjects with type 2 diabetes: systematic review and meta-analysis. Nutr Metab (Lond). 2017;14:60.
218. Youl E, Bardy G, Magous R, Cros G, Sejalon F, et al. Quercetin potentiates insulin secretion and protects INS-1 pancreatic beta-cells against oxidative damage via the ERK1/2 pathway. Br J Pharmacol. 2010;161:799–814.
219. Chen S, Jiang HM, Wu XS, Fang J. Therapeutic effects of quercetin on inflammation, obesity, and type 2 diabetes. Mediat Inflamm. 2016;2016:9340637.
220. Oboh G, Ademosun AO, Ogunsuyi OB. Quercetin and its role in chronic diseases. Adv Exp Med Biol. 2016;929:377–87.
221. Ali MM, Agha FG. Amelioration of streptozotocin-induced diabetes mellitus, oxidative stress and dyslipidemia in rats by tomato extract lycopene. Scand J Clin Lab Invest. 2009;69:371–9.
222. Wang J, Zhang X, Lan H, Wang W. Effect of garlic supplement in the management of type 2 diabetes mellitus (T2DM): a meta-analysis of randomized controlled trials. Food Nutr Res. 2017;61:1377571.
223. Padiya R, Khatua TN, Bagul PK, Kuncha M, Banerjee SK. Garlic improves insulin sensitivity and associated metabolic syndromes in fructose fed rats. Nutr Metab (Lond). 2011;8:53.
224. Gu X, Zhu YZ. Therapeutic applications of organosulfur compounds as novel hydrogen sulfide donors and/or mediators. Expert Rev Clin Pharmacol. 2011;4:123–33.
225. Yun HM, Ban JO, Park KR, Lee CK, Jeong HS, et al. Potential therapeutic effects of functionally active compounds isolated from garlic. Pharmacol Ther. 2014;142:183–95.
226. Liu CT, Hse H, Lii CK, Chen PS, Sheen LY. Effects of garlic oil and diallyl trisulfide on glycemic control in diabetic rats. Eur J Pharmacol. 2005;516:165–73.
227. Axelsson AS, Tubbs E, Mecham B, Chacko S, Nenonen HA, et al. Sulforaphane reduces hepatic glucose production and improves glucose control in patients with type 2 diabetes. Sci Transl Med. 2017;9:eaah4477.
228. Xu Y, Fu JF, Chen JH, Zhang ZW, Zou ZQ, et al. Sulforaphane ameliorates glucose intolerance in obese mice via the upregulation of the insulin signaling pathway. Food Funct. 2018;9:4695–701.
229. Sekikawa A, Ihara M, Lopez O, Kakuta C, Lopresti B, et al. Effect of S-equol and soy isoflavones on heart and brain. Curr Cardiol Rev. 2019;15:114–35.
230. Ding M, Pan A, Manson JE, Willett WC, Malik V, et al. Consumption of soy foods and isoflavones and risk of type 2 diabetes: a pooled analysis of three US cohorts. Eur J Clin Nutr. 2016;70:1381–7.
231. Xu Z, Ju J, Wang K, Gu C, Feng Y. Evaluation of hypoglycemic activity of total lignans from *Fructus Arctii* in the spontaneously diabetic Goto-Kakizaki rats. J Ethnopharmacol. 2014;151: 548–55.
232. Kwon DY, Kim DS, Yang HJ, Park S. The lignan-rich fractions of Fructus Schisandrae improve insulin sensitivity via the PPAR-gamma pathways in in vitro and in vivo studies. J Ethnopharmacol. 2011;135:455–62.
233. Misawa E, Tanaka M, Nomaguchi K, Nabeshima K, Yamada M, et al. Oral ingestion of aloe vera phytosterols alters hepatic gene expression profiles and ameliorates obesity-associated metabolic disorders in zucker diabetic fatty rats. J Agric Food Chem. 2012;60:2799–806.
234. Ponnulakshmi R, Shyamaladevi B, Vijayalakshmi P, Selvaraj J. In silico and in vivo analysis to identify the antidiabetic activity of beta sitosterol in adipose tissue of high fat diet and sucrose induced type-2 diabetic experimental rats. Toxicol Mech Methods. 2019;29:276–90.
235. Kruger MJ, Davies N, Myburgh KH, Lecour S. Proanthocyanidins, anthocyanins and cardiovascular diseases. Food Res Int. 2014;59:41–52.
236. Xu M, Xue W, Ma Z, Bai J, Wu S. Resveratrol reduces the incidence of portal vein system thrombosis after splenectomy in a rat fibrosis model. Oxidative Med Cell Longev. 2016;2016: 7453849.
237. Lannan KL, Refaai MA, Ture SK, Morrell CN, Blumberg N, et al. Resveratrol preserves the function of human platelets stored for transfusion. Br J Haematol. 2016;172:794–806.

238. Kirimlioglu V, Sozen H, Turkoglu S, Haberal M. Protective effect of resveratrol, a red wine constituent polyphenol, on rats subjected to portal vein thrombosis. Transplant Proc. 2008;40: 290–2.
239. Prakash P, Misra A, Surin WR, Jain M, Bhatta RS, et al. Anti-platelet effects of Curcuma oil in experimental models of myocardial ischemia-reperfusion and thrombosis. Thromb Res. 2011;127:111–8.
240. Mayanglambam A, Dangelmaier CA, Thomas D, Damodar Reddy C, Daniel JL, et al. Curcumin inhibits GPVI-mediated platelet activation by interfering with the kinase activity of Syk and the subsequent activation of PLCgamma2. Platelets. 2010;21:211–20.
241. Pellegrino D. Antioxidants and cardiovascular risk factors. Diseases. 2016;4:11.
242. Sawardekar SB, Patel TC, Uchil D. Comparative evaluation of antiplatelet effect of lycopene with aspirin and the effect of their combination on platelet aggregation: an in vitro study. Indian J Pharmacol. 2016;48:26–31.
243. Endale M, Lee WM, Kamruzzaman SM, Kim SD, Park JY, et al. Ginsenoside-Rp1 inhibits platelet activation and thrombus formation via impaired glycoprotein VI signalling pathway, tyrosine phosphorylation and MAPK activation. Br J Pharmacol. 2012;167:109–27.
244. Khin M, Jones AM, Cech NB, Caesar LK. Phytochemical analysis and antimicrobial efficacy of *Macleaya cordata* against extensively drug-resistant *Staphylococcus aureus*. Nat Prod Commun. 2018;13:1479–83.
245. Abachi S, Lee S, Rupasinghe HP. Molecular mechanisms of inhibition of Streptococcus species by phytochemicals. Molecules. 2016;21:215.
246. Wei LL, Yang M, Huang L, Li JL. Antibacterial and antioxidant flavonoid derivatives from the fruits of *Metaplexis japonica*. Food Chem. 2019;289:308–12.
247. Xie Y, Yang W, Tang F, Chen X, Ren L. Antibacterial activities of flavonoids: structure-activity relationship and mechanism. Curr Med Chem. 2015;22:132–49.
248. Dai XQ, Cai WT, Wu X, Chen Y, Han FM. Protocatechuic acid inhibits hepatitis B virus replication by activating ERK1/2 pathway and down-regulating HNF4alpha and HNF1alpha in vitro. Life Sci. 2017;180:68–74.
249. Ajiboye TO, Habibu RS, Saidu K, Haliru FZ, Ajiboye HO, et al. Involvement of oxidative stress in protocatechuic acid-mediated bacterial lethality. Microbiology. 2017;6:e00472.
250. Bostanghadiri N, Pormohammad A, Chirani AS, Pouriran R, Erfanimanesh S, et al. Comprehensive review on the antimicrobial potency of the plant polyphenol resveratrol. Biomed Pharmacother. 2017;95:1588–95.
251. Liu T, Zang N, Zhou N, Li W, Xie X, et al. Resveratrol inhibits the TRIF-dependent pathway by upregulating sterile alpha and armadillo motif protein, contributing to anti-inflammatory effects after respiratory syncytial virus infection. J Virol. 2014;88:4229–36.
252. Clouser CL, Chauhan J, Bess MA, van Oploo JL, Zhou D, et al. Anti-HIV-1 activity of resveratrol derivatives and synergistic inhibition of HIV-1 by the combination of resveratrol and decitabine. Bioorg Med Chem Lett. 2012;22:6642–6.
253. Xu J, Yin Z, Li L, Cheng A, Jia R, et al. Inhibitory effect of resveratrol against duck enteritis virus in vitro. PLoS One. 2013;8:e65213.
254. Vaughn AR, Haas KN, Burney W, Andersen E, Clark AK, et al. Potential role of curcumin against biofilm-producing organisms on the skin: a review. Phytother Res. 2017;31:1807–16.
255. Gutierrez-del-Rio I, Fernandez J, Lombo F. Plant nutraceuticals as antimicrobial agents in food preservation: terpenoids, polyphenols and thiols. Int J Antimicrob Agents. 2018;52:309–15.
256. Amoussa AM, Lagnika L, Bourjot M, Vonthron-Senecheau C, Sanni A. Triterpenoids from *Acacia ataxacantha* DC: antimicrobial and antioxidant activities. BMC Complement Altern Med. 2016;16:284.
257. Roncero AM, Tobal IE, Moro RF, Diez D, Marcos IS. Halimane diterpenoids: sources, structures, nomenclature and biological activities. Nat Prod Rep. 2018;35:955–91.

258. Kurekci C, Padmanabha J, Bishop-Hurley SL, Hassan E, Al Jassim RAM, et al. Antimicrobial activity of essential oils and five terpenoid compounds against *Campylobacter jejuni* in pure and mixed culture experiments. Int J Food Microbiol. 2013;166:450–7.
259. Putnik P, Gabric D, Roohinejad S, Barba FJ, Granato D, et al. An overview of organosulfur compounds from *Allium* spp.: from processing and preservation to evaluation of their bioavailability, antimicrobial, and anti-inflammatory properties. Food Chem. 2019;276:680–91.
260. Gularte MS, Anghinoni JM, Abenante L, Voss GT, de Oliveira RL, et al. Synthesis of chitosan derivatives with organoselenium and organosulfur compounds: characterization, antimicrobial properties and application as biomaterials. Carbohydr Polym. 2019;219:240–50.
261. Romeo L, Iori R, Rollin P, Bramanti P, Mazzon E. Isothiocyanates: an overview of their antimicrobial activity against human infections. Molecules. 2018;23:624.
262. Aires A, Mota VR, Saavedra MJ, Rosa EAS, Bennett RN. The antimicrobial effects of glucosinolates and their respective enzymatic hydrolysis products on bacteria isolated from the human intestinal tract. J Appl Microbiol. 2009;106:2086–95.
263. Hussain H, Green IR. A patent review of the therapeutic potential of isoflavones (2012–2016). Expert Opin Ther Pat. 2017;27:1135–46.
264. Braga Ribeiro AM, Sousa JN, Costa LM, Oliveira FAA, Dos Santos RC, et al. Antimicrobial activity of *Phyllanthus amarus* Schumach. & Thonn and inhibition of the NorA efflux pump of *Staphylococcus aureus* by Phyllanthin. Microb Pathog. 2019;130:242–6.
265. Lee SH, Choi BY, Kho AR, Jeong JH, Hong DK, et al. Protective effects of protocatechuic acid on seizure-induced neuronal death. Int J Mol Sci. 2018;19:187.
266. Winter AN, Brenner MC, Punessen N, Snodgrass M, Byars C, et al. Comparison of the neuroprotective and anti-inflammatory effects of the anthocyanin metabolites, protocatechuic acid and 4-hydroxybenzoic acid. Oxidative Med Cell Longev. 2017;2017:6297080.
267. Wang Z, Pan X, Wang D, Sun H, Han F, et al. Protective effects of protocatechuic acid on retinal ganglion cells from oxidative damage induced by H2O2. Neurol Res. 2015;37:159–66.
268. Zhang HN, An CN, Xu M, Guo DA, Li M, et al. Protocatechuic acid inhibits rat pheochromocytoma cell damage induced by a dopaminergic neurotoxin. Biol Pharm Bull. 2009;32:1866–9.
269. Hodis HN, Mack WJ. The timing hypothesis and hormone replacement therapy: a paradigm shift in the primary prevention of coronary heart disease in women. Part 2: comparative risks. J Am Geriatr Soc. 2013;61:1011–8.
270. Thorp AA, Sinn N, Buckley JD, Coates AM, Howe PR. Soya isoflavone supplementation enhances spatial working memory in men. Br J Nutr. 2009;102:1348–54.
271. Chinta SJ, Ganesan A, Reis-Rodrigues P, Lithgow GJ, Andersen JK. Anti-inflammatory role of the isoflavone diadzein in lipopolysaccharide-stimulated microglia: implications for Parkinson's disease. Neurotox Res. 2013;23:145–53.
272. Arbabi E, Hamidi G, Talaei SA, Salami M. Estrogen agonist genistein differentially influences the cognitive and motor disorders in an ovariectomized animal model of Parkinsonism. Iran J Basic Med Sci. 2016;19:1285–90.
273. Wu HC, Hu QL, Zhang SJ, Wang YM, Jin ZK, et al. Neuroprotective effects of genistein on SH-SY5Y cells overexpressing A53T mutant alpha-synuclein. Neural Regen Res. 2018;13:1375–83.
274. Arbabi E, Hamidi G, Talaei SA, Salami M. Estrogen agonist genistein differentially influences the cognitive and motor disorders in an ovariectomized animal model of Parkinsonism. Iran J Basic Med Sci. 2016;19:1285–91.

# Chapter 13
# Adverse Effects of Phytochemicals

**Yina Huang and Qian Bu**

**Abstract** Phytochemicals are a wide variety of chemical compounds that are naturally produced by plants. Some phytochemicals are known as phytotoxins that are toxic to animals, including humans. These phytotoxins include anticholinergic, severe gastrointestinal irritants, cardiac glycosides, central nervous system stimulants/hallucinogens, and cyanogens. Some of these toxic phytochemicals have also been harnessed for some therapeutic uses. This chapter focuses on plants that produce toxic phytochemicals which can be harnessed for destruction and implementation of death to mankind and livestock, which include natural goitrogens, environmental antithyroid substances, favism, neurolathyrism, cyanogenic glycosides, lectins, caffeine, curare, strychnine, and atropine. We have also provided brief discussions on phytoalexins and the interactions of certain phytochemical compounds in the diet with established drugs, which can produce adverse diet and drug interactions. Finally, we give a brief introduction to how to gain a clear understanding of the health beneficial effects of dietary phytochemicals.

**Keywords** Phytochemicals · Toxicity · Phytotoxins · Natural chemicals · Adverse effects

## 13.1 Toxicity of Phytochemicals

Members of the plant kingdom are extremely important sources of rich products for members of the animal kingdom. Plants provide oxygen, energy, vitamins, and shelter, which are essential to animal lives. Animals play an important role in plant reproduction, seed dispersal, and they also assist in the breakdown of dead plant materials. Plants also have rich sources of phytochemicals that are nonnutritive natural products with small molecular weight. Phytochemicals can have beneficial (drugs and hormones) or harmful (phytotoxins) effects on animals and humans. Of

Y. Huang (✉) · Q. Bu
West China School of Public Health and West China Fourth Hospital, Sichuan University, Chengdu, People's Republic of China
e-mail: hyn427@scu.edu.cn; buqian7978@scu.edu.cn


L. Zhang (ed.), *Nutritional Toxicology*,
https://doi.org/10.1007/978-981-19-0872-9_13

about 30,000 North American plants, approximately 700 species are considered to be toxic [1]. Poisonous species are found throughout the plant kingdom, including algae, ferns, gymnosperms, and angiosperms. In some cases, groups within a single-family exhibit similar toxicity; however, some plants from the same genera can have vastly different types of toxin and there are no easy ways to classify a plant-derived toxin in food ingredients.

Some of the phytotoxins are endogenous toxins with low molecular weight while others are products of secondary metabolism. Primary metabolism is bioprocess involved in energy metabolisms, such as photosynthesis, growth, and reproduction. Macronutrients and micronutrients are the products of plant primary metabolism. On the other hand, secondary metabolism is species-specific and includes carotenoids, phytosterols, saponins, polyphenols etc. Some of the secondary metabolic products are known as growth inhibitors, neurotoxins, carcinogens, mutagens, and teratogens. Many have not been tested because no government regulations require such testing, and the cost of these tests is very high.

In this chapter, we discuss the selected groups of phytotoxins that are important as examples of natural chemicals with an adverse effect on human health, and that have been widely used as approved therapeutic agents. We have also provided brief discussions on phytoalexins and the interactions of certain phytochemical compounds in the diet with established drugs, which can produce adverse diet and drug interactions.

### *13.1.1 Natural Goitrogens*

Human goiter is still a considerable problem, and iodine deficiency in certain food components may lead to endemic goiter. On the other hand, about 4% of human goiter is caused other than iodine deficiency [2]. Dietary cruciferous plants can be one of the contributing factors in some areas of the world.

Over the years, the methods of evaluating the goitrogenic activity of a substance have developed greatly. Visual inspection and weighing of the thyroid glands of experimental animals fed with tested substances were early applied. More recently, histological examinations of the glands have been applied to add information on the nature of the tested drugs and the conditions that could cause them. Other criteria to evaluate goitrogenic activity also contain the growth rate of the tested animals, basal metabolic rates, and iodine contents in the thyroid as well as blood. The present testing includes the measurement of thyroid uptake of radioactive iodine followed by the feeding of test material [3]. This procedure has been used for rats, chickens, and humans, and it has been shown that the anti-thyroid response to substances varies with the different species. The advantage of this testing over previous tests is its increased speed and sensitivity. A disadvantage is that the test does not provide information on the cumulative anti-thyroid effects of feeding natural products containing goitrogen at low levels. Examination of these problems requires prolonged administration of the test material, usually in a feed of known iodine content following an assessment of the thyroid gland. When some seeds of Brassica species are included in the feed, goiters can be induced consistently in animals. However, the thyroid enlargement is varied and does not occur when the leafy part of the vegetable is contained in the feed. Brassica plants, known as a normal part of an adequate diet, are unlikely to cause thyroid enlargement. Nevertheless, it is noted that the consumption of unusually large amounts of these plants like cabbage can induce thyroid abnormalities. Especially, the frequent consumption of Brassica could lead to a relatively high incidence of goiter in regions with low dietary iodine intake in the world. Although the incidence of goiter in the United States has decreased dramatically since the initiation of iodide supplementation, goiter and hypothyroidism continue to be a serious problem in some regions, even if iodine is supplemented. Research conducted in Maine, Kentucky, and many other states found high rates of goiter even with adequate iodine consumption. Furthermore, in recent years, the problem of iodine deficiency has become increasingly serious [4]. For example, a US national health survey conducted in the early 1970s found that the incidence of moderate to severe iodine deficiency in the population was only about 2.5% and a similar survey conducted in the early 1990s found that at least 11.1% of the population had a moderate iodine deficiency rate. This dramatic increase in iodine deficiency is due to an aggressive effort by public health agencies to restrict sodium intake as a means of reducing the risk of hypertension and cardiovascular diseases. These concerns about excessive sodium consumption have also led to a significant increase in goiter prevalence in many other countries [5]. According to the data from World Health Organization between 1993 and 2004, the prevalence of goiter increased by 81.4% in Africa, by 80.7% in Europe, by 62.9% in the Eastern Mediterranean, and by 31.7% globally, respectively. During the increase in iodine supplements, public health agencies were concerned that the goiter rate would increase.

Goiter is an adaptive response to deficiencies in thyroid hormone levels. In adults, it is usually a nonlethal but acceptable diagnostic symptom of hypothyroidism. Other more serious effects of hypothyroidism include thyroid dysfunction, loss of energy, which can increase the risk of cardiovascular disease, impaired mental, physical development (cretinism), prenatal, and infant mortalities [6]. The adverse effects of hypothyroidism on fetal development are of particular concern because even minor maternal iodine deficiency can lead to severe intellectual, motor, and hearing impairments [7]. There is increasing evidence that iodine deficiency does not combine with other anti-thyroid agents to produce endemic goiter. Goitrogens in low-concentration environments may normally be ineffective but may become significant when iodine supply is limited. In some cases, despite a large amount of iodine, goitrogen might be sufficiently potent by itself to cause goiter or thyroid insufficiency. Environmental goitrogens discussed in the following section are classified according to the pattern of hypothyroidism. As shown in Fig. 13.1, thyroid hormone metabolism is regulated by the hypothalamic/pituitary axis at several levels.

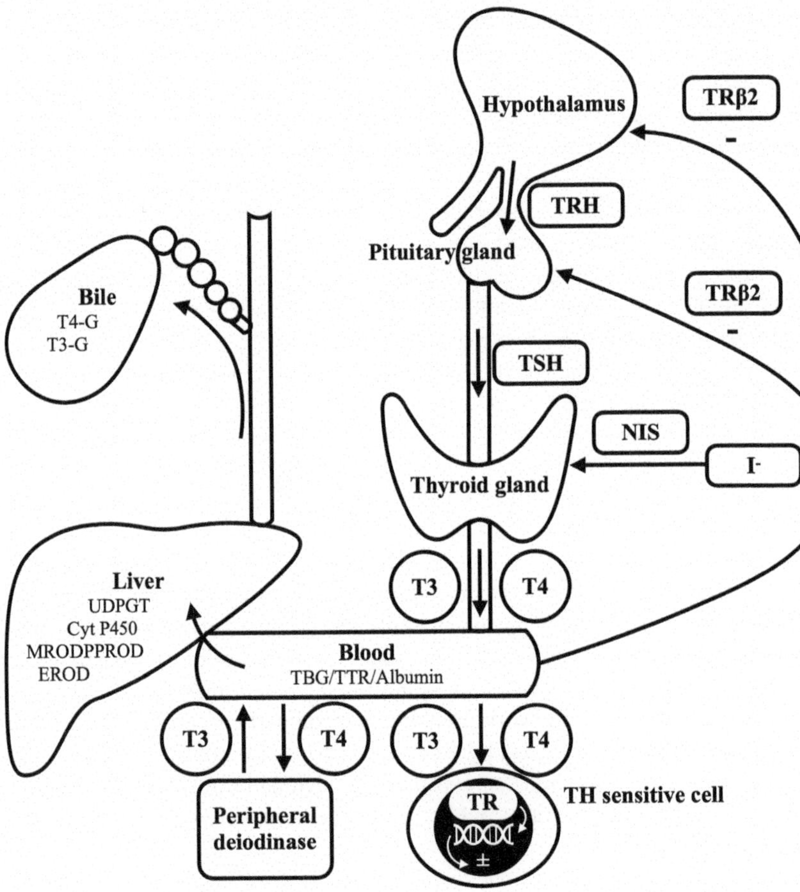


**Fig. 13.1** Metabolic pathway of thyroid hormone

## *13.1.2 Environmental Antithyroid Substances*

Many plants in the mustard family (Cruciferae) contain thyrotoxic substances such as goitrin, allylthiourea, and thiocyanate. However, these substances are not present in undamaged plants but are produced by the enzymatic conversion of thioglucoside precursors known as glucosinolates. Glucosinolates include more than 100 thioglucosides found up to a concentration of 60 mg/g in a wide range of plants of the order Brassicales (especially in the family Brassicaceae), which include cabbage, broccoli, brussels sprout, cauliflower, turnip (rutabaga), radish, black radish root, horseradish, mustard, rape, and others.

Upon enzymatic hydrolysis, the glucosinolate called progoitrin is rapidly converted to goitrin (Fig. 13.2). Progoitrin is present most notably in swedes, turnips, and at low levels in plants of the Brassica genus, such as cabbage,

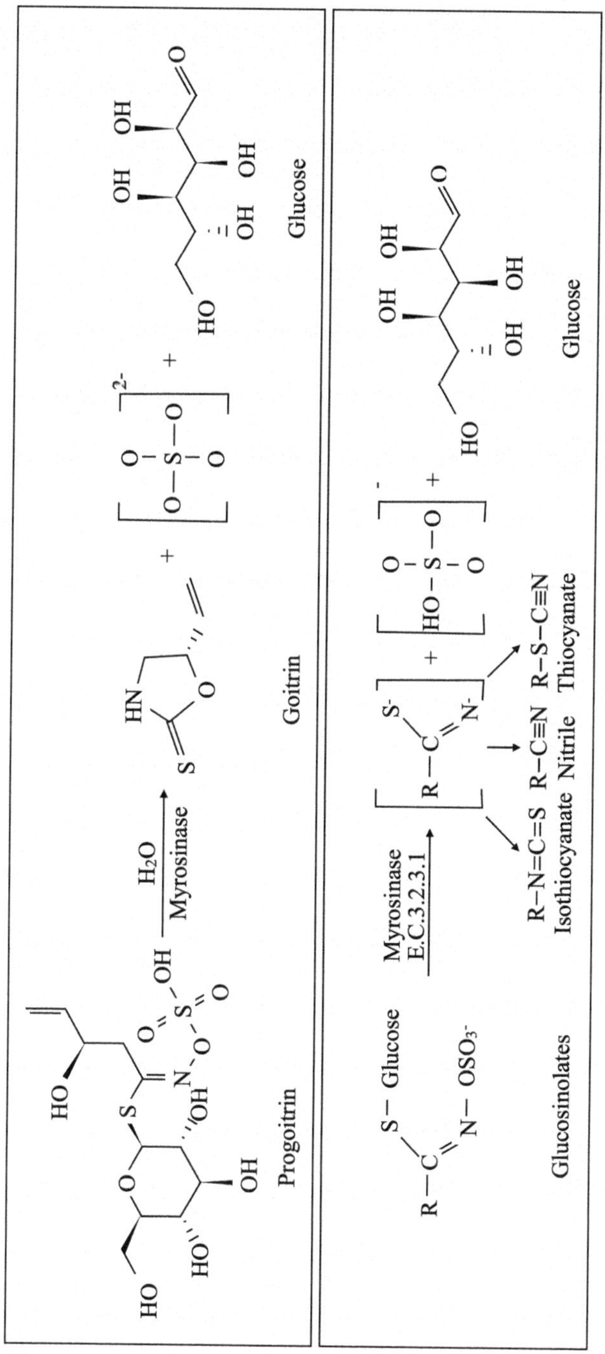


**Fig. 13.2** Goitrin production in Brassica plants

cauliflower, broccoli, kale, Brussels sprouts, kohlrabi, and rutabaga. The seeds of these plants contain the highest levels of progoitrin. Other sources of goitrin-like compounds have been detected in various species of herbs and shrubs of the Barbarea and Residea families.

The goitrogenic activities of Brassica plant components have been widely studied. Studies conducted as early as 1928 showed that a high level of cabbage consumption produced goiter in rabbits. Subsequent studies of the effects of dietary vegetables on thyroid iodine intake in human subjects revealed that raw rutabaga was the most active food plant [8]. That activity was lost by normal cooking, suggesting the requirement for the enzymatic conversion of a precursor in the plant (progoitrin) to the active substance(s) (goitrin). Furthermore, this enzymatic activity was not present in the gastrointestinal tract. When the feed contains Brassica seeds, especially rapeseed, goiter is consistently induced in the animal. The goitrogenic effects of the leafy plants usually vary and are less pronounced than the effects of seeds, especially if the diet is deficient in iodide. Therefore, as a normal part of a nutritionally adequate diet, the consumption of Brassica plants is unlikely to induce goiter. However, it seems plausible that if iodide intake is marginal, the consumption of high levels of Brassica vegetables may produce some symptoms of hypothyroidism.

The goitrogenic effects of goitrin are well established. Twenty days of thyroid administration to rats results in goiters, reduces glandular absorption of iodine, and decreases thyroxine synthesis [9]. Study in Tasmania schoolchildren revealed that goitrin is a causative factor in the development of endemic goiter. Iodine supplementation does not prevent goiter [10]. Goitrin also participates in environmental goitrogen in Finland [11]. Goitrin is secreted in the milk of cows fed on goitrin-rich rapeseed meal at a level of about 0.1% the level in the feed. Studies in rodents have shown that goitrin given to the dam produces symptoms of hypothyroidism in nursing pups.

Thiocyanate is an additional dietary goitrogen of considerable importance. Thiocyanate is produced as a by-product of glucosinolate hydrolysis and as the principal detoxification product of cyanide. Thiocyanate interferes with active uptake and concentration of inorganic iodide by the thyroid and inhibits the enzyme thyroperoxidase, thus preventing the entry of iodine into thyroglobulin. Several staple foods in the tropics contain large amounts of cyanide detoxified in the form of thiocyanate. Poorly refined cassava is a major source of cyanide in parts of Africa. The combined effects of thiocyanate toxicity from cassava consumption plus iodine and selenium deficiency are believed to cause endemic goiter and local cretinism when observed in parts of Africa. Another important group of natural antithyroid substances is plant polyphenols. Chemicals in this category of natural products play a different role in many plants ranging from protection against ultraviolet light and pathogens to providing colors that attract pollinators. Plant polyphenols such as resveratrol, quercetin, and epigallocatechin have many positive effects in the human body and will be discussed in detail later as part of in-depth remedies for a range of diseases including cancer and cardiovascular diseases [12]. Besides, many polyphenols exhibit antithyroid activities primarily due to their inhibitory effects on organification [10].

Inhibitors in this group include genistein and daidzein from soybean, quercetin found in apples, catechin from traditional tea, red grapes, citrus fruits, onions, broccoli, cherries and berries, kaempherol found in grapefruit, rutin from buckwheat, tannins from a nut, and apigenin and luteolin in millet. The latter substances, together with thiocyanate, are thought to contribute to be responsible for the high incidence of goiter in an iodine-deficient population in West Africa for whom millet is a dietary staple. These substances may exacerbate the effects of endemic or sporadic low iodine-induced hypothyroidism in other populations.

A less well-defined source of hypothyroid substances is contaminated drinking water. Studies in India and elsewhere showed that goiters are associated with contaminated water supplies. By simply providing uncontaminated water supply in these regions, the incidence of hypothyroidism in schoolchildren was significantly reduced. Although no specific antithyroid substances were found in these studies, many possible contaminants have this activity. Therefore, powerful antithyroid products such as resorcinol, phthalate esters, methoxyl anthracenes, and phthalic acid are known to contaminate water in areas rich in coal and shale. Some antithyroid disulfides resemble substances formed on onions and garlic and are present in high concentrations in the aqueous wastewater from coal conversion processes. Moreover, 60% of all herbicides, notably, 2,4-dichlorophenoxyacetic acid (2,4-D) and thioureas, exhibit antithyroid activities. Other antithyroid agents include polychlorinated biphenyls (PCBs), mercury, and perchlorate. Thus, normal thyroid function may be compromised by any of a host of pollutants found in the water supplies of some regions.

### *13.1.3 Favism*

Favism, a syndrome of acute hemolytic anemia, is induced by the consumption of raw or cooked Vicia beans, well-known as broad beans or fava beans. Favism occurs to people near the Chinese or Mediterranean people. The disease is limited to a greater extent in males than in females and is severer in infants and young children than in adults. Although adult deaths from favism are rare, modalities have been reported in infants and children. The clinical symptoms of favism include pallor, shortness of breath, nausea, fatigue, fever, abdominal pain, and chills. Renal failure occurs in serious cases [13]. The onset of symptoms usually takes up 24 h after the bean is ingested and lasts for 2 days. Recovery in most individuals is spontaneous and abrupt.

Due to the inability to obtain a suitable animal model, the etiology of the disease has been hindered. Nevertheless, some epidemiologic studies indicated that susceptible individuals have reduced the levels of glucose-6-phosphate-dehydrogenase (G-6-PD) and decreased glutathione (GSH) in red blood cells [14]. G-6-PD catalyzes a reaction in glucose metabolism to produce NADPH [15]. Glutathione reductase-mediated oxidized glutathione (GSSG) reacts with NADPH to maintain adequate levels of GSH [16]. Thus, reducing levels of G-6-PD can lead to a diminished capacity of cells to maintain normal levels of GSH. Adequate levels of GSH, an antioxidant, are required to maintain the stability of the cell membrane, especially the red blood cell.

In experiments with human erythrocyte suspensions, it was found that the GSH levels in cells from individuals susceptible to favism are influenced by components of the fava bean [17]. The GSH levels in cells from normal individuals do not express this sensitivity. The active compounds in fava bean belong to pyrimidine derivatives, isouramil and divicine, those correspond with aglycones of vicine and convicine. These aglycones are readily oxidized in air and rapidly promote the non-enzymatic conversion of GSH to GSSG in solution. So, it is suggested that these pyrimidine derivatives due to enzymatic action of the corresponding glycosides through plants or in the intestine may be causative agents of pathogens. Confirmation of this assumption must await further development of a suitable animal model for the disease or proper human trials.

The fava bean is one of the earliest domesticated food plants. Favas was discovered in Spain with its 6000-year-old archaeological site and the 4000-year-old Egyptian tomb. The bean was cultivated in Massachusetts by some of the first European settlers in the New World. Fava beans, also known as broad beans or horse beans, are ubiquitous in the Mediterranean diet and are the most common food in the Egyptian diet.

The action of fava bean-induced hemolytic anemia has been an important research topic. Analysis of the bean for substances that decrease the levels of reduced glutathione in isolated erythrocytes led to the identification of two related glycosides, vicine and convincing, which occur at very high levels in fresh beans. Cooking does not significantly reduce the glycoside content of the beans. Hydrolysis of the glycosides by exposure to simulated gastric juice or to β-glucosidase, which is present at high levels in ripe beans, produces the active aglycones, isouramil, and divicine. Another phenolic substance that has been implicated in the cause of favism is L-DOPA (3,4-dihydroxy-L-phenylalanine). Although L-DOPA is apparently inactive in producing the effects associated with favism, its combination with divicine and isouramil produces a synergistic increase in hemolytic activity. The main effect of these substances is to produce an oxidative stress in cells, leading to the loss of reduction of glutathione in treated erythrocytes from G-6-PD-deficient subjects. The mechanism of induction of oxidative stress involves the production of quinones and quinone imines from the parent hydroquinones [18]. As shown in Fig. 13.3, redox

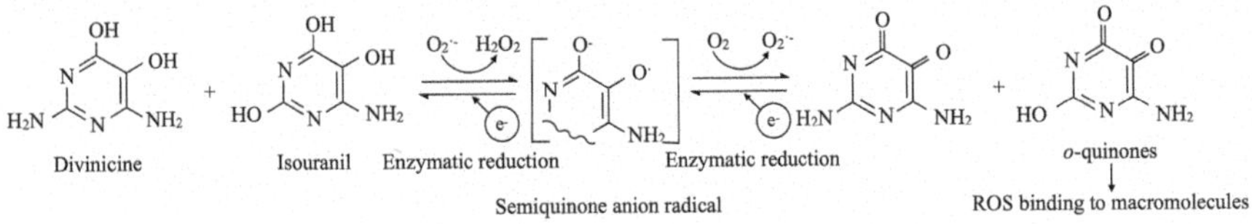


**Fig. 13.3** The redox cycle of oxidant products through high-capacity single electron reductive processes

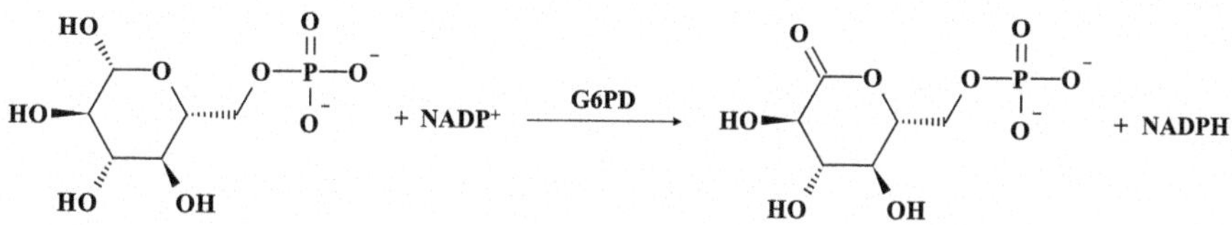


**Fig. 13.4** G-6-PD reaction

cycling of the oxidized products through high-capacity single electron reductive processes (e.g., NADPH-CYP450 reductase) can result in the catalytic generation of reactive oxygen species (ROS). The increased levels of ROS, in turn, induce levels of oxidized glutathione, protein-bound mixed disulfides, cellular protein aggregates, and cross-linked structural proteins in the erythrocyte. These and other abnormalities of the erythrocytes, including the disruption of cellular calcium homeostasis, lead eventually to their destruction by phagocytosis in the spleen.

The G-6-PD-deficient cells can't resist the action of pro-oxidant substances due to the central role of G-6-PD in the control of cellular redox status [19]. As illustrated in Fig. 13.4, G-6-PD is responsible for mediating the oxidation of G-6-P to 6-phosphogluconolactone, which is the first reaction in the pentose phosphate pathway of glucose metabolism. A major function of this pathway is to provide reduced cellular power in the form of NADPH. In turn, NADPH is critical for the cellular metabolic defenses against oxidative stress [20], particularly as a cofactor for reductases, such as catalase, glutathione reductase, glutathione peroxidase, and thioredoxine reductase. Under conditions of oxidative stress in normal cells, cell defense capacity increases by increasing the expression of G-6-PD. Therefore, the inability of G-6-PD-deficient cells to increase the supply of NADPH makes them highly vulnerable to induced oxidative stresses.

### 13.1.4 Neurolathyrism

Neurolathyrism, an ancient disease, is caused by the consumption of peas of the genus Lathyrus (*L. sativus*), commonly known as vetch pea, or grass pea, chickling pea, and by many other names in various languages. The disease was apparently described for the first time in the Hippocratic era (about 400 BC), and an outbreak occurred in parts of North Africa, the Middle East, Asia, and India until the twentieth century. During the Second World War in Europe, prisoners and farmers also experienced extremely high neurotoxicity. Chronic Lathyrus poisoning continues to affect large populations in Ethiopia, India, and Bangladesh. The toxicity of Lathyrus is well known to regulatory agencies and it is illegal to sell peas in most areas. However, Lathyrus is a protein-rich, hardy crop that grows under conditions of poor soil, drought, and even flooding, and thus is used as an emergency crop for food in many areas. As expected, after the drought or flood, the neurolathyrism had the highest incidence. Several aspects of the etiology of neurolathyrism are quite well understood. It is known that neurolathyrism appears in humans after at least 3 months of taking *L. sativus* peas of at least 300 g/day.

Neurolathyrism is a progressive, neurodegenerative condition of motor neurons [21]. It is characterized in humans, primarily men, by strengthening the leg muscles that predominantly target a type of neuron called a Betz cell in the spinal tract. The initial symptoms are increased stiffness of the calf muscles and loss of control of the legs. The toxin in peas that causes the disorder is the amino acid derivative, β-*N*-oxalyl-α,β-diaminopropionic acid (ODAP), which occurs in 0.1–2.5% of the dry weight of the peas. The dose of ODAP administered to male mice for 40 days that would likely be due to the consumption of the toxic peas (i.e., 5 mg/kg body weight) produced decreases in neuronal activities characteristic of human neurolathyrism. Studies of multiple model systems have revealed the mechanism of the toxic action of ODAP [22]. Early studies demonstrated that the toxic effects of ODAP in rodents could be inhibited by the administration of antagonists of the AMPA subclass of glutamate receptors [21]. ODAP has also been shown to promote glutamate release from presynaptic elements and to inhibit glutamate uptake, resulting in a dramatic increase in glutamate concentration at the synapse [23]. This increase in glutamate concentration produces toxic effects in the neurons, mainly due to the release of ROS from inhibition of complex I of mitochondria electron transport chain by glutamate. This role of ROS in the effect of ODAP is consistent with the reported protective effects against neurolathyrism of antioxidants such as vitamin C, as well as a nutritionally adequate diet. Presumably, there are rare cases of neurolathyrism in women because of the protective effect of estrogen on ROS-induced cell damage [24].

The toxic effects of high concentrations of glutamate in neurons involve both receptor-mediated and receptor-independent mechanisms [25]. Under normal neuronal activation conditions, glutamate activation induces the flux of calcium glutamate receptors, which is coupled to a variety of biological responses, including the release of neurotransmitters and growth factors and modulation of neurotransmission. This activation of glutamate-activated neurons is followed by a synergistic inactivation pathway that includes a process in which glutamate concentrations are reduced in synapses. However, uncontrolled activation of the receptor leads to a long-term increase in the intracellular calcium ion concentration, which causes ROS production and activates the cell death pathway [26]. Glutamate also acts through a receptor-independent mechanism to induce cellular oxidative stress known as glutamate-induced oxidative stress. Cells are sensitive to this type of toxicity because high levels of extracellular glutamate can block the glutamate-cysteine antiporter in the plasma membrane, which leads to the inhibition of glutathione synthesis. Glutamic acid-induced oxidation in cultured cells followed by depletion of GSH, increased the production of ROS, inhibition of complex I of the mitochondrial electron transport chain, increased calcium influx, and subsequent cell death [27]. It was observed that antioxidants can block the lathyrogenic effects of ODAP, suggesting that oxidative stress induced by receptor-dependent and receptor-independent mechanisms is essential for the full development of the disorder.

### 13.1.5 Cyanogenic Glycosides

Cyanogenic glycosides are cyanide-containing compounds naturally present in over 2500 plant species. Cyanogenic glycosides, a group of widely occurring natural products, produce ketone or aldehyde, sugar, and the highly toxic cyanide ion by hydrolysis. Toxicity of cyanogenic glycosides is because of the liberated cyanide. The acid or base hydrolysis of cyanogenic glycosides can produce cyanide. The release of hydrogen cyanide is not appreciable in the stomach although of acidic nature of its contents. Hydrogen cyanide is produced from cyanogenic glycosides in chewed or chopped plants or by enzymatic ingestion processes involving two enzymes.

The first step is the cleavage of the sugar to produce a cyanohydrin and a sugar by β-glucosidase. Most of cyanohydrins are not stable and spontaneously decompose into the corresponding ketone or aldehyde and hydrogen cyanide. However, this decomposition is accelerated by hydroxy-nitrile-lyase. When fresh plant material is soaked as in chewing, the cell structure is reasonably decomposed to contact the enzyme and cyanide to produce hydrogen cyanide. This is thought to be the principal mechanism of cyanide poisoning from the consumption of fresh plant material (Fig. 13.5). The two-step therapy is initiated with sodium nitrite, which induces methemoglobinemia that allows the release of cyanide from hemeproteins, followed by sodium thiosulfate, which acts as a substrate for rhodanese, an endogenous hepatic enzyme that catalyzes the conversion of free cyanide to the less toxic thiocyanate.

The development of simple methods to detoxify cyanide-rich plant products has allowed their use as important food sources. For example, cassava roots are the major source of dietary carbohydrates for hundreds of millions of people in South America and Africa [28]. The available cyanide content in unprocessed cassava is high enough that cyanide poisoning has a serious impact on chronic consumption. The detoxification process of cassava has traditionally been a multi-step process that

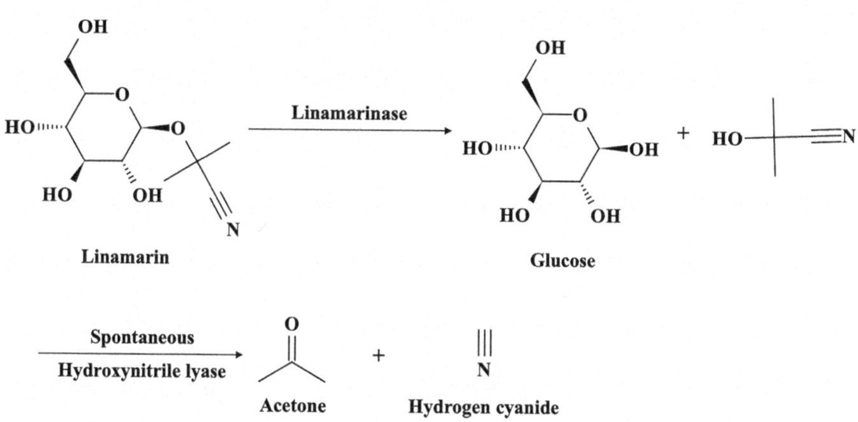


**Fig. 13.5** Cyanide liberation from linamarin

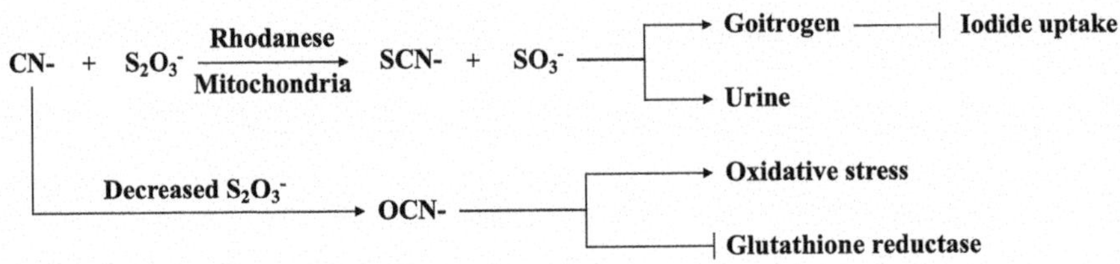


**Fig. 13.6** Metabolic pathways for cyanide ion

includes chopping and grinding in running water, which induces the production of hydrogen cyanide from the enzyme and removes free cyanide and glycosides from the waste water. Fermentation and boiling processes are also used in some traditional procedures for the production of cassava flour. However, despite the fact that well-developed processing programs can greatly reduce the content of hydrogen cyanide in cassava products, the hazard of chronic cyanide poisoning from cassava remains significant in some areas. A major problem remains to be the increased cost of more highly processed (and detoxified) flour. Public health and government organizations are working to promote the production and consumption of adequately processed and safe cassava products.

Cyanide is considered to be a highly toxic substance with acute and chronic effects [29]. Cyanide produces multiple immediate biochemical effects in biological systems, such as inhibition of the antioxidant defense, alteration of critical cellular ion homeostasis, and suppression of cellular respiration [30]. Symptoms of typical acute poisoning include mental confusion, muscle spasms, and respiratory distress. The minimal lethal oral dose of hydrogen cyanide is estimated to be 0.5–3.5 mg/kg body weight. Cyanide exerts its acute toxicity through its binding to the ferric ion of cytochrome oxidase, which is the oxygen-reducing component of the mitochondrial electron transport chain [31]. Recent studies have revealed that this inhibition of cyanide involves a low level of nitric oxide production, which enhances the effect of cyanide by binding to copper ions present in cytochrome oxidase [32]. The overall result is the cessation of cellular respiration.

Cyanide ion can be metabolized to several products as shown in Fig. 13.6. Thiocyanate is the major excretion product of cyanide, which is catalyzed by the mitochondrial enzyme, thiosulfate sulfurtransferase (also known as rhodanese), and mercaptopyruvate sulfur-transferase (MST). Rhodanese is concentrated in the liver but not in mammalian targets of cyanide lethality, including the heart and central nervous system [33]. However, MST occurs ubiquitously in the cytoplasm and mitochondria of cells in all tissues. The cosubstrate for the rhodanese reaction is thiosulfate, which is produced in a multistep process from cysteine that involves desulfuration of cysteine and oxidation of the released hydrogen sulfide. The cosubstrate for MST, mercaptopyruvate, is also produced from cysteine by a direct deamination reaction. Cyanide may become an important metabolite of cyanide in the presence of decreased levels of thiosulfate which may be related to nutritional deficiencies. Cyanate can increase the toxicity of cyanide, as discussed next. Another metabolic pathway for cyanide involves complexation with hydroxocobalamin which is derived from vitamin $B_{12}$. This pathway may be important in low-level cyanide exposures with cyanocobalamin as an end product.

Antidotes for acute cyanide poisoning are an important addition to first aid kits, especially in laboratories or industrial situations that may be exposed to high levels of cyanide. Currently used antidotes include the cobalt complexes, hydroxocobalamin and dicobaltededate, and the oxidants that form methemoglobin, nitrites, and 4-dimethylaminophenol (DMAP). The cobalt atoms of the former substances form strong complexes with cyanide and are then excreted in the urine. The heme oxidant converts heme to iron form. If the concentration is sufficient, cyanide is removed from the ferric ion of cytochrome-c oxidase and allows respiration to resume. A volatile form of nitrite is amylnitrite, which can be taken by inhalation to an unconscious victim. Amylnitrite is a comparatively weak heme oxidant; however, its activity as a vasodilator is now considered to be a major contributor to its well-established antidotal properties. DMAP, a potent heme oxidant, is administered to patients by intravenous injection. The cobalt compounds are administered by intravenous injection, as well, in some cases along with thiosulfate to promote the conversion of cyanide to thiosulfate.

The importance of cyanide poisoning, in terms of the number of infected people, is the effect of chronic low-level cyanide exposure. In parts of Africa and South America, large intakes of cassava in a nutritionally marginal diet, are associated with at least two disorders. These diseases occur with very low incidence in areas where cassava consumption is low and is part of a nutritionally adequate diet, or where cassava is cyanide free. Tropical ataxic neuropathy (TAN) is a neurological syndrome characterized by optic atrophy, ataxia, and deafness. A related neurological disorder caused by long-term consumption of cyanide-rich cassava is known as tropical amblyopia, which is characterized by atrophy of the optic nerves and blindness. A very similar syndrome, called tobacco amblyopia, has also been found in heavy smokers who also consume a nutritionally deficient diet. In this case, tobacco smoke is the source of cyanide. Individuals suffering from these disorders have very low concentrations of sulfur-containing amino acids in the blood and elevated levels of plasma thiocyanate. When the patient is on a cyanide-free and nutritious diet, the symptoms in the early stages of the disease are lessened, especially if the diet contains sufficient amounts of sulfur-containing amino acids and vitamin $B_{12}$. Regardless of the nutritional quality of the cassava diet, goiter may be ubiquitous due to the goitrogenic effect of prolonged exposure to thiocyanate in these populations. Similarly, like the incidence of goiter, the incidence of TAN and tropical amblyopia depends on economic factors. It was evidenced by that the incidence of TAN increased during Nigeria's economic downturn in the 1990s, compared to the more prosperous 1960s for this region.

The mechanism of the chronic toxicity of cyanide is an extension of the acute effects of this toxin [33]. Thus, incomplete inhibition of cytochrome-c oxidase in mitochondria results in the increased generation of reaction oxygen species (ROS) and decreased production of ATP. Under these conditions, cells continue to respire but with decreased efficiency and with increased exposure to oxidative stress. Since cells need sufficient ATP production to resist oxidative stress, this combination of effects of cyanide is particularly hazardous to the cell. A further exacerbating factor is the production of cyanate from cyanide under low-sulfur nutrient conditions. Cyanate is an inhibitor of glutathione reductase, the inactivation of which further reduces cellular defenses against ROS. Although cyanide is undoubtedly toxic to all cells, the optic neurons are particularly sensitive. It is believed that small caliber axons, especially long axons of the optic nerve, are particularly sensitive to cyanide poisoning because their respiratory frequency is usually high, and the apparently large lipid membrane surface area is susceptible to ROS.

### *13.1.6 Lectins*

Lectins are non-enzymatic thermolabile proteins, glycoproteins, or lipoproteins that selectively bind to saccharide groups. Lectins are a rather prominent group of proteins and glycoproteins that possess the ability to bind certain carbohydrates. Lectins will cause agglutination of cells whose cell walls contain these carbohydrates as the components. The ability of lectins agglutinating red blood cells is used as a basis for blood type analysis. Lectins, which bind to carbohydrate components of intestinal epithelial cells, may decrease absorption of nutrients from the digestive tract.

Lectins are widely present in nature. Extracts from over 800 plant species and from numerous animal species possess agglutinating activity. Lectins present in various legumes used as feed or food sources are of particular interest. Lectin activity occurs in a wide variety of legumes used for food, including black beans, kidney beans, lima beans, soybeans, peas, and lentils. Some plant lectins may exert adverse effects if one person only eats the raw plant. Their adverse effect is generally caused by the binding of the lectin molecules to the membranes of the intestinal cells. Furthermore, this binding non-specifically inhibits the active and passive transport of vital nutrients such as amino acids, fats, minerals vitamins, and intestinal necrosis through the cell wall. Some of lectins are highly toxic to animals, although they are a class of substances recognized for their ability to agglutinate or clump red blood cells. For example, lectins isolated from black beans produce growth retardation when fed to rats on a diet of 0.5% [34]. Moreover, lectins from kidney beans produce death in rats fed on lectin at 0.5% of the diet for 2 weeks [35]. Soybean lectin, a less toxic lectin, produces only growth retardation in rats fed at 1% of the diet. The $LD_{50}$ of soybean lectin is detected to be 50 mg/kg.BW. Ricin, a lectin from the castor bean, is one of the most toxic natural compounds with an $LD_{50}$ value of 0.05 mg/kg.BW by injection. Castor beans (not a legume) must be thoroughly heated to deactivate the ricin before used as animal feed due to their high toxicity.

The exact toxic role of lectins in various beans and legumes is a controversial subject and seems to depend on the specific legume in question. Un-cooked beans as the main ingredient of the diet usually do not provide the growth of animals. Thoroughly heated beans do support growth. Symptoms of toxicity are exhibited when the lectin fractions of black beans and kidney beans are fed to animals along with the heated bean material. With respect to soybeans, about half of the growth inhibition resulting from raw soy meal can be attributed to lectin. In addition, the nutritional quality of the soybean meal from which the lectin component had been removed was observed to be slightly improved. Hence, lectins and other substances such as inhibitors of digestive enzymes seem to contribute to the growth inhibitory effects of raw beans.

The final toxic mechanism by lectins is controversial as well. It is noted that lectins from various sources are present in the intestinal epithelium adsorb nutrients, thereby reducing intestinal absorption of these nutrients. The resulting inefficient use of nutrients may itself lead to poor growth due to a diet rich in uncooked legumes. This effect may also amplify the loss of protein because of pancreatic hypersecretion caused by trypsin inhibitors also present in legumes. However, the intestinal microflora seems to play a role in legume- and lectin-induced toxicity. Germfree birds (i.e., birds free of intestinal bacteria) used as test species exhibited less growth depression than conventional birds when fed raw legumes or isolated lectins. For instance, diets containing raw jack beans meal produce high mortality in Japanese quails. However, germfree birds showed no toxic effects under exactly the same experimental conditions. It was suggested that lectins may damage the defense of body's system against bacterial infection, resulting in an increased tendency for an invasion by gut and other bacterial flora.

Although the toxicities of most plant lectins are characterized as chronic antinutritional disorders, a few lectins are very effective acute toxins. Most notable in this group is ricin, which is a lectin from castor bean. The seeds of castor bean have been considered to be toxic for thousands of years, and its use in Indian folk medicine in the sixth century BCE was explicitly mentioned. As few as five castor beans are said to contain enough ricin to kill an adult human. The toxic effects of ricin, including nausea, diarrhea, abdominal cramps, vomiting, internal bleeding, liver, and kidney failure, can be seen within 2–3 h after consumption of these castor beans or ricin-contaminated material. Rapid heartbeat can also occur. If the entire castor beans are swallowed whole, the severity of poisoning will be less than if the beans are chewed because ricin is rarely released from the whole bean. Breathing dust that contains ricin causes cough, fever, weakness, nausea, muscle aches, difficulty breathing, and cyanosis. Breathing dust can also cause respiratory and circulatory failure. The oral $LD_{50}$ for ricin in mice is 30 mg/kg.BW, which is approximately 1000-fold higher than by injection or inhalation [36]. The estimated $LD_{50}$ after oral administration for humans is in the range of 1–20 mg of ricin/kg of body weight, and presumably much lower following inhalation [37].

In the late nineteenth century, ricin was isolated from castor beans in a purified form with a yield of 1–5%, as a relatively thermostable glycoprotein with a molecular weight of approximately 60 kDa. The steam distillation process used to produce castor oil is sufficient to inactivate this protein, making this commonly used folk remedy safe to consume. Subsequent studies revealed that ricin is resistant to digestion. As shown in Fig. 13.7, ricin is composed of two protein chains, A and B, linked by a disulfide bond. The B chain is a lectin that binds to galactose-containing glycoproteins and glycolipids expressed on the cell surface, facilitating

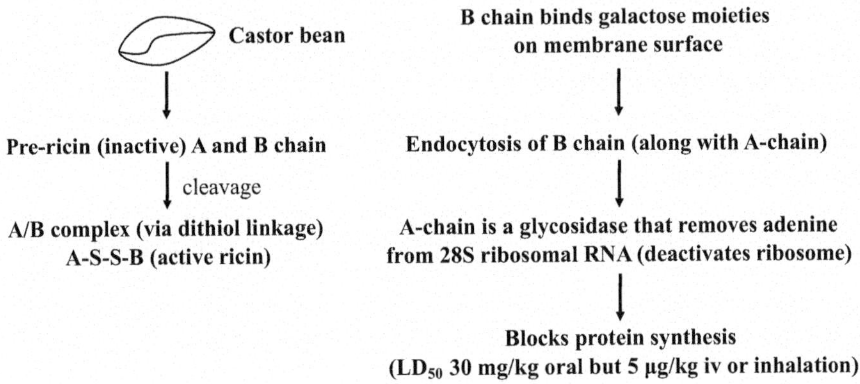


**Fig. 13.7** Mechanism of toxic action by ricin

the entry of ricin into the cytosol. The A-chain inhibits protein synthesis by irreversibly inactivating eukaryotic ribosomes by removing a single adenine residue from the 28S ribosomal RNA loop contained within the 60S subunit. This process prevents chain elongation of polypeptides and causes cell death due to inhibition of protein synthesis. Additional mechanisms of toxicity have also been reported, such as activation of apoptosis pathways, alteration in membrane structure and function, damage on cell membrane, and release of cytokine inflammatory mediators. A broad group of bacterial and plant toxins have similar A- and B-chain protein components such as diphtheria toxin.

Since ricin is a readily available and highly potent toxin, it has been put to a range of negative and positive uses in modern times. The properties of ricin as a chemical warfare agent were tested and developed in the United States and elsewhere in the early twentieth century. Recently, ricin is thought to have been used to assassinate international journalists and attempt to assassinate US politicians. On the other hand, ricin is positively being studied for cancer chemotherapy, bone marrow transplantation, and cell-based research. Clearly, malignant cells are more susceptible to the toxicity of ricin than nonmalignant cells because the former express more carbohydrate-containing surface lectin binding sites than do nonmalignant cells. The antibody-conjugated ricin can target cancer cells and has been studied as an immunotherapeutic agent.

### *13.1.7 Vasoactive Amines*

Vasoactive amines are low molecular weight basic organic compounds that can cause the sympathetic, so-called "flight-or-fight" response in higher organisms. In response to an appropriate stimulus, neurons in this system release primarily norepinephrine and epinephrine, also known as noradrenaline and adrenaline, respectively, which bind to adrenergic receptors in peripheral tissues. As shown in

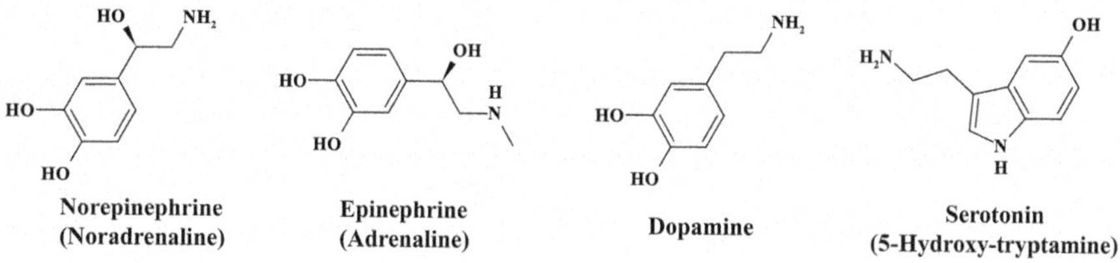


**Fig. 13.8** Chemical structures of vasoactive amines

Fig. 13.8, together with dopamine, these substances are structurally related catecholamines. Also included in the group of vasoactive amines are serotonin and tyramine. The activation of the adrenergic pathways by these substances induces the characteristic responses, including dilated pupil, increased heart rate, increased sweating, occasional vomiting, and increased blood pressure. If this response is not controlled properly, it can lead to life-threatening symptoms, such as heart arrhythmia, hypertensive crisis, and death. Since exogenous vasoactive amines, such as those that occur in certain foods, may be highly toxic, an effective method for detoxification of these substances has developed. This fact can be clearly seen from the observation that these substances are far more toxic by intravenous administration than by ingestion. In fact, circulating levels of exogenous vasoactive amines are carefully controlled by the action of monoamine oxidases (MAO), MAO-A, and MAO-B. MAO-A deaminates serotonin in the central nervous system and dietary monoamines in the gastrointestinal system. MAO-B is mainly present in liver and muscles and deaminates dopamine and phenethylamine. Bioactive amines can also be deaminated by diamine oxidase (DAO) activity in the intestine, which can provide protection from small amounts of amines that are normally in foods. Due to the rapid metabolic conversion of MAO and DAO to amines, oral administration of vasoactive amines to normal mammals has little effect on blood pressure. However, significant effects were observed when patients or experimental animals were also treated with MAO inhibitors. With the development of MAO inhibitors for therapeutic purposes, it is clear that dietary vasoactive amines may pose a major threat to human health. The MAO inhibitors are used to inhibit the action of monoamine oxidases, especially as antidepressants or anti-Parkinsonian drugs in the central nervous system.

The first generation of MAO inhibitor drugs, including isocarboxazid, nardil, marplan, and parnate, is non-specific, inhibiting both MAO isoforms, and the inhibition is considered irreversible. The second generation of reversible inhibitors of monoamines known as RIMA, including L-deprenyl, selegiline, and rasagiline, is selective for the inhibition of MAO-B. They are almost nontoxic at low doses and carry little risk of a hypertensive effect in the treatment of Parkinson's disease. Nevertheless, in order to be effective in the treatment of depression, the required higher dose starts to inhibit all isomers, and the potential of hypertensive crisis increases. The levels of the vasoactive amines have been examined in a wide range of food items. The results of these analyses suggest that the levels of several of these substances such as serotonin, dopamine, tyramine, and norepinephrine are low (0 ~ 28 mg/g) in banana pulp, potato, avocado, tomato, spinach, and orange. In particular, high levels of tyramine (20 ~ 2170 mg/g) are mainly detected in foods, which are intentionally or due to spoilage resulting in microbial action. The most consistent food in this group is therefore considered to be a major risk factor for patients treated with MAO inhibitors, including aged cheese, aged and cured meats, poultry and fish, improperly stored or spoiled meats, sauerkraut, marmite (yeast extract), soy sauce and other soy sauce condiments, and tap beer. Lower levels of vasoactive amines and lower hazard are associated with red and white wine, and bottled or canned beers.

Any foods containing free amino acids, especially tyrosine and phenylalanine, are subject to vasoactive amine formation if poor sanitation and low-quality foods are used, or if the food is subjected to storage conditions suitable for bacterial growth. It is estimated that 80% of all hypertensive events are associated with cheese consumption in patients treated with MAO inhibitors. These patients should not eat a few foods such as aged cheeses, especially English Stilton, Danish bleu, and Cheshire, because a clinically significant level of tyramine may be present in one ounce or less.

### 13.1.8 *Caffeine*

Caffeine, a methylated xanthine derivative, naturally occurs at highest levels in coffee, tea, and chocolate products. As shown in Table 13.1, the levels of caffeine in coffee range from 40 to 180 mg/5-oz. cup and from 20 to 110 mg/cup for traditional teas made from leaves of Camellia sinensis. Baker's chocolate and dark chocolate contain caffeine levels of 5–35 mg/oz. with levels ranging from 2 to 30 mg/cup for chocolate drinks. A large amount of caffeine is also added to certain soft drinks and over-the-counter headache medications. It is also worth noting that the caffeine content of the very large coffee portion sold at the outlet is almost equal to the caffeine content of the three over-the-counter caffeine pills. Because of the widespread availability and use of caffeine and products containing caffeine, caffeine is considered to be the most widely self-prescribed and used drug in modern society.

The discoveries of caffeine-containing plants are ancient and the related legends are part of many cultural folklore. Archaeological evidence shows that our human ancestors may have consumed caffeine-containing plants as early as 3500 BCE. Early peoples discovered that chewing the seeds, bark, or leaves of certain plants had the effects of relieving fatigue, stimulating awareness, and improving mood. A popular Chinese legend claims that an emperor who reigned around 3000 BCE accidentally discovered that when some leaves (maybe tea leaves) fell into boiling water, they produce a fragrant and restorative drink. As early as the ninth century CE, the history of coffee has been recorded. A popular legend traces its discovery to an Ethiopian goat herder who observed that goats became ecstatic and sleepless at

**Table 13.1** Caffeine levels in various beverages

| Item | Average (mg) | Range |
|---|---|---|
| Coffee (5-oz. cup) | | |
| Brewed, drip method | 115 | 60–180 |
| Brewed, precolator | 80 | 40–170 |
| Instant | 65 | 30–120 |
| Decaffeinated, brewed | 3 | 2–5 |
| Decaffeinated, instant | 2 | 1–5 |
| Teas (5-oz. cup) | | |
| Brewed, major U.S. brands | 40 | 20–90 |
| Brewed, imported brands | 60 | 25–110 |
| Instant | 30 | 25–50 |
| Iced (12-oz. glass) | 70 | 67–76 |
| Some soft drinks (6 oz.) | 18 | 15–30 |
| Cocoa beverage (5 oz.) | 4 | 2–20 |
| Chocolate milk beverage (8 oz.) | 5 | 2–7 |
| Milk chocolate (1 oz.) | 6 | 1–15 |
| Dark chocolate, semi-sweet (1 oz.) | 20 | 5–35 |
| Baker's chocolate (1 oz.) | 26 | 26 |
| Chocolate-flavored syrup (1 oz.) | 4 | 4 |
| NoDoz max strength (1 pill) | 200 | |
| Mountain dew (12 oz.) | 55 | |
| Excedrin (1 pill) | 35 | |
| Espresso (1 oz. Starbucks) | 35 | |
| Coffee (8 oz. Starbucks) | 250 | |
| Coffee, grande (16 oz. Starbucks) | 550 | |

night after browsing on coffee shrubs. Goat herders experienced the same vitality when trying berries that the goats had been eating. The legend further suggests that the wife of the goat herder inadvertently baked berries and used them to make an aqueous decoction that we call coffee. The use of cola nuts and cocoa is also ancient, and there is evidence of Maya culture using the latter dating to 600 BCE.

Caffeine is a neurological stimulant which produces biological effects in almost all organs of the body. The major cellular action of caffeine is to block adenosine receptors, which act as negative regulators of overall body function. These receptors are blocked by caffeine, related purine and theophylline, and then stimulate the activity of related organs and tissues, including the central nervous system. At a low adult dose of about 200 mg, caffeine produces central nervous system stimulation, diuresis, cardiac muscle stimulation, relaxation of smooth muscles, and increased gastric secretion. Many studies have confirmed the centuries-old belief that caffeine can improve the physical performance of fatigued individuals, but caffeine does not improve the physical performance of the rested subject.

The authors of one such study measured the effects of caffeine on the performance of members of the US Navy Seals Special Forces, noting that "even in the most unfavorable circumstances, moderate doses of caffeine could improve cognitive function, including vigilance, memory, learning, and mood state" [38]. The authors of another study of the effects of caffeine on students' test pressures suggest that "caffeine had significant effects on blood pressure and heart rate of habitual coffee drinkers and persisted for many hours during the activities of daily living" and that "caffeine may exaggerate responses to normal daily stress events [39]. There is currently considerable interest in the observation that caffeine intake is negatively correlated with the incidence of Alzheimer's disease. Studies using rodent models of this disease also show that caffeine has a strong protective effect against this disorder.

Excessive consumption of caffeine may also produce many adverse reactions. The most common observed toxic effects of caffeine include nervousness, irritability, and cardiac arrhythmias. Adverse effects can also include vomiting, agitation, abdominal pain, and seizures. The oral $LD_{50}$ for caffeine is approximately 200 mg/kg.BW, approximately 12 g/60-kg person. Therefore, the adult lethal dose of this moderately toxic substance would require rapid consumption of about 50 cups of coffee or 50 caffeine tablets. Death occurs from ventricular fibrillation with blood caffeine levels over 100 mg/ml.

Caffeine intake by pregnant and nursing women has been receiving attention recently. Until recently it was widely believed that low caffeine intake (<150 mg/day) during pregnancy could not harm the fetus, and only high caffeine intake (>300 mg/day, equivalent to more than three cups of coffee/day) should be avoided during pregnancy due to an association with increased rates of birth defects. The results of an important recent study of a large number of patients related to major health institutions and controlled for many variables resulting from pregnancy indicate that the allowance for caffeine intake during pregnancy may be too generous. This recent study showed that women who consumed only 200 mg or more of caffeine per day had fully twice the risk of miscarriage as women who do not consume caffeine [40]. The results further showed that even women who consumed less than 200 mg of caffeine daily had an increase in the risk of miscarriage by more than 40% compared with women who consumed no caffeine. Caffeine intake from non-coffee sources such as caffeinated tea, soda, and hot chocolate showed a similar increased miscarriage risk [41]. Thus, the increased risk of miscarriage seems to be due to the caffeine itself rather than other possible chemicals in coffee [42]. If these findings are reproduced by other researchers in other settings, they may recommend greater restrictions on caffeine consumption by pregnant women. At the same time, it is prudent for pregnant or imminent pregnant women to reduce or stop taking caffeine.

### *13.1.9 Curare*

Since prehistoric times, indigenous peoples around the world have used natural plant materials as hunting and war poisons [43]. In the centuries before the arrival of

**Fig. 13.9** Chemical structure of D-tubocurarine

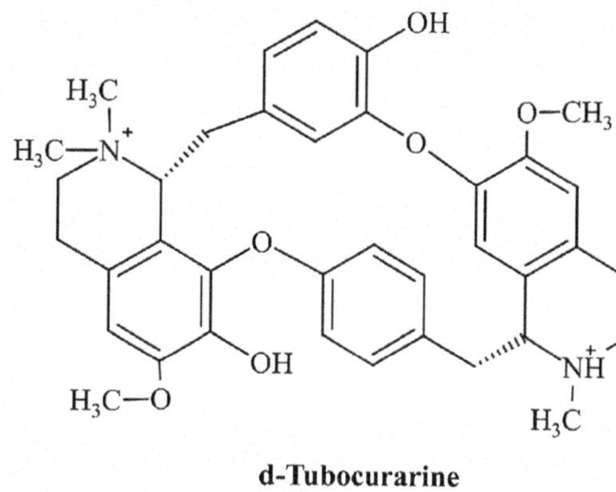

Europeans, the indigenous populations of the Amazon basin in South America used extracts primarily of the climbing shrubs or vines *Chondrodendrontomentosum* and *Strychnostoxifera* as sources of effective arrows and dart poisons known as curare.

According to reports on curare that first published in 1516, tribes in the Amazon region had developed a variety of methods for preparing poison. All the methods involved crushing and cooking the roots, stems, and bark of the selected plant, followed by adding other plants and poisonous animals. The light syrup was prepared by repeatedly boiling the mixture, and then applied to the tips of arrows and darts. The latter were carefully heated near a flame to produce a hardened, dark brown, or black, tarred coating that was resistant to damage during storage or use.

Continuous studies since the discovery of curare active compounds have established their mode of toxic action. In 1935, Harold King isolated crystalline D-tubocurarine in classic studies, which is the primary active compound in curare (Fig. 13.9). D-Tubocurarine was shown subsequently to block nicotinic acetylcholine receptors at neuromuscular junctions to produce paralysis in muscles. This alkaloid first affects the muscles of the toes, eyes, and ears, and then muscles of the neck, arms, and legs. Finally, when the dose is high enough, the lungs are paralyzed and the victim dies of asphyxiation. The victim keeps awake and realizes that he cannot move and is losing the ability to breathe. The $LD_{50}$ for D-tubocurarine was found to be only 0.5 mg/kg.BW by intravenous injection into rabbits, thereby classifying it as an extremely toxic agent. Recovery is possible, however, if respiration is maintained, which allows for the metabolism, detoxification, and excretion of the toxin. Curare alkaloids are not toxic after ingestion, apparently because their high polarity inhibits their absorption of the gastrointestinal tract. This is why natives can safely test the potency of their poison preparations by tasting the bitterness of the mixture.

There have been many unproven uses of curare in the folk medicine of indigenous tribes. Thus, curare has been used as a diuretic and for madness, bruises, dropsy, edema, fever, and kidney stones. D-Tubocurarine has been used very effectively in Western medicine, and it has been used as a muscle relaxant during surgical anesthesia since the 1940s. However, the medical use of curare compounds has been replaced by a variety of curare-like synthetic drugs that have similar pharmacodynamic characteristics but with fewer side effects.

### 13.1.10 Strychnine

Like curare, the first uses of strychnine were as a mixture in plant extract preparations in combination with other natural products that have been used by native hunters for arrows and darts tips since the most recent record. The main source of strychnine is StrychnosNux vomica, an evergreen tree native to Southeast Asia, especially Myanmar and India, and cultivated elsewhere. Its dried seeds or beans, sometimes its bark (called nuxvomica), were used as ordeal poisons by natives and continue to be used by some groups in herbal remedies. The seeds were brought to Europe for the first time in the fifteenth century and could become poisons for games and rodents. In 1640, the seeds were first used as stimulants in European.

Due to its powerful biological effects, studies of the active principles of *S. nuxvomica* were initiated very early in the history of the development of organic chemistry. The potent emetic activity of this plant extract is well known in Western sciences by the time the species was named with the Latin designation for "vomiting nut" by Linnaeus in the eighteenth century. Follow-up studies conducted by French chemists Pelletier and Caventou in 1818 were guided by this activity and led to the separation of strychnine as an active ingredient. However, the structural complexity of the compound exceeded the analytical capacity of the time, and more than 100 years later Sir Charles Robertson and his colleagues elucidated the structure of strychnine in 1947. The total synthesis of the compound was achieved in 1963 by the American chemist, R. B. Woodward, and coworkers. Their successful research on strychnine was the crowning achievement of careers for Robinson and Woodward, and they won Nobel Prize in Chemistry shortly after their publication of this famous natural product.

The biological activity and mechanism of action of strychnine have been the subjects of a large number of studies since the compound was purified in the early nineteenth century. Strychnine is considered to be one of the bitterest known substances with a detectable limit of as low as 1 ppm by weight in water. In sharp contrast, the taste threshold of sucrose approaches 2000 ppm for humans. This very intense bitterness is believed to contribute to its emetic properties and may account for its use in ordeal trials. A fearless innocent individual quickly consumes a nuxvomica formulation, resulting in the rapid and effective discharge of gastric contents before the toxin is absorbed in sufficient quantities to cause death. However, a slower consumption with exploratory rates by a person would prevent vomiting reflections with fatal consequences. The primary toxic effects of strychnine are from flexed muscles to painful immobility, convulsions, and eventually muscle paralysis. Spasms of the chest muscles and diaphragm lead to hypoxia and respiratory failure. Lethal doses for adults are only in the range of 0.2–0.4 mg/kg body weight, which makes strychnine as an extremely toxic agent.

Mechanism studies of strychnine suggest that this alkaloid is a highly selective neurotoxin. Strychnine is readily absorbed following ingestion or inhalation and can accumulate in lipid storage sites in the body. So, chronic exposure can eventually produce a toxic effect. Strychnine blocks postsynaptic receptors of the inhibitory neurotransmitter, glycine, in the spinal cord and motor neurons [44]. Studies have shown that glycine and strychnine bind to partially overlapping sites on this receptor [45]. The glycine receptor is the main carrier of rapid inhibitory transmission of synapses in the vertebrate spinal cord and brainstem. These receptors belong to the family of ligand-gated ion channels and also include acetylcholine and GABA receptors. Activation of this kind of receptor by glycine or other agonists induces the opening of the anion-selective channel of the receptor, allowing chloride flow into the cytoplasm. The resulting hyperpolarization of the postsynaptic membrane stabilizes the resting potential of the cell, thereby inhibiting neuronal firing.

### *13.1.11 Atropine*

Atropine, a neuroactive natural product, has a long and interesting history of both illicit and licit uses. Atropine and several other tropine alkaloids occur mainly in solanaceous plants, primarily *Mandragora officinarum*, previously known as *Atropa mandragora* or *Atropa belladonna*. *M. officinarum* is a low-growing, large-leafed, flowering plant that is native to the Mediterranean area and southern Europe, and is widely distributed throughout the Middle East. *A. belladonna*, also called belladonna, deadly nightshade, witch's berry, or a Circe herb, is a flowering bush that is native to Europe, Western Asia, and North Africa. The term belladonna, meaning beautiful lady in Italian, is usually considered to be derived from the practice of fashionable women, dating back to Cleopatra, the first century BCE in Egypt using plant extracts to dilate their pupils for cosmetic purposes. Other interpretations imply that the name may have referred to a magical or beautiful lady of the forest, and this term be used instead of the witch, suggesting that it is related to the use of herbs and mysterious potions.

The use of these plants is thought to date back to prehistoric times, and written records show that plants were used in ancient Greece and Egypt. For instance, the use of mandrake for promoting conception was described in Book of Genesis in the Bible (circa 1400 BCE). Artifacts from ancient Egypt around 1370 BCE suggest that mandrake was used as aphrodisiac and analgesic. The early Greek scholar, Theophrastus (370–285 BCE), stated that mandrake could be useful for skin infection, wounds, gout, and sleeplessness. It was reported that the mixture of mandala and wine was used by General Carthage to immobilize an invading army around 200 BCE. In 184 BCE, Hannibal's army used belladonna plants to induce enemy forces to lose their way. Around 35 BCE, Mark Anthony's army is said to have been poisoned by *A. belladonna*, and Cleopatra used atropine containing plant extracts to dilate her pupils. The Greek herbalist Dioscorides (about 40–90 CE) was apparently the first to record the use of *A. belladonna* as an anesthetic agent.

In medieval Europe, a large amount of belladonna was used to produce hallucinations by witchcraft and demon worship cults. Belladonna was also used for psychological torture in order to obtain true confessions from stubborn victims. In 1860, the preparation of the root of *A. belladonna* began to be used as a topical pain reliever. Over the next few decades, belladonna became a major product of world commerce for pain relief.

The separation of the active ingredients and modes of action of the belladonna and mandrake were quickly performed, and these plants were established for real medical use. The primary active ingredient, atropine, was first isolated in pure crystalline form from dried belladonna root by a German pharmacist, A. Mein, with a mydriatic activity-guided method. In 1901, the substance was first synthesized by Nobel Prize-winning German chemist Richard Willstatter. Subsequently, the natural product was found to be a racemic mixture of D- and L-carnitine and that most of its physiological effects were ascribed to L-caroline. Many dramatic effects of atropine were observed in humans and experimental animals. The toxic symptoms of atropine include dry mouth, dizziness, drowsiness, nausea, and constipation. Other effects include pupil dilation, fever, blurred vision, heart arrhythmia, inability to urinate, and excessively dry mouth and eyes. Higher doses may cause sensations of a burning throat, restlessness and mania, delirium, hallucinations, hot and dry skin flushing, and difficulty breathing. Death will ensue with constriction of the airways and suffocation. Although the therapeutic dose of atropine is only in the range of about 30 mg/kg.BW or less, lethal effects have been reported in humans at doses of about 1.5 mg/kg.BW or higher [46]. Atropine has a relatively minor impact on most domestic animals and birds, while it is quite toxic to dogs and cats. The effects are attributed to atropine's ability to lower the "rest and digest" activity of all muscles and glands regulated by the parasympathetic nervous system. Atropine is an effective competitive antagonist of human muscarinic acetylcholine receptors. The various recognized activities of atropine include narcotic, sedative, diuretic, mydriatic, and antispasmodic. The most important uses of atropine are in the treatment of eye diseases and as antidotes for anticholinesterase nerve agents, which recently have been used in times in chemical warfare and terrorism.

### *13.1.12 Phytoalexins*

Phytoalexins are plant metabolites produced by a plant in response to environmental stresses. Invasive organisms such as bacteria, fungi, viruses, and nematodes can induce the production of phytoalexins in plants, often resulting in the inhibition of the growth of invading organisms. However, this phenomenon is a general response to stress because exposure to cold, physical damage, ultraviolet light, and certain chemical compounds (such as polyamines, metal salts, and certain pesticides) also trigger the production of the same phytoalexins in a given plant. Typical examples of production of these so-called stress metabolites or natural pesticides occur in potatoes inoculated with the blight fungus *Phytophthora infestans*. When applied to the

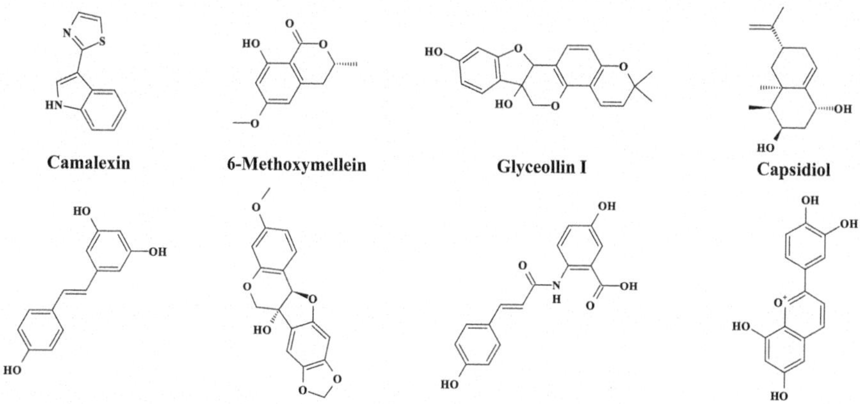


**Fig. 13.10** Chemical structures of several phytoalexins

surface of potato slices, certain strains of the fungus will initially grow rapidly and then the growth rates will gradually decrease. If the extract of the infected potato slice comes in contact with a pure culture of the same fungus, the fungus does not grow. Many other plants such as peas, broad beans, soybeans, green beans, carrots, cabbages, beets, and broccoli could be observed in this response to various fungi. In many of these situations, the plant response is triggered by certain polysaccharide components of the fungal cell wall. The amount of phytoalexins produced by a plant can be very significant. For example, soybeans infected with the fungus *Phytophthora megasperma* can produce glyceolin I at levels up to 10% of the dry weight of the infected tissue in a matter of days.

Generally, the chemical structures of the phytoalexins are modifications of those of metabolites produced in the unstressed plants, implying that phytoalexins are produced by modifications of the plant's normal metabolism. The chemical structures of several well-studied phytoalexins are presented in Fig. 13.10.

Phytoalexins have been the focus of considerable attention due to their abundance, chemical diversity, insecticide-like activity, and presence in the feed and food chain. In addition, phytoalexins possess well-demonstrated effects on livestock and potential effects on humans. Sweet potato phytoalexins indicate that these metabolites may have considerable toxic potential. It is known that the consumption of moldy sweet potatoes in feed can cause severe respiratory distress, congestion, pulmonary edema, and death of cattle. The toxic substances that are present in the moldy sweet potatoes, but not in the uninfected sweet potatoes, are a group of structurally related furans (Fig. 13.11). Two of the compounds, ipomeamaronol and ipomeamarone, cause hepatic degeneration in experimental animals with $LD_{50}$ of 230 mg/kg.BW [47]. The lung edema factors from infected tubers include 4-ipomeanol, ipomeanine, 1-ipomeanol, and 1,4-ipomeadiol, with ipomeanine

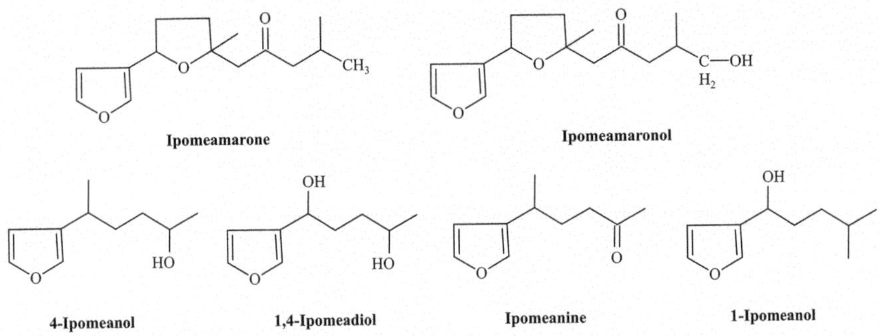


**Fig. 13.11** Toxins from damaged sweet potatoes

being the most toxic of the group to mice with an oral $LD_{50}$ of 26 mg/kg.BW. All these substances can induce an acute toxic response in mice, which is indistinguishable from the acute response produced by the administration of crude extract from infected sweet potato tubers.

The toxic terpenes in damaged sweet potatoes can occur only in slightly damaged tubers used for food. The presence of these substances is related to the darkened lateral rings below the tuber skin. The liver toxin, ipomeamarone, was shown to be present in commercial sweet potatoes in amounts ranging from 0.1 to 7.8 mg/g of sweet potato. Therefore, it is likely that humans are commonly exposed to toxic phytoalexins from sweet potatoes and other plant products. Fortunately, the level of exposure is generally lower than the toxic range because there appears to be no documented case of human poisoning from metabolites of stressed plants of any genus. In contrast, some phytoalexins such as resveratrol in grapes have potentially important beneficial effects on humans.

## 13.2 Effects of Phytochemicals on Nutrients/Nutrition

The consumption of fruits and vegetables is inversely correlated with the risk of many pathological diseases, and the beneficial effects are attributed to various protective phytonutrients. The mechanism of this correlation has not yet been fully elucidated. In any cases, it is generally believed that diets rich in fruits and vegetables are beneficial for health and with preventing coronary heart disease and even some cancers. The nutrients responsible for the protective action are unknown, but antioxidants, vitamins, and flavonoids are among the possible candidates. Dietary supplements are often applied to increase the plasma levels of these nutrients, which can also be achieved by increasing the proportion of vegetables and fruit in the diet. Prevention of disease through dietary factors or small amounts of consumed foods and exotic plants may be one of the strategies to reduce the risk of development of malignancy in humans.

The identification of new molecules that can reduce metastatic and proliferative potential of cancer cells is the goal of new differentiation therapies. Noteworthy examples of diet-derived substances that have been shown to reduce experimental carcinogenesis are isothiocyanate (ITC) from cruciferous vegetables, epigallocatechin gallate (EGCG) from tea, curcumin from the root of curcuma, and resveratrol (RSV) from red wine. Vegetables and fruits contain fiber, minerals, vitamins, and various bioactive compounds, such as flavonoids, carotenoids, indoles, and sterols, all of which could be responsible for the protective effect. In addition, herbs, spices, and rare plants are gaining wide uses as pharmacologic and therapeutic agents. A better understanding of the actions of phytochemicals will promote their use in more specific clinical trials. It may allow for the design for the synthetic derivatives that are equally effective but less toxic, and potentially offer insight into additional therapeutic uses for their antioxidant potential.

The chemopreventive effects of dietary phytochemicals on malignant tumors have been extensively studied due to their relative low toxicity. However, in order to achieve a desirable effect, high doses are generally required for treatment with a single drug. Thus, studies on effective combinations of phytochemicals at lower concentrations may contribute to chemopreventive strategies. The field of chemoprevention has been greatly developed. Nanotechnology was used as a novel approach to deliver packaged chemopreventive agents, which allows drugs to be delivered selectively to the target tissues.

For instance, the bioavailability of EGCG that had been packaged into nanoparticles has been reported by Mukhtar and colleagues. These results suggested that nano-chemoprevention could provide a new approach to avoid systemic toxicity and increase bioavailability. Similarly, it was reported that packaging RSV into solid lipid nanoparticles could improve the intracellular delivery of RSV and reduce the toxicity of RSV. The application of lipid- or polymer-based nanoparticles or nanoshells for improved delivery of chemopreventive or chemotherapeutic agents has facilitated the delivery of agents to selective tissues. Moreover, these applications can lessen systemic toxicity by reducing the amount of required agent and/or limiting the exposure to the body. Of particular importance is that heyneanol, a tetramer of RSV, has comparable or better anti-tumor efficacy than RSV in the mouse model of lung cancer. The current interest in phytochemicals has been driven mainly by epidemiological studies. In order to establish conclusive evidence for the effectiveness of dietary phytochemicals in disease prevention, however, it is useful to better define the bioavailability of these compounds so that their biological activity can be assessed. The bioavailability seems to differ greatly among the different plant compounds, and the most abundant of our diet is not necessarily those that have the best bioavailability characteristics. The evaluation of the bioavailability of phytochemicals has recently received increasing attention because the food industry is continually involved in the development of new products, defined as functional food, due to the presence of specific compounds. Definitive conclusions on the bioavailability of most bioactive compounds are difficult to obtain although the amount of available data is increasing. Hence, further studies on the bioavailability of bioactive compounds are necessary. At least four critical lines of research should be explored to gain a clear understanding of the health beneficial effects of dietary phytochemicals:

1. The potential biological activity of many dietary phytochemicals metabolites needs to be better studied. Indeed, metabolomic research, including the identification and the quantification of metabolites, is currently on behalf of a significant and growing field of research.
2. Strategies to improve the bioavailability of phytochemicals are required to be developed. Moreover, it is necessary to determine whether these methods increase biological activity.
3. The findings from in vitro studies need to be supported by in vivo experiments, although they have revealed the mode of action of individual dietary phytochemicals. The health benefits of dietary phytonutrients must be verified in appropriate animal models of disease and in humans at appropriate doses.
4. Novel technologies, such as nanotechnology, and a better understanding of stem cells, are certain to continue to promote the development of chemoprevention of cerebrovascular disease and cancer in the coming years.

## References

1. Berenbaum MR, Zangerl AR. Phytochemical diversity. In: Phytochemical diversity and redundancy in ecological interactions. New York: Springer; 1996. p. 1–24.
2. Derwahl M, Studer H. Nodular goiter and goiter nodules: where iodine deficiency falls short of explaining the facts. Exp Clin Endocrinol Diabetes. 2001;109(05):250–60.
3. Meier DA, Kaplan MM. Radioiodine uptake and thyroid scintiscanning. Endocrinol Metab Clin North Am. 2001;30(2):291–313.
4. Zimmermann MB, Jooste PL, Pandav CS. Iodine-deficiency disorders. Lancet. 2008;372 (9645):1251–62.
5. Zimmermann M, Saad A, Hess S, Torresani T, Chaouki N. Thyroid ultrasound compared with World Health Organization 1960 and 1994 palpation criteria for determination of goiter prevalence in regions of mild and severe iodine deficiency. Eur J Endocrinol. 2000;143(6): 727–31.
6. Biondi B. Thyroid and obesity: an intriguing relationship. Oxford: Oxford University Press; 2010.
7. Grüters A, Krude H. Detection and treatment of congenital hypothyroidism. Nat Rev Endocrinol. 2012;8(2):104.
8. Truong T, Baron-Dubourdieu D, Rougier Y, Guénel P. Role of dietary iodine and cruciferous vegetables in thyroid cancer: a countrywide case–control study in New Caledonia. Cancer Causes Control. 2010;21(8):1183–92.
9. Gonçalves CFL, de Souza dos Santos MC, Ginabreda MG, Fortunato RS, de Carvalho DP, Ferreira ACF. Flavonoid rutin increases thyroid iodide uptake in rats. PLoS One. 2013;8(9): e73908.
10. Gaitan E. Goitrogens in food and water. Annu Rev Nutr. 1990;10(1):21–37.
11. Sarne D. Effects of the environment, chemicals and drugs on thyroid function. In: Endotext. South Dartmouth: MDText. com; 2016.
12. Iacopini P, Baldi M, Storchi P, Sebastiani L. Catechin, epicatechin, quercetin, rutin and resveratrol in red grape: content, in vitro antioxidant activity and interactions. J Food Compos Anal. 2008;21(8):589–98.

13. Torres D, Chandía M. Favism presenting as an acute renal failure: report of one case. Rev Med Chil. 2012;140(8):1043–5.
14. Bilmen S, Aksu TA, Gümüşlü S, Korgun DK, Canatan D. Antioxidant capacity of G-6-PD-deficient erythrocytes. Clin Chim Acta. 2001;303(1–2):83–6.
15. Luzzatto L, Arese P. Favism and glucose-6-phosphate dehydrogenase deficiency. N Engl J Med. 2018;378(1):60–71.
16. Youngster I, Arcavi L, Schechmaster R, Akayzen Y, Popliski H, Shimonov J, Beig S, Berkovitch M. Medications and glucose-6-phosphate dehydrogenase deficiency. Drug Saf. 2010;33(9):713–26.
17. Cappellini MD, Fiorelli G. Glucose-6-phosphate dehydrogenase deficiency. Lancet. 2008;371(9606):64–74.
18. Yoshio N, Moldéust P. Cytotoxic effects of phenyl-hydroquinone and some hydroquinones on isolated rat hepatocytes. Biochem Pharmacol. 1992;44(6):1059–65.
19. Schafer FQ, Buettner GR. Redox environment of the cell as viewed through the redox state of the glutathione disulfide/glutathione couple. Free Radic Biol Med. 2001;30(11):1191–212.
20. Birben E, Sahiner UM, Sackesen C, Erzurum S, Kalayci O. Oxidative stress and antioxidant defense. World Allergy Organ J. 2012;5(1):9.
21. Spencer PS. Food toxins, AMPA receptors, and motor neuron diseases. Drug Metab Rev. 1999;31(3):561–87.
22. Kusama-Eguchi K, Miyano T, Yamamoto M, Suda A, Ito Y, Ishige K, Ishii M, Ogawa Y, Watanabe K, Ikegami F. New insights into the mechanism of neurolathyrism: L-β-ODAP triggers [Ca2+] i accumulation and cell death in primary motor neurons through transient receptor potential channels and metabotropic glutamate receptors. Food Chem Toxicol. 2014;67:113–22.
23. Van Moorhem M, Lambein F, Leybaert L. Unraveling the mechanism of β-N-oxalyl-α, β-diaminopropionic acid (β-ODAP) induced excitotoxicity and oxidative stress, relevance for neurolathyrism prevention. Food Chem Toxicol. 2011;49(3):550–5.
24. Khandare AL, Ankulu M, Aparna N. Role of glutamate and nitric oxide in onset of motor neuron degeneration in neurolathyrism. Neurotoxicology. 2013;34:269–74.
25. Zhang Y, Bhavnani BR. Glutamate-induced apoptosis in neuronal cells is mediated via caspase-dependent and independent mechanisms involving calpain and caspase-3 proteases as well as apoptosis inducing factor (AIF) and this process is inhibited by equine estrogens. BMC Neurosci. 2006;7(1):49.
26. Szydlowska K, Tymianski M. Calcium, ischemia and excitotoxicity. Cell Calcium. 2010;47(2): 122–9.
27. Radi R, Rodriguez M, Castro L, Telleri R. Inhibition of mitochondrial electron transport by peroxynitrite. Arch Biochem Biophys. 1994;308(1):89–95.
28. Montagnac JA, Davis CR, Tanumihardjo SA. Nutritional value of cassava for use as a staple food and recent advances for improvement. Compr Rev Food Sci Food Saf. 2009;8(3):181–94.
29. Baskin SI, Kelly JB, Maliner BI, Rockwood GA, Zoltani C. Cyanide poisoning. Med Aspects Chem Warfare. 2008;11:372–410.
30. Nelson L. Acute cyanide toxicity: mechanisms and manifestations. J Emerg Nurs. 2006;32(4): S8–S11.
31. Siedow JN, Umbach AL. The mitochondrial cyanide-resistant oxidase: structural conservation amid regulatory diversity. Biochim Biophys Acta. 2000;1459(2–3):432–9.
32. Pearce LL, Bominaar EL, Hill BC, Peterson J. Reversal of cyanide inhibition of cytochrome c oxidase by the auxiliary substrate nitric oxide an endogenous antidote to cyanide poisoning? J Biol Chem. 2003;278(52):52139–45.
33. Brenner M, Kim JG, Lee J, Mahon SB, Lemor D, Ahdout R, Boss GR, Blackledge W, Jann L, Nagasawa HT. Sulfanegen sodium treatment in a rabbit model of sub-lethal cyanide toxicity. Toxicol Appl Pharmacol. 2010;248(3):269–76.
34. Drickamer K, Taylor ME. Biology of animal lectins. Annu Rev Cell Biol. 1993;9(1):237–64.

35. Herzig K-H, Bardocz S, Grant G, Nustede R, Fölsch U, Pusztai A. Red kidney bean lectin is a potent cholecystokinin releasing stimulus in the rat inducing pancreatic growth. Gut. 1997;41 (3):333–8.
36. Winder C. Toxicity of ricin. Toxin Rev. 2004;23(1):97–103.
37. Garber EA, Eppley RM, Stack ME, McLaughlin MA, Park DL. Feasibility of immunodiagnostic devices for the detection of ricin, amanitin, and T-2 toxin in food. J Food Prot. 2005;68(6): 1294–301.
38. Ullrich S, de Vries YC, Kühn S, Repantis D, Dresler M, Ohla K. Feeling smart: effects of caffeine and glucose on cognition, mood and self-judgment. Physiol Behav. 2015;151:629–37.
39. Lane JD, Adcock RA, Williams RB, Kuhn CM. Caffeine effects on cardiovascular and neuroendocrine responses to acute psychosocial stress and their relationship to level of habitual caffeine consumption. Psychosom Med. 1990;52(3):320–36.
40. Nawrot P, Jordan S, Eastwood J, Rotstein J, Hugenholtz A, Feeley M. Effects of caffeine on human health. Food Addit Contam. 2003;20(1):1–30.
41. Savitz DA, Chan RL, Herring AH, Howards PP, Hartmann KE. Caffeine and miscarriage risk. Epidemiology. 2008;19:55–62.
42. Weng X, Odouli R, Li D-K. Maternal caffeine consumption during pregnancy and the risk of miscarriage: a prospective cohort study. Am J Obstetr Gynecol. 2008;198(3):279.e1–8.
43. Lee MR. Curare: the South American arrow poison. J R Coll Physicians Edinb. 2005;35(1): 83–92.
44. Kanthasamy AG, Kanthasamy A, Matsumoto RR, Vu TQ, Truong DD. Neuroprotective effects of the strychnine-insensitive glycine site NMDA antagonist (R)-HA-966 in an experimental model of Parkinson's disease. Brain Res. 1997;759(1):1–8.
45. Zhang Y, Wu S, Eger EI, Sonner JM. Neither GABAA nor strychnine-sensitive glycine receptors are the sole mediators of MAC for isoflurane. Anesth Analg. 2001;92(1):123–7.
46. Koplovitz I, Menton R, Matthews C, Shutz M, Nails C, Kelly S. Dose-response effects of atropine and HI-6 treatment of organophosphorus poisoning in guinea pigs. Drug Chem Toxicol. 1995;18(2–3):119–36.
47. Takayuki LSB. Introduction to food toxicology. 2nd ed. 2009; In Academic Press.

The manufacturer's authorised representative in the EU is Springer Nature Customer Service Centre GmbH, Europaplatz 3, 69115 Heidelberg, Germany. If you have any concerns regarding our products, please contact ProductSafety@springernature.com

Printed and bound by CPI Group (UK) Ltd, Croydon, CR0 4YY
30/09/2026
02225316-0001